Experimental Particle Physics

Understanding the measurements and searches at the Large Hadron Collider

Experimental Particle Physics

Understanding the measurements and searches at the Large Hadron Collider

Deepak Kar

School of Physics, University of Witwatersrand, Johannesburg, South Africa

IOP Publishing, Bristol, UK

Cover image shows a visualization of the highest-mass dijet event, (Event 4144227629, Run 305777) recorded in 2016 by the ATLAS Experiment. The two central high-pT jets each have transverse momenta of 3.74 TeV, they have a y* of 0.38 and their invariant mass is 8.02 TeV. © CERN, reproduced with permission.

Permission to make use of IOP Publishing content other than as set out above may be sought at permissions@ioppublishing.org.

ISBN   978-0-7503-2112-9 (ebook)
ISBN   978-0-7503-2110-5 (print)
ISBN   978-0-7503-2111-2 (mobi)

DOI   10.1088/2053-2563/ab1be6

Version: 20190801

IOP Expanding Physics
ISSN 2053-2563 (online)
ISSN 2054-7315 (print)

British Library Cataloguing-in-Publication Data: A catalogue record for this book is available from the British Library.

Published by IOP Publishing, wholly owned by The Institute of Physics, London

IOP Publishing, Temple Circus, Temple Way, Bristol, BS1 6HG, UK

US Office: IOP Publishing, Inc., 190 North Independence Mall West, Suite 601, Philadelphia, PA 19106, USA

To all my CDF and ATLAS collaborators over the years,
from whom I have learned everything!

Contents

Preface

*If you can't explain something to a first year student, then you haven't really
understood it.*

—Richard Feynman

My interactions with new students in experimental particle physics have led me to
believe that many of them lack a broad overview of what they are doing, although
they are often proficient in coding, quick in obtaining results, and fearless in
presenting in meetings. Also, while there are excellent books covering theoretical
aspects of collider physics, a practical guide to terminology and techniques used in
the field is missing. Newcomers in the field mostly end up learning from their peers,
seniors or from week-long schools, where many of the topics are covered at an
insanely fast pace.

This book is an attempt to address that shortcoming. It originated from teaching
a one semester-long, advanced undergraduate introduction to experimental particle
physics course at the School of Physics at the University of Witwatersrand. The aim
of the course was that at the end, the student should be able to digest an
experimental paper, and the book follows the same philosophy.

Although in some sense a compilation, my hope is that a single book will fill the
void for a one-stop resource. This can be either used as a textbook for a semester
course (with the last two chapters as optional reading), or simply as required reading
for a student starting in this field. Many of the exercises are designed to trigger
thinking, or point to common pitfalls a beginner encounters.

Of course there are obvious shortcomings in such an effort. Many of the topics are
not covered in as much detail as they merit, which is somewhat on purpose. The idea
is to make beginners aware of concepts, point them to the resources, and let them
take it from there. This will, hence, not be a replacement for a theoretical textbook
introducing the Standard Model, but rather be complementary to it. Keeping that in
mind, elaborate calculations have been avoided as much as practicable, and focus
has been on introducing the commonly used analysis methods. Although I have
strived hard to keep everything as general as possible, there is an overload of
examples from ATLAS, and this is solely due to the fact that it was easier for me to
find those.

I am heavily indebted to numerous colleagues over the years, both experimen-
talists and theorists, from whom I learned everything I discuss here. I would also
love to hear from the community, so that subsequent editions can be improved.

First and foremost, Nabanita Mukherjee (AbbVie pharmaceuticals, formerly at
Duke University, PhD in Statistics from University of Florida) helped me to write
the statistics part. I cannot thank her enough for her time and patience, especially
when she was going through a difficult time.

Many colleagues read the whole book or some chapters and provided invaluable
suggestions. I am immensely grateful to all of them. Andy Buckley, Rick Field,
Marvin Flores, Andrew Larkoski, Jong Soo Kim, Swagata Mukherjee, Tuhin Roy

and Michael Spannowsky I cannot thank you enough. Danielle Wilson spent her end of second year undergraduate break reading the book, and pointed out the parts where I was not clear enough. Her review, from a non-expert perspective, helped massively. The same goes to my PHYS4029A classes of 2017 and 2018, as a large fraction of the material was fine-tuned in response to discussions in class.

During the course of writing the book, the conversations I had with Stefan von Buddenbrock, Jon Butterworth, Tasnuva Chowdhury, Valentina Cairo, Kaustuv Datta, Bruce Mellado, Debarati Roy, Xifeng Ruan, Seema Sharma and Sukanya Sinha have contributed significantly to both the content and presentation. In fact, many examples I use are motivated from the discussions we had in ATLAS, so it is only fair that I thank the whole ATLAS community. I sincerely thank Heather Russell and Frederic Dreyer for letting me use their figures. Sukanya proofread the entire book, and Stefan also helped in drawing a number of figures.

I received useful advice from many friends, and I would specifically mention Debashree Dattaray. The constant support of Nikhliesh Kar, Ramala Kar, and Nandini Kar must be acknowledged, as well as of Saswati Roy.

Finally, it would be a travesty not to thank Dan Heatley from IoP Publishing, without whose persistent push, this book would have probably never seen the light of day.

Author biography

Deepak Kar

Deepak Kar is currently an associate professor at the School of Physics, at the University of Witwatersrand, in Johannesburg, South Africa. He is an active of the ATLAS collaboration at the Large Hadron Collider at CERN. Previously, he was a post-doctoral researcher at the University of Glasgow (2012–15), and at Technische Universität Dresden (2009–11). He completed his PhD from University of Florida in 2008, working on CDF experiment at Tevatron in Fermilab. He likes travelling, follows current affairs and sports avidly, and is a big fan of South African wines.

IOP Publishing

Experimental Particle Physics
Understanding the measurements and searches at the Large Hadron Collider
Deepak Kar

Chapter 1

Groundwork

Physics is really nothing more than a search for ultimate simplicity, but so far all we have is a kind of elegant messiness.

—Bill Bryson

We start by briefly reviewing natural units in section 1.1, followed by a discussion of the particles discovered so far and open questions in section 1.2. Relativistic kinematics is used extensively in particle physics, so we include a brief overview and some examples in section 1.3.

A full theoretical overview of Quantum Field Theory (QFT), Quantum Chromodynamics (QCD) and the Standard Model (SM) of particle physics is beyond the scope of this book. The interested reader is referred to books [1–7], which the author has found helpful. Application of statistical methods to interpret the results is also of critical importance. Again, for a comprehensive overview, dedicated resources [8–10] should be consulted. These references are by no means complete, and additionally there are excellent lecture notes and slides from numerous summer schools, which the author always finds extremely helpful.

1.1 Natural units

We are familiar with the three base units, [Length], [Mass] and [Time]. Along with them, [Electric Current], [Temperature], [Luminous intensity], and [Amount of substance] complete the basic units (square brackets is the usual way of denoting units or dimensions). In the most commonly used *Systéme International* (SI) units, the first three are represented by meter, kilogram and second. That is sufficient for everyday instances, but when dealing with quantities which are orders-of-magnitude larger or smaller, as encountered in particle physics, they prove inadequate. Let us see how we can derive units suitable for our purpose.

In electromagnetism, we have the force $\vec{F}$ experienced by a charge q, moving with a velocity $\vec{v}$, in an electric field $\vec{E}$ and magnetic field $\vec{B}$, given by:

$$\vec{F} = q(\vec{E} + (\vec{v}/c) \times \vec{B}),$$

where c is the speed of light. In Quantum Mechanics, Planck's constant h (actually the reduced Planck's constant, $\hbar = h/2\pi$, which has the dimension of energy multiplied by time) is the fundamental constant. Energy, E, is expressed as:

$$E = \hbar\omega$$

with ω being the angular frequency which is quantised.

We adopt the convention where $c = \hbar = 1$. That implies that the velocities can be expressed as dimensionless fractions of c, and the energies (and masses, which now have the same unit) can be expressed as reciprocals of time. This is termed *natural units*. The advantage is that we end up with reasonable numbers for the quantities of interest in the subatomic domain, as c and $\hbar$ are respectively very large and small.

This imposes two constraints on three of the basic units, leaving us the freedom to define one of them as per our convenience. That is usually taken to be energy, and expressed in terms of electron-volts (eV, with prefixes as needed), which is defined in terms of the rest mass of the electron (0.511 keV or 9.11×10^{-31} kg). This choice is arbitrary. To give a sense of this measure, one TeV is about the energy caused by the motion of a flying mosquito. At the Large Hadron Collider (LHC), energy of the order of TeV is produced at a much smaller length scale in each collision, essentially within the radius of a proton.

In particle physics, the quantities are almost always presented in natural units. Mass, momentum and energies will all be expressed in terms of energy, using some prefix of eV. The conversion to ordinary units merely involves multiplying or dividing the formula or the value by some combination(s) of c and $\hbar$.

For completeness, the values of second, meter and kg in natural unit are given by:
1 m = 5.076×10^{15} GeV^{-1}.
1 s = 1.519×10^{24} GeV^{-1}.
1 kg = 5.625×10^{26} GeV.

When we refer to the mass of the Higgs boson as 125 GeV, this actually means the mass is 125×10^9 c^{-2} eV = 2.228×10^{-25} kg.

1.2 Particle content

1.2.1 What we have

The elementary particles discovered so far are depicted in figure 1.1, which is almost like the periodic table of elements, but with elementary particles instead. The Standard Model [11] of particle physics is the theoretical framework which encapsulates our best understanding of the interactions of the elementary particles discovered so far.

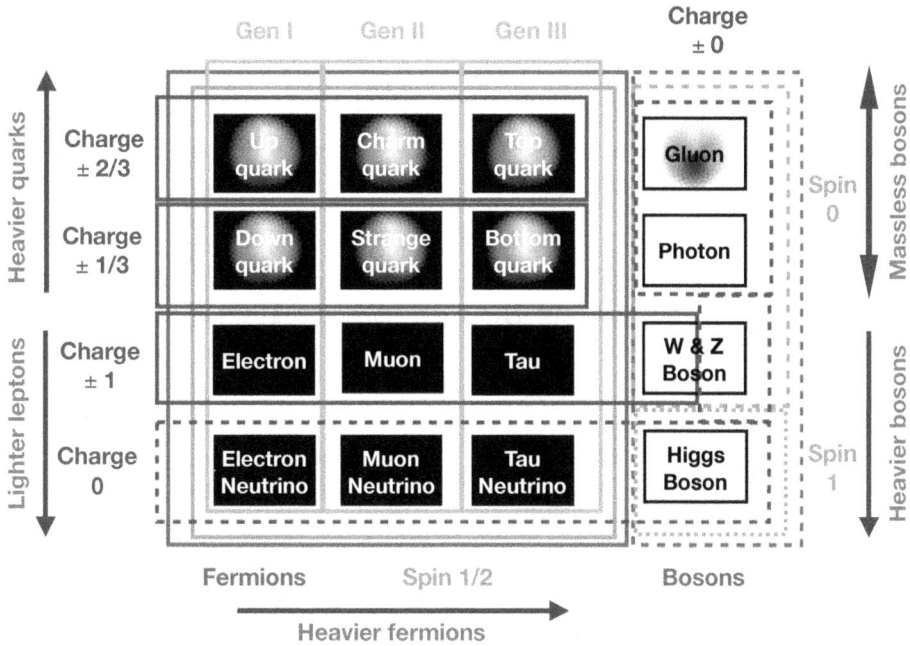

Figure 1.1. The 'periodic table' of elementary particles, adapted from [12]. The fermions are in the black boxes, while the bosons are in the white boxes. The colour wheels in the boxes indicate that the corresponding particles have colour charge, and the absence of those indicate that the corresponding particles do not have colour charge. The green lines group together particles by spin. The purple lines group together particles by charge. The orange lines group together fermions by generation.

The elementary particles are classified as *fermions* (named after Enrico Fermi) or *bosons* (named after Satyendranath Bose), depending on whether they have half integral or integral spin quantum numbers. The so-called matter particles are fermions, while the bosons are the force carriers. Fermions are divided into leptons and quarks, based on how they interact. Three of the leptons: electrons, muons and the tau leptons (τ) carry electromagnetic charge, and each has a corresponding neutral neutrino. The electron and muon are stable at the LHC detectors, while the τ leptons decay before reaching the detector. They can decay to an electron or a muon with the corresponding anti-neutrino or to one or two pions. Therefore, experimentally the term lepton refers only to electrons and muons in most cases, and we will follow the same convention here. Their (electromagnetic and weak) interactions are described by quantum electrodynamics (QED).

The quarks in addition carry so-called colour charge, and their (strong) interactions are described by quantum chromodynamics (QCD) (the name quark comes from James Joyce's novel *Finnegans Wake*). They can also interact with the leptons via electromagnetic and weak interactions. The six quarks are named up (u), down (d), charm (c), strange (s), top (t) and bottom (b). Each of the quarks can have three possible colour charges, which are termed as red (R), blue (B) and green (G), with corresponding anti-colours. A combination of red, blue and green quarks, as well as of one coloured quark and its anti-coloured quark results in a white (or uncoloured, alternatively termed colour-neutral) particle. It must be kept in mind that the colours are just labels for us to visualise abstract quantum numbers, and have no connection to the everyday colours from the light spectrum. Gluons carry a combination of colour and anti-colour charge in order to participate in strong interactions (i.e. the glue that holds the quarks together). That can lead to nine possible combinations. However, combinations like $R\bar{R}$, $G\bar{G}$ or $B\bar{B}$, that correspond to colourless states cannot exist, neither can a linear combination ($R\bar{R} + G\bar{G} + B\bar{B}$), as that cannot change the colour of a quark in an interaction. So colour-neutral states are not possible for gluons. However, unbalanced linear combinations are possible, but only two combinations can be shown to be linearly independent from QCD. So it leads to eight gluons, designated as $R\bar{G}$, $R\bar{B}$, $G\bar{B}$, $G\bar{R}$, $B\bar{R}$, $B\bar{G}$, $\frac{1}{\sqrt{2}}(R\bar{R} - G\bar{G})$, and $\frac{1}{\sqrt{6}}(R\bar{R} + G\bar{G} - 2B\bar{B})$, where the last two combinations can be formed differently as well. This is referred to as the colour *octet*[1].

Each fermion has a corresponding anti-particle, which is identical to the specific fermion, except it has opposite charge. For the leptons, the anti-particles carry opposite electromagnetic charge, and the anti-quarks (denoted by, for example, $\bar{t}$, corresponding to t), carry anti-colour charge as discussed above (red and anti-red, for example, can also combine to form a white). The fermions are divided into three generations (or flavours), with members of subsequent generations having higher masses than the preceding ones. Also, only the first generation is considered stable, the rest can decay with different lifetimes. In experiments, sometimes the second or third generation quarks are referred to as *heavy flavours*. The top quark is the heaviest known elementary particle (roughly as heavy as a tungsten atom), and it has a lifetime of 10^{-23} s, which means it does not even get to decay to hadrons directly.

QCD dictates that the colour-charged particles cannot be free in nature, and the properties of strong forces results in their *confinement*. Quarks can create colour-neutral objects in two ways, either by forming a quark and anti-quark pair, termed mesons, (integral spin of zero and $\pm$ unity) or by the appropriate colour (or anti-colour) combination of three quarks (or anti-quarks), termed baryons (half integral spin of $\pm$ 1/2 or $\pm$ 3/2). The baryons and mesons are collectively called the *hadrons*.

Some examples of mesons are pions, kaons, ρ, Ω etc. They all include two different types of quarks and anti-quarks (like neutral pion is made from u and $\bar{d}$, and the charged pions from combinations of $u\bar{u}$ and $d\bar{d}$). The exceptions are ϕ, J/ψ (it is believed that the name was chosen to look like the Chinese name of Samuel Ting, the discoverer) and Υ mesons, that are composed of $s\bar{s}$, $b\bar{b}$ and $c\bar{c}$ combinations, respectively. When a meson includes the strange quark s, it is sometimes termed as a strange meson. Protons and neutrons are examples of baryons (consisting of u, u, d and u, d, d respectively). That explains the naming of the LHC as well, with the word large characterising the collider. Tetraquarks containing two heavy (bottom) quarks and two light anti-quarks have been

---

[1] Formally QCD is expressed in terms of non-abelian symmetry group SU(3), where quarks are in fundamental representation of SU(3) while gluons are in adjoint representation of SU(3), which leads to the colour octet gluons, but such a discussion is beyond the scope of this book.

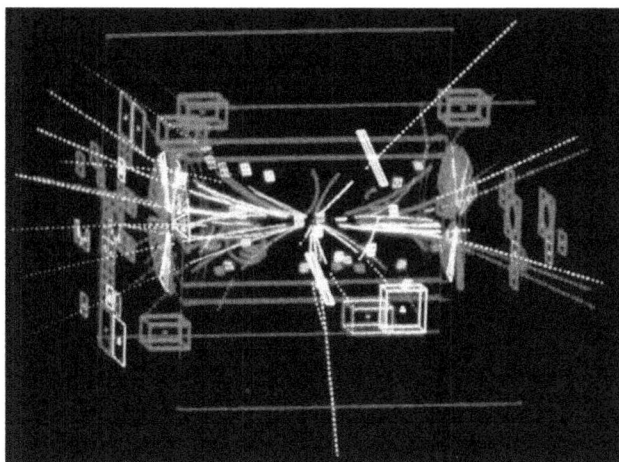

Figure 1.2. An actual event display from one of the first events recorded with a Z boson [31]. The white dashed lines indicate the two muons coming from the decay of the Z boson (© 2018 CERN).

proposed, and pentaquarks have been observed by the LHCb experiment [13] in Λ_b^0 baryon decay (the Λ baryons contain one up quark, one down quark, and a third quark from a higher generation, in this case a b-quark).

The bosons mediate in all the interactions. The bosons with non-zero spin are termed *vector or gauge* bosons, which include photon, W/Z bosons, and gluons. The latter participates in the strong interaction. The only *scalar* boson, i.e. a boson with a zero spin, is the Higgs boson. The Higgs boson was postulated in order to theoretically reconcile the existence of the massless photon, which carries the long-range electromagnetic force, and the massive W/Z bosons, which carry the short-range weak force. It was based on the independent work of Peter Higgs [14, 15], Francois Englert and Robert Brout [16], and Gerald Guralnik, Carl Hagen and Tom Kibble [17]. It is also closely related to the concept of superconductivity generated by the Anderson mechanism [18]. It must be noted that the Higgs mechanism only gives mass to fermions (quarks and leptons) when they interact with the Higgs field generated by so-called spontaneous symmetry breaking. The mass of everyday objects comes from protons and neutrons, which get their masses from the binding energy of quarks and gluons, governed by QCD. The Higgs boson is often referred to as the 'God particle' in popular culture, which came about when Leon Lederman apparently referred to it as the 'Goddamn Particle', which was (unfortunately) shortened [19]!

A brief comment about the discovery of new particles in colliders is in order here. The W and Z bosons were discovered in 1983 at the CERN Super Proton Synchrotron (SPS) collider, by the UA1 and UA2 experiments [20–23]. The discoveries were made essentially by looking at event displays (figure 1.2 shows an example) where the Z boson decayed to an electron–positron pair. Interested readers can refer to the book *Nobel Dreams* [24] to read the fascinating background stories about this discovery. The Υ (and, hence, the bottom quark) was discovered earlier at Fermilab in 1977, but it was at a fixed target experiment [25]. The top quark was discovered in 1995 at the Tevatron [26, 27] by CDF (Collider Detector at Fermilab) and D Ø (which is just named after the quadrant position of the detector in the ring; in fact CDF was supposed to be named BØ before it chose its name) experiments. The Higgs boson was discovered in 2012 by the ATLAS (A Toroidal Lhc ApparatuS) and CMS (Compact Muon Solenoid) experiments at the LHC [28, 29], and the book *Smashing Physics* [30] gives behind the scene stories from the effort.

1.2.2 What we do not have

It is impossible to say what we do not have, in fact that is why the field of particle physics is so exciting. We are always searching for the *known–unknowns*, based on our best guess of what should be there, but there always remains the possibility that we can find a true *unknown–unknown*.

The Standard Model (SM) has been remarkably successful in predicting and explaining most experimental observations so far, but it can be a *low-energy* approximation of a broader framework, just like classical mechanics is perfectly suitable to explain most everyday phenomena, but it is valid at a limit of the underlying quantum mechanical principles governing the world. The SM contains 24 free parameters that are simply not calculable and have to be taken from experiment. Some of the significant open questions are:

- Why do particles have these particular mass values? Why are there three generations of fermions, with increasing masses? Why are the neutrinos not massless (or so light)? How do neutrinos transform between the flavours (which is known as *neutrino oscillation*)? Is there a fourth type of neutrino, called a *sterile neutrino* that we have not discovered yet?

- Planck (after Max Planck) mass is defined to be $\sqrt{c\hbar/G}$, where G is the gravitational constant. It roughly corresponds to the mass of the smallest possible black hole. An obvious question arises, why is the mass of everything we see, including that of elementary particles, so small compared to Planck mass? In other words, why is the gravitational mass scale so large? This is called the *hierarchy problem*. This is tied to the mass of the Higgs boson, which determines the electroweak symmetry breaking. The quantum corrections to the Higgs field almost magically cancel each other out, making the Higgs boson stable. This is called the little hierarchy problem, with an explanation still eluding us. I find an analogy from reference [32] particularly helpful. *This is a bit like a bank account which has huge inputs and outputs during the month, but balances at exactly zero on the last day of every month. Looking at this behaviour, an accountant would be unlikely to think this was a coincidence. There is probably some principle underlying the operation of the account which forces this to happen.*

- Why is gravity so weak, and why is it the only force that has not been explained by a quantum field theory? The hypothetical particle *graviton* has been proposed as the force carrier of gravity, without any experimental evidence so far. This leads to the idea that all forces will be unified at some scale (termed grand unification).

- The observation in 1998 by the Hubble Space Telescope, that the Universe is expanding now at a much faster rate than before, raised an important question. Current models predicted a slowing down of the expansion, and the proposed solution was to postulate the existence of *dark energy*. Other cosmological observations indicated that the galaxies are spinning much faster than they should be, based on the gravitational pull of their visible matter. The existence of *dark matter* was postulated to explain this. The Universe is estimated to consist of 68% dark energy and about 27% dark matter. Not much is known about the origin and composition of dark energy and dark matter.

- The Universe is dominated by matter, whereas during its creation in the Big Bang, matter and anti-matter should have been produced in equal parts. The SM has no mechanism to explain the imbalance. The combination of charge conjugation (particles and anti-particles behave the same way) and parity symmetry is referred to as CP. Strong and electromagnetic interactions obey CP symmetry, while CP violation is seen to occur in some weak decays. The first observation was from Fermilab's KTeV collaboration [33], where kaons were seen to preferentially decay to electrons rather than to positrons (along with a charged pion and the corresponding neutrino). The kaons consist of strange quarks, and later CP violation was also observed in neutral B mesons, containing bottom quarks. Recently, the LHCb collaboration at the LHC has observed CP violation in the D^0 meson, which contains a pair of charm quarks [34]. However, the observed CP violation is not enough to explain the observed matter–anti-matter asymmetry in the Universe.

So far experimental particle physics has been mostly driven by theoretical predictions, in the same way that predictions of the Higgs boson in the middle of the last century dictated experimental programmes. Now we are perhaps at a unique position, as we do not have a specific answer to any of the unanswered questions, so in some sense experimental results have to show the way forward.

1.3 Relativistic kinematics

1.3.1 Basic ideas

Particles participating in or emerging from collisions at colliders move at speeds close to the speed of light. Therefore, they follow relativistic kinematics. We can define the usual dimensionless quantities, $\beta = v/c$ and $\gamma = 1/\sqrt{1 - \beta^2}$, where v denotes the velocity and γ is referred to as the *boost* factor, as it represents the (relativistic) increase of the energy of the colliding particle. In natural units, $\beta = v$.

The collision is best visualised at the centre-of-mass (c.m.) frame, where colliding particles have zero combined momentum, and the collision energy is simply twice the energy of one proton. However, the result of the collision is measured in the inertial frame of LHC detectors (which will be referred to as the *detector frame*), in which we would see two equal energy particles of equal masses colliding. We will discuss how to connect these two frames.

In relativistic kinematics, three-spatial-component vectors need to be expanded with a time-like component, and they are termed four-vectors. A generic four-vector can be denoted by $X = (x_0, \vec{X}) = (x_0, x_1, x_2, x_3)$, where the convention of having the 0th component as time-like is chosen, and $\vec{X}$ is the ordinary three-vector. A four-vector (and in general a tensor), which is represented with lower indices or subscripts as above, is called a covariant vector. It will have a corresponding contravariant vector, which inverts the sign of the space-like components, and will be denoted by upper indices or superscripts, such as $(x^0, -x^1, -x^2, -x^3)$. Equivalently, the covariant and the contravariant vectors can be denoted by X_μ and X^μ, where μ runs from 0 to 3.

Then a dot product (mathematically called an inner product), analogous to the three-vector dot product is defined, which must be between a covariant and a contravariant vector. So for two four-vectors X and Y, the dot product is:

$$X \cdot Y \equiv X_\mu Y^\mu = x_0 y^0 - x_1 y^1 - x_2 y^2 - x_3 y^3$$

In three-vector notation, the dot product represents the projection of one vector in the direction of another, multiplied by the length of the second vector. In four-vector notation, the notion of distance or direction does not apply, but the dot product still serves an important role.

A Lorentz transformation of a four-vector X to X', to a frame moving with velocity β along x-axis between is given by:

$$x_0' = \frac{x_0 - \beta x_1}{\sqrt{1 - \beta^2}}$$

$$x_1' = \frac{x_1 - \beta x_0}{\sqrt{1 - \beta^2}}$$

$$x_2' = x_2$$

$$x_3' = x_3$$

This can be rewritten as:

$$x_0' = \frac{x_0}{\sqrt{1 - \beta^2}} - \frac{\beta x_1}{\sqrt{1 - \beta^2}} = x_0 \gamma - \beta x_1 \gamma$$

$$x_1' = \frac{x_1}{\sqrt{1 - \beta^2}} - \frac{\beta x_0}{\sqrt{1 - \beta^2}} = x_1 \gamma - \beta x_0 \gamma$$

In general, this can be written in matrix form:

$$\begin{bmatrix} x_0' \\ x_1' \\ x_2' \\ x_3' \end{bmatrix} = \begin{bmatrix} \gamma & -\gamma\beta & 0 & 0 \\ -\gamma\beta & \gamma & 0 & 0 \\ 0 & 0 & 1 & 0 \\ 0 & 0 & 0 & 1 \end{bmatrix} \begin{bmatrix} x_0 \\ x_1 \\ x_2 \\ x_3 \end{bmatrix}$$

The transformation of another vector Y to Y' can be written similarly. Now it is straightforward to show:

$$X \cdot Y = X' \cdot Y'$$

This is a very important result, which means that the dot products of two four-vectors is frame-independent, or as it is generally called, *Lorentz invariant*. Therefore, any quantity which can be expressed as a four-vector dot product, can be calculated in any frame, yielding the same result. In collider physics, the most commonly encountered four-vectors are the position and momentum four-vectors (often termed *four-momentum*). The former can be represented by $(t, x, y \cdot z)$, and the latter as (E, p_x, p_y, p_z), where the spatial components are the usual three-vectors.

The dot products of the position and momentum four-vectors are Lorentz invariant. For position vectors, it is designated as the space–time interval:

$$s^2 = X' \cdot X' = x_0'^2 - x_1'^2 = x_0^2 - x_1^2 = X \cdot X$$

In general terms, for a spatial separation of Δx_1 and a temporal separation of Δx_0:

$$(\Delta s)^2 = (\Delta x_0)^2 - (\Delta x_1)^2$$

is constant for all observers. The sign of $(\Delta s)^2$ is not positive definite. For space-like intervals, $(\Delta s)^2 < 0$, while for time-like intervals, $(\Delta s)^2 > 0$. Similarly, for momentum four-vectors, the dot product with itself is invariant, which is given by:

$$P \cdot P = E^2 - p^2,$$

where p is the magnitude of three-momentum. This is defined as the *invariant mass* (as this quantity will be same in any inertial frame). For a single particle, invariant mass is equivalent to its rest mass (the mass it has at a frame that is at rest, which is a measure of inertia), however, for a system of particles that participates in an interaction, the appropriate term to use is the invariant mass, which can be thought of as the effective rest mass of the system, accounting for their motion[2]. In the limiting case of a particle at rest, $p = 0$, this can also be thought of as equivalent to Einstein's famous equation: $E = mc^2$ in terms of the rest mass $m = P \cdot P$ (and not using natural units). The original equation involved relativistic mass, a concept which is deprecated now.

---

[2] Occasionally the term relativistic mass is used to denote $m_{rel} = \gamma m$, where m is the rest mass. This definition is useful only to keep the standard form of momentum $p = m_{rel}v$, but it is not used otherwise.

For a particle decaying to multiple particles, then the outgoing daughter energies and momenta can be added to find the invariant mass. As this is the same in all frames, including the centre-of-mass frame, then the particle can be identified in any frame from its decay product energies and momenta.

The momentum four-vector can be obtained from the position four-vector by dividing it by a quantity analogous to time. Proper time, $d\tau = dt\sqrt{(1 - \beta^2)}$, which is equivalent to the invariant space–time interval (this can be considered as the time measured by a clock moving with the particle) is used, so:

$$P \equiv m(dx_0/d\tau, dx_1/d\tau, dx_2/d\tau, dx_3/d\tau),$$

where m is the mass. For $\beta \ll 1$, it reduces to $P_1 = mv$, the usual three-momentum, as can be seen from the first two components, $P = (m/\sqrt{1 - \beta^2}, mv/\sqrt{1 - \beta^2})$. The first term then stands for the rest energy.

1.3.2 Specific examples

- Fixed target experiments: this is equivalent to a situation where particle of mass m_1 and energy E_1 hits a stationary particle of mass m_2. The available energy (at the c.m. frame) to create a new particle (or a particle pair) of certain invariant mass can be expressed as (from the definition of invariant mass):

$$s = (E_1 + E_2)^2 - (p_1 + p_2)^2,$$

where p_1 and p_2 denote the (three) momentum of the particles, and E_2 denotes the energy of the target particle. As the target particle is at rest, $p_2 = 0$, and $E_2 = m_2$. This leads to:

$$s = E_1^2 + m_2^2 + 2E_1m_2 - p_1^2 = m_1^2 + m_2^2 + 2E_1m_2$$

Alternatively, this can be rearranged to:

$$E_1 = \left(s - m_2^2 - m_1^2\right)/2m_2,$$

which gives the required energy of the incoming particle to reach a certain target c.m. energy at the collision.

If the two particles are protons, and a proton anti-proton pair is to be created (each having a mass m_p), then $s = (4m_p)^2$, where $2m_p$ comes from the beam and target protons and other $2m_p$ from a newly created pair. The total momentum is zero at the c.m. frame. That gives: $E_1 \approx 7m_p = 6.6$ GeV. If the target is a nucleus instead of a bare proton, then protons due to their uncertainty in position within the nucleus, will necessarily have some momentum as governed by the uncertainty principle. That will make the three-momentum in the detector frame somewhat smaller.

- Collider experiment: when colliding particles with masses m_1 and m_2 have energies E_1 and E_2, the c.m. energy is:

$$\sqrt{s} = E_1 + E_2$$

Equating the momentum magnitudes in the c.m. frame,

$$E_1 - m_1^2 = E_2 - m_2^2$$

or

$$(E_1 + E_2)(E_1 - E_2) = m_1^2 - m_2^2$$

it follows that:

$$E_1^2 - E_2^2 = \left(m_1^2 - m_2^2\right)\big/\sqrt{s}$$

Adding the two,

$$E_1 = \sqrt{s}\big/2 + \left(m_1^2 - m_2^2\right)\big/2\sqrt{s}$$

Now for the fixed target case before, if $\sqrt{s} \gg m_1, m_2$, we get $E_1 \approx s/2m_2$, and for the colliding case here, we get, $E_1 \approx \sqrt{s}/2$. That means the available energy in the collider set-up is more, as energy is wasted as the kinetic energy of the created particle in the c.m. frame for fixed target, where it is at rest in the c.m. frame.

- Decays:

For a heavy particle $(M, 0)$ (in the c.m. frame, therefore the momentum is zero) decaying into two particles with energies and three-momenta given by $(E_{a/b}, \pm\vec{p}\,)$:

$$M^2 = E_a + E_b$$

$$E_a = \sqrt{m_a^2 + p^2} = \frac{M^2 - \left(m_a^2 - m_b^2\right)}{2M}$$

$$E_b = \sqrt{m_b^2 + p^2} = \frac{M^2 - \left(m_b^2 - m_b^2\right)}{2M}$$

From this, we can get:

$$p = \frac{\sqrt{M^2 - \left(m_a^2 - m_b^2\right)}\sqrt{M^2 - \left(m_a^2 + m_b^2\right)}}{2M}$$

Which shows, $M > m_a + m_b$. A heavy particle can decay only if the mass exceeds the sum of decay product's masses. In other words, if a particle has mass exceeding the masses of two other particles, that particle is unstable and decays unless decay is forbidden by some conservation law, i.e. conservation of charge, momentum, or angular momentum.

Here we work out an example, the decay of neutral pion to two photons. In the c.m. frame of the pion, the photons are back-to-back. We work in the x–z

plane, and assume the photons are at angles of ϕ and $-\phi$ with the z-axis. As the photons are massless, their four-momentum can be written as: $(M_\pi/2, (M_\pi/2) \sin \phi, 0, (M_\pi/2) \cos \phi)$ and $(M_\pi/2, -(M_\pi/2) \sin \phi, 0, -(M_\pi/2) \cos \phi)$, where M_π is the mass of the pion, and the energy of each photon is just half of it.

Now in the detector frame, the pion can be considered to be moving with a velocity β along the $+z$-direction. The two reference frames are shown in figure 1.3, where θ_1 and θ_2 are the angles the photons make with the z-axis in the detector frame.

If in the detector frame, the photons have energies E_1 and E_2, then their four-momentum can be written as: $(E_1, E_1 \sin \theta_1, 0, E_1 \cos \theta_1)$ and $(E_2, -E_2 \sin \theta_2, 0, E_2 \cos \theta_2)$.

Then the four-momentum in the detector frame and in c.m. frame are related by:

$$
\begin{bmatrix} E_1 \\ E_1 \sin \theta_1 \\ 0 \\ E_1 \cos \theta_1 \end{bmatrix} = \begin{bmatrix} \gamma & 0 & 0 & \gamma\beta \\ 0 & 1 & 0 & 0 \\ 0 & 0 & 1 & 0 \\ \gamma\beta & 0 & 0 & \gamma \end{bmatrix} \begin{bmatrix} M_\pi/2 \\ (M_\pi/2) \sin \phi \\ 0 \\ (M_\pi/2) \cos \phi \end{bmatrix}
$$

and

$$
\begin{bmatrix} E_2 \\ -E_2 \sin \theta_2 \\ 0 \\ E_2 \cos \theta_2 \end{bmatrix} = \begin{bmatrix} \gamma & 0 & 0 & \gamma\beta \\ 0 & 1 & 0 & 0 \\ 0 & 0 & 1 & 0 \\ \gamma\beta & 0 & 0 & \gamma \end{bmatrix} \begin{bmatrix} M_\pi/2 \\ -(M_\pi/2) \sin \phi \\ 0 \\ -(M_\pi/2) \cos \phi \end{bmatrix}
$$

corresponding to the boost of β along the z-axis, with $1 - \beta^2 = 1/\gamma^2$.

A few useful results can be obtained, derivations of which are left to the reader as an exercise. The invariant mass can be calculated in either frame of reference, and they must be identical. The invariant mass, m in the detector frame can be shown to be:

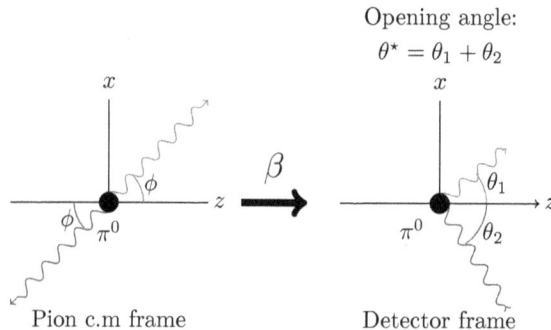

Figure 1.3. The center-of-mass and the detector reference frames for the neutral pion to diphoton decay.

$$m^2 = 2E_1E_2(1 - \cos(\theta_1 + \theta_2)) = E_1E_2 \sin^2(\theta^\star/2)$$

where $\theta^\star = \theta_1 + \theta_2$ is the opening angle between the two photons.

For large boost, and using Taylor expansion in powers of $(2/\gamma^2)$, we can show that $\theta^\star \approx (2/\gamma) \sin\phi$. Since the pion has zero spin, the angle ϕ is isotropic, resulting in no preferred direction for decay products in the detector frame. This is an important general relationship. It shows that the decay products of a highly boosted particle tend to be collimated with an opening angle that decreases as the boost factor γ increases, as shown in figure 1.4. The minimum opening angle occurs for $\phi = \pi/2$ corresponding to a symmetric decay in the detector frame and the maximum opening angle is π, corresponding to $\phi = 0$ (one photon going forward, the other going backward). The distribution of the opening angle can be seen to peak at the minimum corresponding to $\sin(\theta^\star/2) = 1/\gamma$, and vanish at the maximum value of π.

The distribution of the energy, E_γ of one of photons, dN/dE_γ can be expressed as:

$$dN/dE_\gamma = (dN/d\cos\theta)(d\cos\theta/dE_\gamma) = (1/2)(d\cos\theta/dE_\gamma)$$

where θ is taken as the angle of the photon under consideration with the z-axis in the detector frame, and the first term is half by normalising it to unity over $-1 \geqslant \cos\theta \leqslant 1$. Again, using the above mentioned Lorentz transformation, it can be shown that $dE_\gamma/d\cos\theta = (M_\pi/2)\gamma\beta$, which leads to:

$$dN/dE_\gamma = 1/P_\pi$$

where P_π is the momentum of the pion in the detector frame. This distribution is flat, with limiting values of $(E_\pi \pm P_\pi)$, where E_π is the energy of the pion in the detector frame.

The decay rate at the c.m. frame for a general two-body decay is given by $W = 1/\tau$, where τ is the mean lifetime. For short-lived particles, a related quantity is the width Γ. It can be interpreted from quantum field theory (QFT), or from experimental observations. In QFT, the probability amplitude for the decay of a particle with energy E and mass M is given by the so-called Breit–Wigner relation [35]:

Decay of a particle:

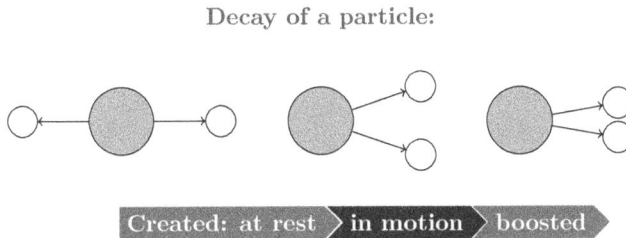

Created: at rest ⟩ in motion ⟩ boosted

Figure 1.4. Illustration of orientation of the decay products from a two-body decay. The more the energy of the initial particle is, the less the angular separation of the decay products will be. Boost refers to the transverse energy of the initial particle.

$$P = \frac{1}{(E^2 - M^2)^2 - M^2\Gamma^2}$$

This means that we expect to see a narrow spike in the mass distribution of a particle constructed from the invariant mass of the decay products. Experimentally, however, a bell-shaped curve is obtained, as shown in figure 1.5. This distribution is often referred to as the *lineshape*. The width not only comes from potential experimental mis-measurements, but from the Breit–Wigner width of the decay. Another way of saying this is: it is the manifestation of uncertainty principle, $\Delta E \Delta t > \hbar$, where mass is related to the energy E, and the width is related to the time t. The width of the distribution at half height is Γ, which also represents the uncertainty in mass. Measuring the lineshape of a resonance therefore corresponds to measuring the mass (i.e. the value for which the production rate is maximum), the width of the Breit–Wigner around that mass (according to the above formula), and the production rate at the peak itself.

> Probability of a particle to decay to a particular channel is called the branching fraction.

For a particle with several different modes of decay, one can also define partial decay rates or widths or individual modes, termed *branching fraction*, $BF = \Gamma_{process}/\Gamma$. The total width is the sum of all partial widths.

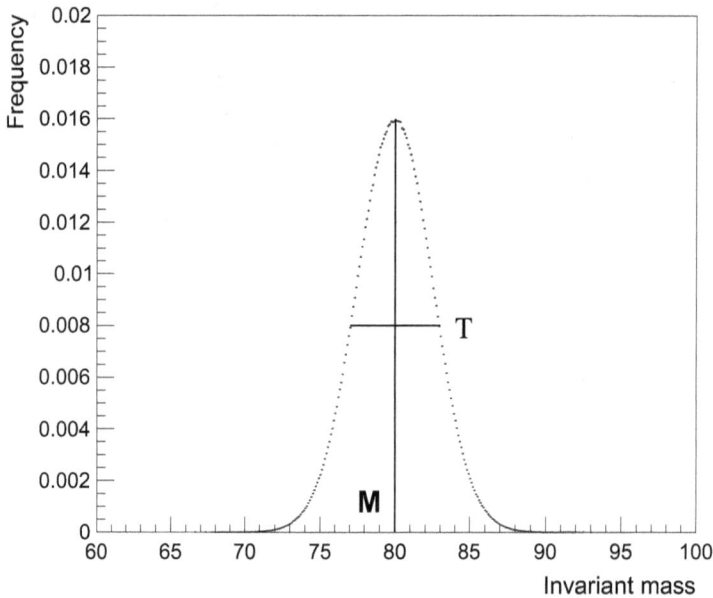

Figure 1.5. Illustration of the finite width of the invariant mass distribution around the actual mass M of a particle, constructed from its decay products. The width of the distribution at half the height is denoted by Γ.

Exercises

1. While defining natural units, we expressed energy in terms of electron-volt. What would the problem be if we tried to express energy in terms of $\hbar$, and derive the other units? Express length and mass in terms of natural units.

2. It can be seen from section 1.2, that all the colliders mentioned had (at least) two experiments (corresponding to recoding data independently in two detectors). SPS at CERN had UA1 and U1, Tevatron at Fermilab had CDF and DØ, and LHC has ATLAS and CMS (focussing on general purpose ones). The HERA experiment at DESY, which we have not talked about, made fundamental contributions to the field as well, and the main experiments were H1 and ZEUS. The Relativistic Heavy Ion Collider (RHIC) at Brookhaven National Laboratory (BNL) in the USA also has STAR and PHENIX experiments. The question is, why?

3. What prevents anti-matter from being dark matter?

4. We have two photons with (E, p_x, p_y, p_z) given by:
   Photon1: 109.38, −41.85, −9.29, −100.74 GeV.
   Photon2: 40.21, 34.10, 8.35, 19.49 GeV.
   Find the invariant mass of the combined object. What do you think was the original particle that decayed to these two photons?

5. If the energy of a proton when it undergoes a collision at the LHC is 6.5 TeV, what is its speed in terms of the speed of light? What is the energy seen by one proton for another?

6. If the Z boson mass and decay width are measured to be 90.1 GeV and 2.5 GeV, what is the lifetime in seconds?

7. Why can a single photon not produce an electron–positron pair? How does this process, termed pair production, take place?

8. Show that the decay $e^- \to e^- + \gamma$ is kinematically not allowed.

9. The Ω^{-1} baryon was predicted by Murray Gell-Mann. It is produced by the strong interaction process:

$$K^- p \to \Omega^{-1} K^+ K^0$$

The *strangeness* of a particle can be defined by number of strange quarks making up a hadron, with anti-strange quarks resulting in negative strangeness. In order for this process not to violate strangeness conservation, what is the strangeness of Ω^{-1}? Can it decay via strong interaction to the following?

$$\Omega^{-1} \to \overline{K^0} + \Xi$$

Check all the conservation laws you can think of. PDG [36] is your friend.

10. Is the decay of a neutral pion to one or three photons possible?
    Is the decay: $\Lambda^0 \to p + \pi^-$ possible (where the p stands for proton)?

References

[1] Halzen F and Martin A 1984 *Quarks & Leptons: An Introductory Course in Modern Particle Physics* (New York: Wiley)

[2] Keith Ellis R, James Stirling W and Webber B R 1996 QCD and collider physics *Camb. Monogr. Part. Phys. Nucl. Phys. Cosmol.* **8** 1–435

[3] Peskin M E and Schroeder D V 1995 *An Introduction to Quantum Field Theory* (Reading, MA: Addison-Wesley)

[4] Schwartz M D 2018 *TASI Lectures on Collider Physics* 65–100 ch 2

[5] Pruneau C A 2017 *Data Analysis Techniques for Physical Scientists* (Cambridge: Cambridge University Press)

[6] Campbell J, Huston J and Krauss F 2018 *The Black Book of Quantum Chromodynamics: A Primer for the LHC Era* (Oxford: Oxford University Press)

[7] Larkoski A J 2017 *Elementary Particle Physics: An Intuitive Introduction* (New York: Cambridge University Press)

[8] Lyons L 1986 *Statistics for Nuclear and Particle Physicists* (Cambridge: Cambridge University Press)

[9] Cowan G 1998 *Statistical Data Analysis* (New York: Oxford University Press)

[10] Behnke O, Kröninger K, Schott G and Schörner-Sadenius T 2013 *Data Analysis in High Energy Physics: A Practical Guide to Statistical Methods* (Weinheim: Wiley)

[11] Gaillard M K, Grannis P D and Sciulli F J 1999 The Standard model of particle physics *Rev. Mod. Phys.* **71** S96–111

[12] *What is the world made of?* http://www.fnal.gov/pub/inquiring/matter/madeof/index.html (accessed: 2018-12-30)

[13] Aaij R *et al* 2016 Evidence for exotic hadron contributions to $\Lambda_b^0 \to J/\psi p \pi^-$ decays *Phys. Rev. Lett.* **117** 082003 [Addendum: 2017 *Phys. Rev. Lett.* **118** 119901].

[14] Higgs P W 1964 Broken symmetries, massless particles and gauge fields *Phys. Lett.* **12** 132–33

[15] Higgs P W 1964 Broken symmetries and the masses of gauge bosons *Phys. Rev. Lett.* **13** 508–9 [160 (1964)]

[16] Englert F and Brout R 1964 Broken symmetry and the mass of gauge vector mesons *Phys. Rev. Lett.* **13** 321–3 [157 (1964)]

[17] Guralnik G S, Hagen C R and Kibble T W B 1964 Global conservation laws and massless particles *Phys. Rev. Lett.* **13** 585–7 [162 (1964)]

[18] Anderson P W 1963 Plasmons, gauge invariance, and mass *Phys. Rev.* **130** 439–42

[19] Lederman L and Teresi D 1993 *The God Particle: If the Universe Is the Answer, What Is the Question?* (New York: Dell Publishing)

[20] Arnison G *et al* 1983 Experimental observation of isolated large transverse energy electrons with associated missing energy at $s^{(1/2)} = 540$ GeV *Phys. Lett. B* **122** 103–16

[21] Arnison G *et al* 1983 Experimental observation of lepton pairs of invariant mass around 95 GeV c^{-2} at the CERN SPS collider *Phys. Lett. B* **126** 398–410

[22] Bagnaia P *et al* 1983 Evidence for Z0 $\to e^+e^-$ at the CERN anti-p p collider *Phys. Lett. B* **129** 130–40

[23] Banner M *et al* 1983 Observation of single isolated electrons of high transverse momentum in events with missing transverse energy at the CERN anti-p p collider *Phys. Lett. B* **122** 476–85

[24] Taubes G 1986 *Nobel Dreams: Power, Deceit, and the Ultimate Experiment* (New York: Random House)

[25] Herb S W *et al* 1977 Observation of a dimuon resonance at 9.5 gev in 400 gev proton-nucleus collisions *Phys. Rev. Lett.* **39** 252–5

[26] Abe F *et al* 1995 Observation of top quark production in $\bar{p}p$ collisions *Phys. Rev. Lett.* **74** 2626–31

[27] Abachi S *et al* 1995 Observation of the top quark *Phys. Rev. Lett.* **74** 2632–37

[28] Aad G *et al* 2012 Observation of a new particle in the search for the Standard Model Higgs boson with the ATLAS detector at the LHC *Phys. Lett. B* **716** 1–29

[29] Chatrchyan S *et al* 2012 Observation of a new boson at a mass of 125 GeV with the CMS experiment at the LHC *Phys. Lett. B* **716** 30–61

[30] Butterworth J 2015 *Smashing Physics: Inside the World's Biggest Experiment* (London: Headline Publishing)

[31] *W and z particles discovered* https://timeline.web.cern.ch/events/w-and-z-particles-discovered (accessed: 2018-12-30).

[32] *Naturalness* http://www.symmetrymagazine.org/article/april-2013/naturalness (accessed: 018-12-30).

[33] Alavi-Harati A *et al* 2003 Measurements of direct CP violation, CPT symmetry, and other parameters in the neutral kaon system *Phys. Rev. D* **67** 012005
Alavi-Harati A *et al* 2004 Measurements of direct CP violation, CPT symmetry, and other parameters in the neutral kaon system *Phys. Rev. D* **70** 079904 Erratum

[34] Aaij R *et al* 2019 Observation of *CP* violation in charm decays *Phys. Rev. Lett.* **122** 211803

[35] Breit G and Wigner E 1936 Capture of slow neutrons *Phys. Rev.* **49** 519–31

[36] *Particle data group* http://pdg.lbl.gov/ (accessed: 2018-12-30).

Chapter 2

Collisions

Despite my resistance to hyperbole, the LHC belongs to a world that can only be described with superlatives. It is not merely large: the LHC is the biggest machine ever built. It is not merely cold: the 1.9 kelvin (1.9 degrees Celsius above absolute zero) temperature necessary for the LHC's supercomputing magnets to operate is the coldest extended region that we know of in the Universe—even colder than outer space. The magnetic field is not merely big: the superconducting dipole magnets generating a magnetic field more than 100,000 times stronger than the Earth's are the strongest magnets in industrial production ever made.

And the extremes don't end there. The vacuum inside the proton-containing tubes, a 10 trillionth of an atmosphere, is the most complete vacuum over the largest region ever produced. The energy of the collisions are the highest ever generated on Earth, allowing us to study the interactions that occurred in the early Universe the furthest back in time.

—Lisa Randall

We start this chapter by briefly describing how collisions are made to happen at the LHC in section 2.1. Then in section 2.2, we introduce the concept of cross-section and luminosity as measures of collisions. In section 2.3 we describe the coordinates used, along with the motivation for that choice. Then in section 2.4, we describe different classes of collisions such as elastic, inelastic, and diffractive. Finally the chapter is wrapped up with the definitions of hard scatter, minimum bias, underlying event and pile-up, and pointing out the connection between them.

2.1 Effecting collisions

Protons are collided against protons at four collisions points along the Large Hadron Collider (LHC) ring (27 km in circumference), as shown in figure 2.1.

doi:10.1088/2053-2563/ab1be6ch2

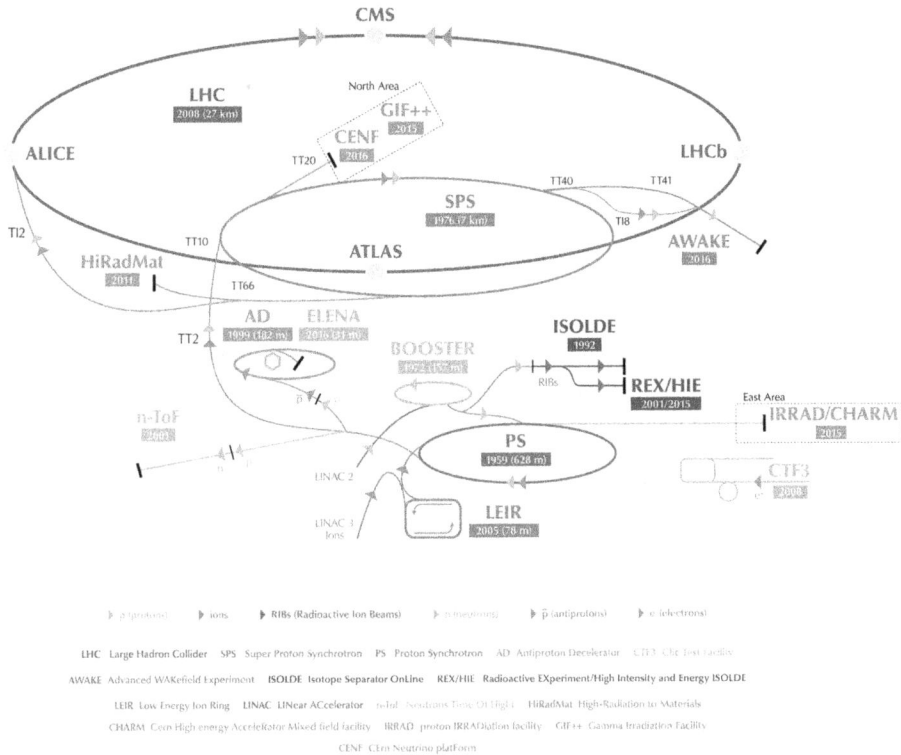

Figure 2.1. An overview of CERN's accelerator complex layout [1] (© 2018 CERN). It consists of a succession of machines that accelerate particles to increasingly higher energies, before injecting them into the LHC. Linac 2, the first accelerator in the chain, accelerates the protons to the energy of 50 MeV. The beam is then injected into the proton synchrotron booster (PSB), which accelerates the protons to 1.4 GeV, followed by the proton synchrotron (PS), which pushes the beam to 25 GeV. Protons are then sent to the super proton synchrotron (SPS) where they are accelerated to 450 GeV.

Protons are obtained by applying an electric field to hydrogen gas contained in a metal cylinder, which disintegrates the hydrogen atoms. The protons are then accelerated in stages by using a series of interconnected linear and circular accelerators, radio frequency quadrupole (QRF), linear accelerator (LINAC), proton synchrotron booster (PSB), proton synchrotron (PS) and super proton synchrotron (SPS), before being injected into the LHC ring. At the LHC, protons are injected both in a clockwise and anticlockwise direction, in bunches. The LHC is designed to run with 2808 bunches per beam (each containing ≈100 000 million protons), separated by a gap of just 25 ns. That means a collision rate of 40 MHz, or about 600 million collisions per second. So far the highest energy of each beam has been 6.5 TeV (as opposed to the design luminosity of 7 TeV per beam).

The beams are kept on their trajectory by superconducting (at a temperature of 1.9 K) dipole magnets placed along the path (of 8.3 T magnetic field), and the quadrupole magnets squeeze the beams to make them narrower as they enter the detectors. After colliding, the particle beams are separated again by dipole magnets.

The beam areas can be represented as ellipses. The beam size then is expressed in terms of two quantities, transverse emittance ϵ, and amplitude function β, with the beam area given by $\sigma = \sqrt{\epsilon\beta}$. Emittance is the area containing a certain fraction of beam particles, so it indicates how confined the beam particles are. It is usually constant along the trajectory. The β reduces to the aspect ratio for upright ellipses. If β is low, the beam is narrower, or 'squeezed'. If β is high, the beam is wide and straight. Typically, β's are different for two perpendicular directions in the plane transverse to the beam direction. The value of the β at an interaction point is termed $\beta^\star$. It is typically adjusted to have a local minimum there, in order to minimise the beam size and thus maximise the interaction rate.

A *fill* at the LHC corresponds to one injection to depletion of the beam. A typical fill usually consists of the following stages:

1. Injection: circulating beams are inserted, corresponding to $\beta^\star$ of 30 cm.
2. Ramp: inside the LHC, the beams continue to accelerate. This takes about 20 min.
3. Flat top: beams have been accelerated to the desired energy.
4. Squeeze: beams are made narrower, corresponding to $\beta^\star$ of 25 cm.
5. Stable beams: ready for physics data-taking, about 10 hours for beams to get exhausted in normal running condition.
6. Dump: beams are brought into a gap out of the LHC (planned or in case of an emergency).

For experiments recording data, a *run* is a discrete interval of data-taking. There can be many runs per fill, corresponding to changes in trigger configuration (the concept of triggering will be introduced in the next chapter) or due to enabling or disabling of a subdetector. Even if there is no beam, but the detector is on for calibration purposes (mostly using cosmic rays), the recorded data is assigned run numbers. However, in the context of the LHC, a *run* is a period between *long shutdowns*. Run 1 corresponds to the start of LHC in 2009–13, followed by long shutdown 1 (LS1), while 2015–18 spans Run 2. Run 3 is scheduled start in 2020, following long shutdown 2 (LS2) in 2019. The long shutdowns are utilised for maintenance and upgrade of the LHC machines and detectors.

2.2 Measure of collisions

One collision is referred to as an event.

We collect a large number of events, and thus we need to know the *rates* of different types of events, i.e. corresponding to the production of different final states which can happen in a collision. Quantum mechanics dictates that the outcome of any specific collision cannot be known with certainty, and in principle we do not have an exhaustive list of possible final states (if we did, we would not expect to see a

Proton-proton overlap:

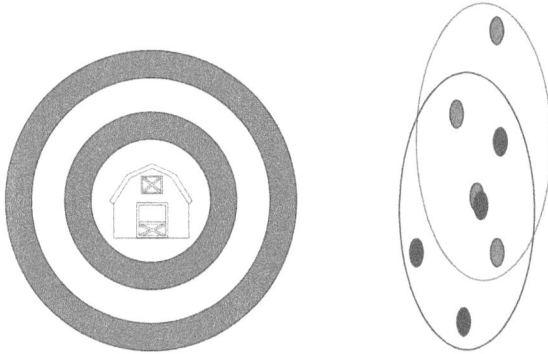

Figure 2.2. Dramatic illustration of cross-section as a target (left) and overlap between two colliding protons (right). The image of a barn is superposed inside the target as a reminder of the origin of the unit of cross-section. The red and blue filled circles indicate the partons inside the red and blue protons, with one actual parton–parton interaction happening in the middle. The proton shapes are elongated because of Lorenz contraction.

new particle!). Instead, we try to find an average rate, or a relative probability of specific processes. The term *cross-section* is introduced as a measure.

In the simplest form, cross-section measures how often a particular process occurs, or the probability of it. It is measured by considering an effective area *available* for that particular process, a concept that came from fixed target experiments. In fixed target experiments (Rutherford's α-particle scattering experiment is an example), a beam particle hits a target particle or an extended object containing many of the target particles. The cross-section is then interpreted as the effective area that a target presents to the incoming particle. A large target with target particles covering a significant fraction of it indicates the interaction is highly probable, whereas if the interaction is very rare, it is thought of as if the target is small.

The cross-section is measured in terms of barns, which is a unit of area, equal to 10^{-28} m^2. During the Manhattan project, physicists who were bouncing neutrons off uranium nuclei described the uranium nucleus as 'as big as a barn' (barns are large buildings in farms to keep livestock or grain, shown in figure 2.2 (left)), which is the origin of this strange name [2]. In particle physics, usually prefixes are used as the cross-section is much smaller. For easy reference, a conversion table is given in table 2.1.

In hadron colliders (or in general in colliding beam experiments), due to their high speed, the protons are Lorentz contracted longitudinally (the transverse component is not affected). That leads to effectively flat, oval-sized protons colliding in the centre-of-mass frame, as shown in figure 2.2 (right). So the same concept of overlapping areas applies for estimating cross-sections. The particles in each beam are stored in *bunches*, which are made to arrive at the collision point with a certain frequency. The collision rate thus depends on the numbers of particles in the bunches, the collision frequency and the size of the bunches. The smaller the bunch,

Table 2.1. Conversion factors between different commonly used prefixes to denote luminosity or cross-section.

Unit		Prefix	
Barn (b)		1	
Unit	Prefix	Unit	Prefix
Mili (mb)	10^{-3}	Pico (pb)	10^{-12}
Micro (μb)	10^{-6}	Femto (fb)	10^{-15}
Nano (nb)	10^{-9}	Atto (ab)	10^{-18}

the higher the rate of collision, since the particles in a small bunch are packed more closely together. Referring to them as beam1 and beam2, the scattering cross-section σ is the effective size of a single particle in beam1, as 'seen' by a particle in beam2. So the probability of a collision, $P_{collision}$ is the fraction of the beam1 area that is actually occupied by the beam1 particles, so in terms of the area of beam1, A_1, and number of particles in the bunch, n_1,

$$P_{collision} = \frac{\sigma}{A_1/n_1}$$

The total number of collisions, $N_{collision}$, is this probability times the number of ways a collision can happen, namely the number of particles, n_2 in the bunch in beam2.

$$N_{collision} = \frac{n_2\sigma}{A_1/n_1} = \sigma\frac{n_1 n_2}{A_1}$$

As this happens every time two bunches cross, $N_{collision}$ is multiplied by the rate of bunch crossing, f to obtain the overall rate at which collisions occur:

$$R_{collision} = \sigma f \frac{n_1 n_2}{A_1}$$

The quantity $f(n_1 n_2)/A_1$ is called instantaneous luminosity, L. It gives a measure of collision rate[1]. It can be thought of as the 'brightness' of the source of the colliding particles. The integrated luminosity is defined to be the instantaneous luminosity integrated over time.

This division is convenient because the first factor depends only on the physical process being considered, and the second factor depends only on the design of the accelerator. The second factor is a good way to characterise the production capacity of a given accelerator. The units of luminosity is inverse of area, and the inverse femtobarn turns out to be roughly the right magnitude for measuring integrated luminosity at current particle accelerators. The luminosity effectively corresponds to the inverse cross-section per second produced in the experiments in hadron collisions.

---

[1] Strictly speaking it measures how many particles are put in a position to collide.

> Cross-section refers to the probability of a specific process occurring, expressed as the number of events. Integrated luminosity is a measure of the total number of events.

To find the actual number of collisions for any particular process, N_{process}, we multiply luminosity by the cross-section for that process, $\sigma_{\mathrm{process}}$, and integrate over time,

$$N_{\mathrm{process}} = \sigma_{\mathrm{process}} \int L \mathrm{d}t$$

Measurement of luminosity is important, as it relates to the measurement of cross-sections. Dedicated instruments can be used to measure it, examples are the BCM (Beam Conditions Monitor) and LUCID (LUminosity measurement using a Cherenkov Integrating Detector) in ATLAS. It is also measured indirectly (which is the most common method at the LHC experiments), where the visible cross-section for a reference process is measured, and scaled appropriately. Another standard way to determine the luminosity is via elastic scattering at small angles, where the optical theorem connects the elastic scattering amplitude in the forward direction to the total cross-section. This method requires very specific beam conditions during special high-β runs.

Perhaps the most common method is *van der Meer (vdM) scans*, named after the inventor [3], where the absolute luminosity is determined using only machine parameters in dedicated runs. The luminosity is obtained from the effective height of the colliding beams, which are determined by observing the counting rate (i.e. the average number of visible interactions per bunch crossing) in a suitable detector while scanning one of the two beams in the vertical plane with respect to the other one (i.e. using the effective widths of the beam overlap region in the x–y plane).

To get a sense of these numbers at the LHC, we can look at some examples. The current bunch crossing rate of 40 MHz (40 million per second) and an average peak luminosity of 2×10^{34} cm^{-2} s^{-1} has been achieved at the LHC. This yields over 1 billion collisions per second, out of which only a tiny fraction can be recorded (discussed in the next chapter), at a rate of about 10–20 kHz, corresponding to about 100–200 potentially interesting events per second. The total integrated luminosity collected by the LHC experiments at the end of Run 2 is about 150 fb^{-1}, and the progression of data accumulation with time is shown in figure 2.3 (left).

The next chapter will describe what information from each event is saved, but in general if an event size is 2–10 MB, then an inverse femtobarn of data means roughly 10^{12} recorded collisions, and over 500–2000 TB in size.

The cross-sections for interesting physics processes tend to be very small. In table 2.2, we list rough values of production cross-sections for some SM processes to give an order of magnitude idea. The production processes of these particles will be discussed in chapter 4. The values are obtained either from the latest LHC measurements [5], or from relevant LHC cross-section working group recommendations [6]. They are presented without measurement or theoretical uncertainties.

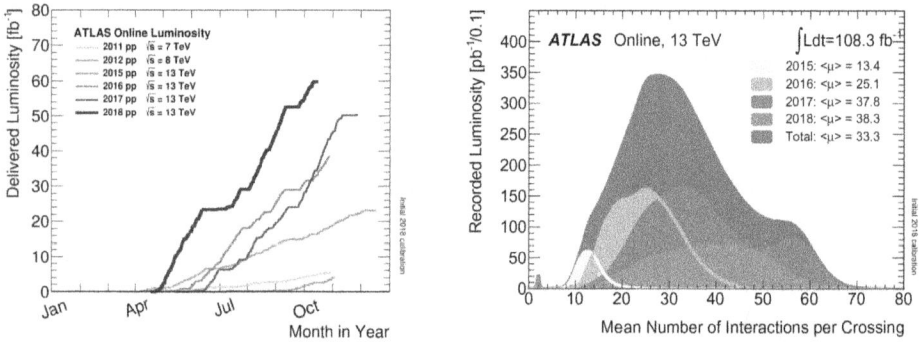

Figure 2.3. The recorded integrated luminosity (left), and the mean number of interactions per bunch crossing (right), recorded [4] by the ATLAS detector in different years (ATLAS Experiment © 2018 CERN).

Table 2.2. Production cross-sections of some commonly occurring or interesting processes at 13 TeV centre-of-mass energy proton–proton collisions. The most dominant channel in each case is shown. The values are obtained either from the latest LHC measurements [5], or from relevant LHC cross-section working group recommendations [6].

Process		Cross-section (in pb)	
Total inelastic		8×10^{10}	
Process	Cross-section (in pb)	Process	Cross-section (in pb)
W^+, W^-	$1.2 \times 10^4, 8.8 \times 10^3$	Wt	94
Z	2×10^3	H	43
$t\bar{t}$	820	WZ	40
Single $t, \bar{t}$	136, 80	ZZ	15
W^+W^-	142	HH	0.03

As mentioned earlier, a single bunch crossing leads to multiple proton–proton collisions in the same bunch crossing. The detector resolution is not often enough to distinguish between these multiple collisions in a single bunch crossing or even between separate collisions (usually the ones just before and after). This contamination is termed pile-up, with the former termed *in-time* pile-up, and the latter *out-of-time* pile-up.

> Pile-up refers to the additional proton–proton collisions in the same interaction, contaminating the event with extra particles.

Out-of-time pile-up is parametrised in terms of the parameter μ, which denotes the average number of additional proton–proton collisions per bunch crossing.

A measure of the in-time pile-up is given by the parameter N_{vtx}, which is the average number of reconstructed primary vertices in the event. Pile-up interactions are basically reflective of inclusive collisions, with high-p_T objects produced in a small fraction of events. In figure 2.3 (right), the distribution of average μ over the years is shown, which indicates the pile-up increases with increased luminosity. At the LHC, we have had mean number of interactions per crossing peaking at 54 already.

2.3 Coordinates

The usual coordinate systems are the Cartesian (x, y, z) or spherical (r, θ, ϕ). In the hadron colliders, the energies and, therefore, the momenta of the incoming hadrons are known but the energies and momentum fractions of the respective constituents that interact are not known *a priori*. We measure the particles in the detector to characterise the collisions. Therefore, it is convenient to use a modified form of spherical coordinate system, suited to the geometry of the detectors.

The positive z-axis lies along the incident proton beam direction, as the beam-axis is uniquely defined. The origin of the coordinate system is fixed where the beams collide, since that is usually a well defined point at the centre of the detector. The angle ϕ is the usual azimuthal angle around the beam-axis. The transverse momentum $p_T \equiv \sqrt{p_x^2 + p_y^2}$ which is the component of momentum in the transverse plane is used, as only components along the x–y plane is measured. This implies $p_x = p_T \cos \phi$ and $p_y = p_T \sin \phi$. Similarly, transverse mass is defined as $m_T \equiv \sqrt{m^2 + p_T^2}$, which is invariant under longitudinal boosts as well.

Then rather than the polar angle θ, a new quantity, rapidity, y is introduced[2], and defined in terms of the particle kinematics:

$$\cosh y \equiv \gamma, \ \sinh y \equiv \beta\gamma \Rightarrow \tanh y = \beta$$

which can be equivalently written as:

$$y = \frac{1}{2} \ln \left(\frac{E + p_z}{E - p_z} \right)$$

where E and p_z are respectively the energy and z-component of the momentum of the particle being considered. The detector geometry is cylindrical, and each detector element covers the same area in y–ϕ space, hence the y–ϕ coordinates exploit the cylindrical symmetry better. So y can be directly mapped to detector geometry.

The difference between the rapidities of two particles is invariant with respect to Lorentz boosts along the z-axis, which corresponds to the transformation between the reference frame connected to the detector (detector-frame) and the centre-of-

---

[2] This actually corresponds to a transformation of the coordinates of the form: $x' = x \cosh y + t \sinh y$ and $t' = x \sinh y + t \cosh y$, corresponding to an imaginary angle of rotation.

mass frame. Any boost is parametrised by a boost parameter γ and by defining an axis. Energy and three-momentum along this axis (here for obvious reasons the z-axis) then change according to:

$$E' = E \cosh \gamma - p_z \sinh \gamma$$
$$p'_z = p_z \cosh \gamma - E \sinh \gamma$$

Hence,

$$y' = \frac{1}{2} \log \frac{E' + p'_z}{E' - p'_z} = \frac{1}{2} \log \frac{(E + p_z)(\cosh \gamma - \sinh \gamma)}{(E - p_z)(\cosh \gamma + \sinh \gamma)}$$
$$= y + \frac{1}{2} \log \frac{e^{-\gamma}}{e^{\gamma}} = y - \gamma$$

The reason that this is important is because it is often the case that one of the individual quarks or gluons involved in the collision may have a lot more momentum than the other, so all the particles produced come out near one end of the detector. However, by moving from detector-frame to centre-of-mass frame, the particles come out symmetrically distributed, which makes our calculations easier. This choice of coordinates results in having the angular information interleaved with the energy information in one single coordinate y.

A way to visualise rapidity is to consider that the hyperbolic-tangent of rapidity yields the ordinary spatial velocity, $v = \tanh y$. The rapidity is zero when $v = 0$, while the rapidity becomes infinite if $v = c$. In other words, the rapidity is zero when a particle is close to transverse to the beam-axis, but tends to infinity when a particle is moving close to the beam-axis in either direction.

However, the rapidity can be hard to measure for highly relativistic particles. The z-component of the momentum is usually not measured in the detectors, as it is along the beam direction. Therefore, another quantity has been introduced, called the pseudorapidity, which is commonly denoted by η. The pseudorapidity can be obtained by measuring only the polar angle θ

$$\eta \equiv -\ln \tan \theta/2$$

The reason for the factor of half is because it scales both the angular coordinates in the same way. Then a jet at $\theta = \pi/2$ (where the resolution is the best) should have the same behaviour in physical space as in η–ϕ space.

This detector coordinate system is shown in figure 2.4, along with the conversion from polar angle to pseudorapidity. The pseudorapidity values directly indicate how far from the centre of the detector (i.e. the detector element surrounding the collision points) a particle is being produced. The pseudorapidity values range from minus to plus infinity within the detector geometry.

For massless particles (or particles with very high energies with $p \gg m$), rapidity and pseudorapidity are identical (where p is the magnitude of three-momentum: $p = \sqrt{p_x^2 + p_y^2 + p_z^2}$). This can be seen from:

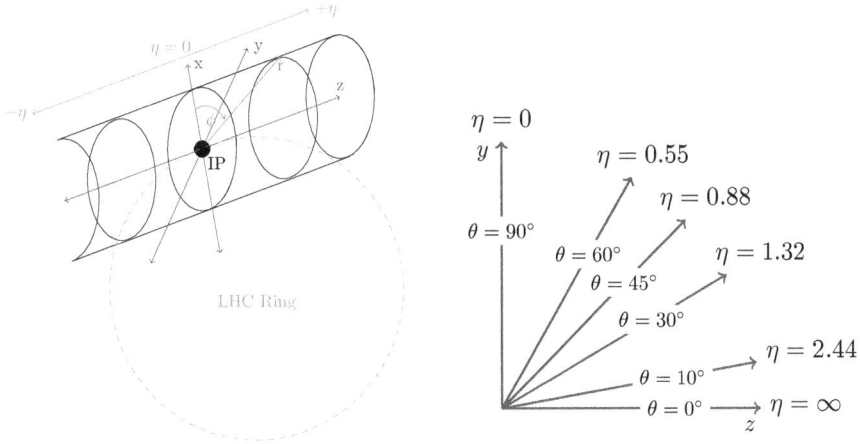

Figure 2.4. Illustration of the coordinate system used in LHC experiments (left), and a conversion between the polar angle and psuedorapidity (right). The z-axis is along the incoming proton beam's direction, and the azimuthal angle ϕ is defined with respect to the x–y plane, as the usual practice. The pseudorapidity η is defined to exploit the cylindrical symmetry, and it increases in value as we go away from the centre of the detector.

$$y = \frac{1}{2} \ln \frac{E + p_z}{E - p_z} = \frac{1}{2} \ln \frac{\sqrt{p^2 + m^2} + p \cos \theta}{\sqrt{p^2 + m^2} - p \cos \theta}$$

$$= \frac{1}{2} \ln \frac{p + p \cos \theta}{p - p \cos \theta} = \ln \frac{(1 + \cos \theta)^{1/2}}{(1 - \cos \theta)^{1/2}}$$

$$= -\ln \tan \theta/2 = \eta$$

Pseudorapidity can be further expressed in terms of momentum:

$$\eta = -\ln \tan \theta \Big/ 2 = \frac{1}{2} \ln \frac{p + p \cos \theta}{p - p \cos \theta}$$

$$= \frac{1}{2} \ln \frac{p + p_z}{p - p_z}$$

Or momentum can be expressed in terms of pseudorapidity:

$$p = p_T \cosh \eta$$

by summing:

$$e^\eta = \sqrt{\frac{p + p_z}{p - p_z}}$$

$$e^{-\eta} = \sqrt{\frac{p - p_z}{p + p_z}}$$

and using $p_T = \sqrt{p^2 - p_z^2}$. Similarly, by subtraction it is easy to show: $p_z = p_T \sinh \eta$. In contrast, for massive particles (or jets), pseudorapidity and rapidity do not coincide (in fact it can be shown that $|\eta| > |y|$), and rapidity should be used.

Rapidity (or pseudorapidity) is often paired with the azimuthal angle ϕ at which a particle is emitted, so that the angle of emission of a particle from an interaction point is often given as the coordinate pair (y, ϕ). This way, the angular separation of two events, $(\eta_2 - \eta_1, \phi_2 - \phi_1)$ is invariant with respect to boosts along the beam-axis. This separation is usually expressed in terms of $\Delta R = \sqrt{\Delta\eta^2 + \Delta\phi^2}$. This is equivalent to the standard opening angle in spherical coordinates:

$$d\Omega^2 = d\theta^2 + \sin^2(\theta)d\phi^2 = (1/\cosh^2\eta)(d\eta^2 + d\phi^2)$$
$$= (1/\cosh^2\eta)dR^2$$

by using $\sin\theta = 1/\cosh\eta$, for massless particles. It also follows, that for massless particles:

$$E = p_T \cosh\eta$$
$$E\Delta\Omega = p_T\Delta R,$$

which establish the previous assertion.

> The position of a particle in the detector is expressed in terms of azimuthal angle ϕ and the rapidity y or the pseudorapidity η.

2.4 Types of collisions

When protons are brought to collision at the centre of the detectors[3], effectively the quarks and gluons from each proton (collectively referred to as partons) interact. The proton consists of two up type and one down type quark, and these are termed the *valence quarks*. Gluons are present to carry the strong interaction force. Additionally, quantum fluctuations lead to creation of other quark–anti-quark pairs inside the proton, which are termed the *virtual* or *sea quarks*.

> Proton–proton collisions are effectively the interaction of partons from each proton.

Two protons can collide elastically or inelastically, analogous to collisions described by classical mechanics. For elastic collisions, which happens most of the time, protons do not dissociate, and there are no new particles created or energy lost. The protons then go down the beam-pipe, and are not detected by ATLAS and CMS. For inelastic collisions, one or both protons have a change in energy and direction,

---

[3] Each of the protons also contain photons, although their energy is usually at sub-percent level. However, that means that the LHC also collides photons against photons, at a very small rate.

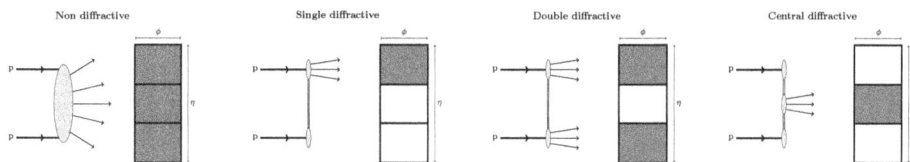

Figure 2.5. Schematic illustration of non-diffractive, single-diffractive, double-diffractive and central-diffractive processes (left to right). The shaded areas in the rectangular representation of the detector in η–ϕ plane indicate if outgoing particles are observed there.

and this leads to their dissociation and the creation of new particles. From a collider physics perspective, we are only interested in inelastic collisions. There is no way for us to look at the collisions directly, but experimental evidence leads us to believe that inelastic collisions can proceed in a few different ways. The non-diffractive collisions are the ones where partons from two protons interact directly, leading to the creation of new particles. The greatest amount of energy exchange is possible in this case, so these are the collisions in which we are usually most interested. The other class of inelastic collisions can be broadly categorised as diffractive. Single or double-diffractive collisions happen when one or both protons dissociate without direct interactions between them. A detailed discussion of diffraction [7, 8] is beyond the scope of this text, but these events are characterised in the detector by the absence of hadronic activity in certain angular regions of the final state phase space. In the Regge theory of strong interaction [9], diffraction is thought to be the result of exchange of a colour singlet object called the *Pomeron* with vacuum quantum numbers. The dissociation is caused by hadrons getting excited to a high mass state by absorbing the Pomeron, and then decaying. For events in which both colliding hadrons remain intact as they each emit a Pomeron, we have the so-called central diffractive events. This is known as the double Pomeron exchange (DPE) process, where both incoming hadrons are quasi-elastically scattered and the particles in the centre region are produced by Pomeron–Pomeron interaction.

Different types of collision events can be characterised in terms of where in the detector the resulting particles are observed. Figure 2.5 schematically shows the types of collisions mentioned. For a non-diffractive collision, the outgoing particles are usually spread over the whole volume of the detector. Particles emitted in diffractive interactions are mainly found at psuedorapidities close to that of the parent proton, resulting in so-called (psuedo)rapidity gaps in the detector. Therefore, it is possible to differentiate between single, double or central diffraction processes depending on the location of the rapidity gaps.

In figure 2.6, the cross-section of these different types of events at different centre-of-mass energies are shown. These estimates are obtained from a commonly used simulation programme (a topic that will be discussed in chapter 4) not from data, as precise measurements of them are extremely difficult. Experimentally it is impossible to separate them cleanly. The total cross-section for the events (i.e. number of collisions) consist of all of these.

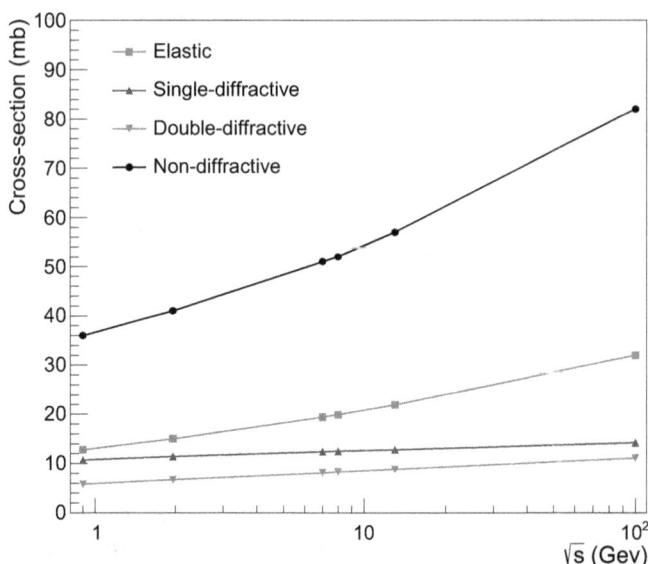

Figure 2.6. The rate of increase of cross-sections corresponding to elastic, single-diffractive, double-diffractive and non-diffractive processes are shown, as a function of centre-of-mass energy. The numbers are obtained from simulation, so they are not precise, but are useful to visualise the general trend.

$$\sigma_{total} = \sigma_{el} + \sigma_{in}$$
$$\sigma_{in} = \sigma_{sd} + \sigma_{dd} + \sigma_{nd}$$

As can be seen, the total cross-section increases gradually, mainly as a result of inelastic cross-section increasing, while diffractive cross-sections become almost flat.

Coming back to a typical inelastic, non-diffractive collision, a full event is thought to consist of not only the hard partons emitted from the collisions, but also of:

- Initial and final state radiation (I/FSR): A parton can emit further partons both before and after the QCD interaction vertex giving rise to initial and final state parton showers, respectively. Photon radiation due to QED radiations are usually also present.
- Multiple parton interactions (MPI): additional *semi-hard* 2-to-2 parton–parton scatterings in the same event. Double parton interaction (DPI) is a subset of MPI.
- Beam–beam remnants (BBR): particles that come from the partons not participating in hard scatter.

Since free partons cannot exist in nature, they need to form colour neutral hadrons. This is accomplished by splitting partons into other partons (termed fragmentation) such that they can combine to form colour neutral hadrons (termed hadronisation).

Everything except the hard scattering is generally referred to as an underlying event (UE). However, experimentally it is impossible to uniquely separate out the different components of a collision on an event-by-event basis. The initial paper

Definition of Impact Parameter (IP):

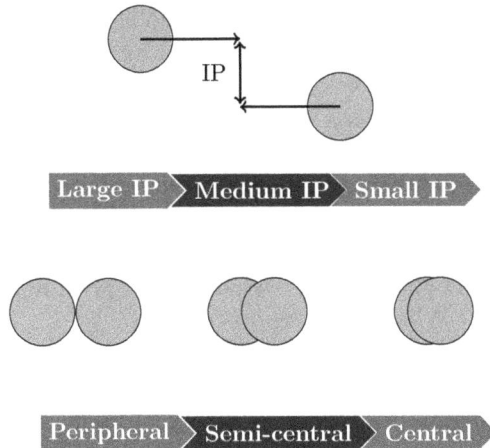

Figure 2.7. Schematic illustration of the definition of impact parameter (top) and centrality (bottom) in hadron–hadron collisions. The impact parameter indicates the amount of overlap the hadrons have, and, depending on that the collisions can be categorised as peripheral, semi-central and central.

from CDF on the subject [10] points this out: 'Of course, from a certain point of view there is no such thing as an *underlying event* in a proton–antiproton collision. There is only an *event* and one cannot say where a given particle in the event originated'.

An underlying event is all the activity in a single proton–proton collision that is not coming from the main parton–parton interaction.

Further, the collisions can either be *soft*, with less momentum transfer or *hard*, with appreciable momentum transfer. These terms are of course subjective. We are mostly interested in the hard scattering part, where significant momentum transfer leads to creation of *new* particles with appreciable p_T or mass (not new as in never observed, rather newly created in the collision). The branching fractions for most of these final states are rather low, leading to a large fraction of collisions producing soft outgoing particles. It is intuitive to understand soft and hard collisions in terms of the *impact parameter*, which is a measure of overlap of the colliding protons, as shown in figure 2.7. The overlap essentially indicates what fractional areas or volumes of protons actually participate in the collisions. This is also referred to as *centrality*. In proton–proton collisions, most hard interactions tend to be fairly central. LHC also collides heavy nuclei like lead or xenon (these runs are collectively termed heavy ion runs), and the collisions there are separated into centrality classes, based on the overlap between the colliding nuclei.

Exercises

1. The LHC is a proton–proton circular collider. Tevatron was a circular proton–antiproton collider. The proposed Superconducting Super Collider (SSC) at Texas, which was cruelly abandoned after the construction got under-way would have been a proton–proton circular collider. The previous experiment at CERN, LEP (Large Electron–Positron Collider), was a circular electron–positron collider, and so is the proposed Circular Electron Positron Collider (CEPC) in China. Another proposed experiment is the International Linear Collider (ILC), which plans to collide electrons and positrons in a linear tunnel. What are the advantages and disadvantages of these designs? What about muon colliders?

2. What limits the centre-of-mass energy of a collider, such as the LHC?

3. We mentioned in section 2.1, that bunches of protons are put in the LHC at one go in opposite direction. Why do we use such bunches, specially since they lead to pile-up?

4. The angular difference between two objects is expressed in terms of $\Delta\phi$ or ΔR, where the latter is given by, $\Delta R = \sqrt{(\Delta\phi)^2 + (\Delta\eta)^2}$. The detector spans 2π, i.e. a full circle, but it can be either expressed as between 0 to 2π, or between $-\pi$ to π. Convince yourself that the difference $\Delta\phi$ (and correspondingly ΔR) is independent of this choice. Let us take the convention of defining ϕ between $-\pi$ to π. When calculating $\Delta\phi$, if you end up with a value greater than 2π, how should you proceed?

5. We will see later that particles are specified with some minimum value of the transverse momentum, p_T and within a range of pseudorapidity, η (mostly due to limitations of our detector, as we will see in the next chapter). Why do we ignore the third coordinate, the azimuthal angle ϕ?

References

[1] *Overview of Cern's accelerator layout* https://lhc-machine-outreach.web.cern.ch/lhc-machine-outreach/lhc_in_pictures.htm (accessed: 2018-12-30).

[2] Holloway M G and Baker C P 1972 How the barn was born *Phys. Today* **25** 7

[3] van der Meer S 1968 Calibration of the effective beam height in the ISR *Technical Report* CERN-ISR-PO-68-31. ISR-PO-68-31, CERN, Geneva. http://inspirehep.net/record/1098817?ln=en

[4] *ATLAS luminosity publc results run2* https://twiki.cern.ch/twiki/bin/view/AtlasPublic/Luminosity PublicResultsRun2 (accessed: 2018-12-30).

[5] *Summary plots from the atlas standard model physics group* https://atlas.web.cern.ch/Atlas/GROUPS/PHYSICS/CombinedSummaryPlots/SM/ (accessed: 2018-12-30).

[6] *LHC Higgs cross section working group* https://twiki.cern.ch/twiki/bin/view/LHCPhysics/LHCHXSWG (accessed: 2018-12-30).

[7] Donnachie S, Dosch G, Landshoff P and Nachtmann O 2002 *Pomeron Physics and QCD Cambridge Monographs on Particle Physics, Nuclear Physics and Cosmology* (Cambridge: Cambridge University Press)

[8] Barone V and Predazzi E 2002 *High-energy Particle Diffraction Texts and Monographs in Physics* vol 565 (Berlin, Heidelberg: Springer)

[9] Collins P D B 1977 *An Introduction to Regge Theory and High Energy Physics Cambridge Monographs on Mathematical Physics* (Cambridge: Cambridge University Press)

[10] Affolder T *et al* 2002 Charged jet evolution and the underlying event in proton-antiproton collisions at 1.8 TeV *Phys. Rev. D* **65** 092002

IOP Publishing

Experimental Particle Physics
Understanding the measurements and searches at the Large Hadron Collider
Deepak Kar

Chapter 3

Analysis objects

I have done a terrible thing: I have postulated a particle that cannot be detected.
—Wolfgang Pauli

Continuing from the previous chapters, now we focus on the particles created from a collision, and how they are identified by the detector, in section 3.1. Then in section 3.2, we cover how the most ubiquitous objects, jets are reconstructed. The principle of triggering, and of calibration of various objects, including the *performance* aspects like *b*-tagging, fakes, overlap removal, isolation requirement are covered in section 3.4.

3.1 Detector objects

Particle physics detectors are designed to act like high resolution cameras (equivalent to 100 Megapixels or more) to take snapshots of the collisions, which happen when accelerated particles are made to collide at the centres of the detectors.

The ATLAS and CMS detectors both have the same physics goal [1–3]. They are referred to as 4π (or *hermetic*) detectors, as the aim is to completely surround the collision point, or cover all 4π steradians of solid angle around the collision. The presence of the beam-line and the mechanical support means there need to be small openings in order to allow them to pass through. The shapes of the detectors are cylindrical, with several layers, each wrapped around the previous one. The detectors can be categorised into two main types: the racking detectors, which reveal the path of electrically charged particles, while the calorimeters stop, absorb and measure the energy that the particles deposit. Different layers of the calorimeters are built with materials chosen and optimised to stop specific types of particles. A comparison of

doi:10.1088/2053-2563/ab1be6ch3

Table 3.1. A comparison of ATLAS and CMS detectors focussing on the barrel regions. The granularity is in pseudorapidity and azimuthal angle, for the second layer (inner three layers) of the ATLAS calorimeters. The symbols, E, p_T and d_0 respectively refer to energy, transverse momentum and transverse impact parameter, the last of which will be defined in the next section. The notation $\oplus$ indicates addition in quadrature (inspired by a similar table from [8]).

	Tracking detector			
	Magnetic field	Minimum p_T	$1/p_T$ resolution	d_0 resolution
ATLAS	2 T	0.1 GeV	$5 \times 10^{-4} \, p_T \oplus 0.01$	100–40 μm
CMS	4 T	0.2 GeV	$1.5 \times 10^{-4} \, p_T \oplus 0.005$	140–80 μm

	EM calorimeter		Hadronic calorimeter		Muon chamber
	E resolution	granularity	E resolution	granularity	p_T resolution
ATLAS	$0.1/\sqrt{E} \oplus 0.007$	0.025×0.025	$0.5/\sqrt{E} \oplus 0.03$	0.1×0.1	2% at 50 GeV, 10% at 1 TeV
CMS	$0.03/\sqrt{E} \oplus 0.003$	0.017×0.017	$1/\sqrt{E} \oplus 0.05$	0.087×0.087	1% at 50 GeV, 5% at 1 TeV

ATLAS and CMS detectors is presented in table 3.1, using information from [4–7]. The relevant terms are explained in the text as we go along.

> The particles emanating from a collision are reconstructed from their interactions and energy deposits in different layers of the detector.

A schematic diagram of a generic detector is shown in figure 3.1. Rather than going into the technical details of each detector's (and sub-detector's) components (which are mostly experiment specific), we will focus on how various particles are detected. We will now schematically depict each of the physics objects in an idealised detector, where four layers correspond to the tracking detector, EM and hadronic calorimeters and the muon chamber, in the cross-sectional view. The real event displays from ATLAS will be shown sometimes in parallel. In subsequent chapters though, we will only use the real event displays.

> The fraction of events or objects satisfying some requirement is referred to as the efficiency of imposing that requirement (called a selection or a cut).

3.1.1 Charged particles

The tracking detectors are usually placed the closest to the beam-line. Their aim is to measure charged particles as they emerge from the interaction point (IP). The basic

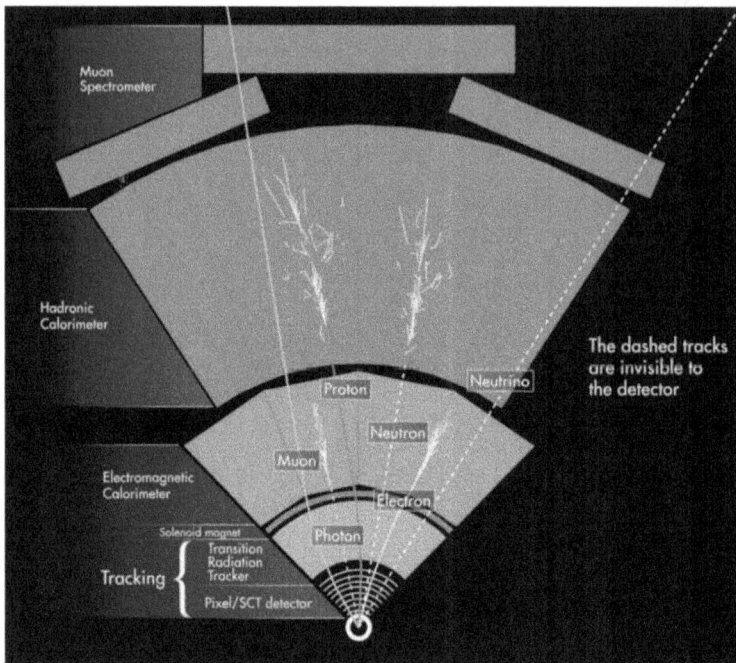

Figure 3.1. Schematic diagram of a generic detector, showing how different sub-detectors detect different particles (© 2018 CERN).

considerations for constructing the tracking detector is to minimise the amount of material the charged (and neutral) particles will pass through to least affect its path, and make the detector radiation insensitive since it resides closest to the interaction point. ATLAS and CMS both use silicon detectors for the innermost part of their tracking systems, and the trackers extend to $|\eta| < 2.5$. Usually tracking detectors contain a solenoidal magnetic field, so that the paths of the charged particles become curved (helical). From the sign of curvature, the charge can be ascertained, and the momentum is estimated from the curvature itself. However, this also implies that for tracks with high p_T, the curvature will be less, thereby introducing more uncertainty on the momentum measurement, and increased possibility of mis-measuring the charge.

> Charged particles are reconstructed from their hits in different layers in the tracking detector.

As charged particles pass through the different layers of the tracking detectors (usually pixel and strips), the position in each layer is recorded with a very high precision. Each of these space points is termed a *hit*, and the hits are categorised into trajectories, termed *tracks* by pattern recognition programmes

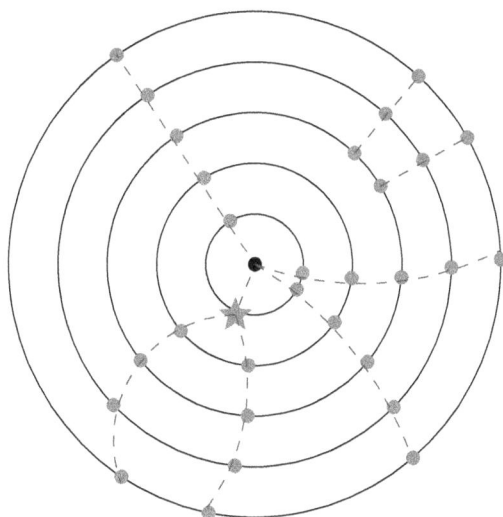

⭐ Photon conversion/Hadronic interactions

Figure 3.2. A schematic diagram to illustrate track fitting from hits in different layers of a tracking detector. The red points denote the hits, the dashed blue lines are the fitted tracks. The red star denotes a photon conversion or a hadronic interaction, resulting in a secondary vertex. Photon conversion means the photon creates an electron–positron pair, while hadronic interaction refers to the situations where a neutral hadron will produce charged particles after interaction with the detector material. Usually the electrons from photon conversion will have a small angle between them, while generally more than two tracks emerge from hadronic interactions, but, for simplicity, that is not shown here. There can be tracks missing hits in innermost layers, but they should still point back toward the PV, as can be seen on the top right corner.

[9], as shown in figure 3.2. These programmes also have to account for Cherenkov effect[1], The most commonly used track-fitting algorithm is the Kalman filter [10], where the full knowledge of the track parameters at each detector layer is used to find compatible measurements in the next detector layer. For a fitted track, the momentum is determined by the radius of curvature of the trajectory and magnetic field strength. The more curved the path, the less momentum the particle had. The tracks are extrapolated back to the point of origin, and by using multiple tracks, the collision vertex is reconstructed.

For the particles originating in the hard collision, this vertex is called the primary vertex (PV), which should coincide with the IP. Particles originating in PV are termed *prompt* particles. However, for particles created from decay of longer lived particles, the extrapolation leads back to a vertex displaced from the PV, called a *secondary vertex*. These can be caused by τ-leptons, b and c-hadrons, which have a

---

[1] When a charged particle travels faster than light does through a given medium, it emits Cherenkov radiation at an angle that depends on its velocity. The particle's velocity can be calculated from this angle. The velocity can then be combined with a measure of the particle's momentum to determine its mass, and, therefore, its identity.

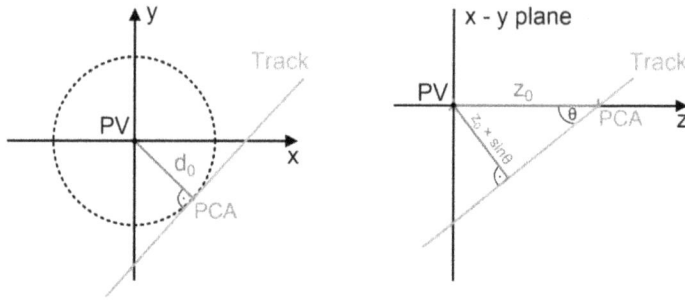

Figure 3.3. The transverse (left) and longitudinal (right) impact parameters of a track are defined in terms of their point of closest approach (PCA) to the PV.

longer lifetime for decay. Photon *conversions*, i.e. creation of an electron–positron pair from a photon can also have displaced vertex, The tracking detectors can also contribute to particle identification by measuring the rate of energy loss (often denoted by dE/dx).

Prompt particles are the ones originating in the actual interaction point.

The tracks are characterised by two impact parameters, transverse, d_0, and longitudinal z_0, shown in figure 3.3. They are measured with respect to the PV. The transverse distance from the position of the PV to the point of closest approach (PCA) of the track in (η–ϕ plane) is defined to be d_0, while the distance along z-axis is termed z_0. Often $z_0 \sin \theta$ is used as well, where θ is the polar angle.

The momentum resolution, σ_{p_T}, of tracking detectors is usually of the form [9]:

$$\frac{\sigma_{p_T}}{p_T} \approx \frac{p_T}{BL^2} \times \sigma_s$$

where B and L denote the magnetic field and the length of the track, and σ_s is the uncertainty on the measurement of *sagitta*, which denotes the deviation from the straight line of the track (this comes from the Latin word for bow). In order to optimise BL^2, ATLAS designed a larger detector, while CMS put in a stronger magnet. The impact parameter resolution depends on a constant term, and another term inversely proportional to the p_T of the track. The former term depends on the geometry of the tracking detector, and it improves according to the rate of increase of the radius of the successive layers. The latter term depends on the material of the tracking detectors.

The most common charged particles are electrons, muons, and pions. Electrons give curved short tracks, while muons cause long tracks as they are heavier. There are many charged hadrons as well, like different types of baryons and mesons, but the fraction of pions are dominant among them, and they usually give spiral tracks. Tracks are usually selected with some minimum p_T requirement (100 or 500 MeV), determined by the fitting algorithm, and they must be within the geometric expanse of the tracking detector ($|\eta| < 2.5$ for ATLAS and CMS). Additionally, requirement on the number of hits in different layers and on the impact parameters is enforced in order to reject spuriously reconstructed tracks, and tracks coming from pile-up interactions.

In figure 3.4 (left) a generic event display with many tracks is shown, while figure 3.4 (right) shows a real event event display from ATLAS, where pile-up interactions created 11 separate vertices.

3.1.2 Electrons

The signature of an electron is a reconstructed track in the tracking detector, associated with a narrow, localised cluster of energy in the electromagnetic (EM) calorimeter. In figure 3.5 (left), a generic event display with two oppositely charged electrons is shown (this can correspond to a $Z \rightarrow e^+e^-$ event). The electrons deposit energy in electromagnetic calorimeters, and the tracks point to where the energy deposits are. The electron energy is determined from the corresponding energy deposits, optimised to reduce pile-up and noise contribution.

Electrons are reconstructed from their tracks matched to their energy deposits in electromagnetic calorimeters.

The challenge is to correctly identify an electron in the detector. Jets (or photons) can be misidentified (called *fakes*) as electrons, and electrons from non-prompt sources (photon conversion, semi-leptonic heavy-flavour decays, decays of charged pions or K-mesons) are also considered background for prompt electrons. Therefore, discriminating variables need to be constructed which will provide a highly efficient electron selection, with large background rejection.

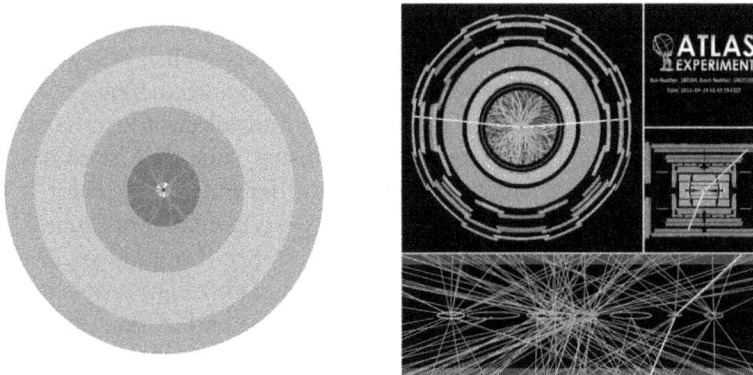

Figure 3.4. A generic detector display with only tracks, indicated as green lines in the tracker, which is the innermost part of the detector (left) and an event display from ATLAS (ATLAS Experiment © 2018 CERN) showing a lot of tracks created from 11 vertices in the same event (right).

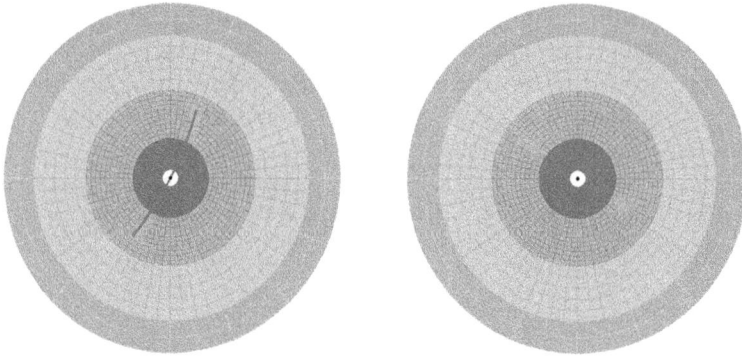

Figure 3.5. A generic detector display showing an electron–positron pair (left) and a photon pair (right). The blue lines indicate the tracks in the tracking detector, and the orange squares indicate energy deposits in the electromagnetic calorimeters. The charged electrons leave tracks roughly consistent with the direction of the energy deposits, while neutral photons do not. Note that the electron and the positron tracks have opposite curvature.

The electron causes a shower (cascade of lower energy particles originating from the electron) in the EM calorimeter. The so-called *shower shape* variables use calorimeter information, exploiting the lateral and longitudinal segmentation. Examples are the fractions of energy deposits in different layers of the calorimeters, the width of the shower, the leakage of energy in the hadronic calorimeter, the number of hits corresponding to the electron track, the impact parameters, and quality of the track fit. Combining information from the tracker and calorimeter provides additional background discrimination. These variables correspond to different angular measures of track-cluster matching, and also E/p ratio (ratio of the electron energy measured in the EM calorimeter, to the track momentum determined from the tracker) is used. Real electrons are expected to lose only a small amount energy via bremsstrahlung radiation, resulting in values close to unity, while hadrons will deposit a large fraction of energy in the hadronic calorimeter.

Optimised thresholds on these variables are combined to come up with the final identification criteria. Depending on the analysis requirements, often three working points are employed. They are referred to, in order of increasing background rejection (or in decreasing order of purity of electron selection) as: loose, medium, and tight. The operating points are usually inclusive, such that loose is a subset of medium, which in turn is a subset of tight. Since the tracking detector is used, the electrons can only be detected within $|\eta| < 2.5$. Although electron candidates can be reconstructed with a few GeV of p_T, usually a higher threshold of 20–30 GeV is required to obtain a better selection efficiency and calibration. Also single-electron triggers (a concept that will be introduced toward the end of this chapter) become fully efficient at those values. Lower p_T electrons can be used for specific analyses, but careful study is needed to protect against inefficiencies and fakes. Triggers requiring two electrons tend to become fully efficient at lower values.

Figure 3.6. Event display from light-by-light scattering observed [11] by the ATLAS experiment (ATLAS Experiment © 2018 CERN). The yellow lines indicate the scattered photons.

3.1.3 Photons

Photons originating from hard interactions do not leave any tracks, but deposit all their energy in an electromagnetic calorimeter. They are reconstructed from energy deposits in the calorimeter systems. In figure 3.5 (right) a generic event display with two photons is shown (this can correspond to a $H \rightarrow \gamma\gamma$ event). The photons are seen to deposit energy in electromagnetic calorimeters, but they have no tracks. In figure 3.6, an event display from a rare photon–photon scattering process observed by ATLAS [11] is shown.

Photons are reconstructed from their energy deposits in electromagnetic calorimeter, and from the absence of any corresponding track.

Similar to electrons, fake photons (mis-reconstruction of a jet or an electron as a photon or photons from a decay of a hadron) are discarded by shower shape variables. The other aspect is to differentiate between unconverted and converted photons, where the latter have interacted with detector material upstream of the calorimeters and produced an electron–positron pair. Converted photons will have at least one track originating from a vertex, whereas unconverted photons do not have such a track. The photons are also detected within the tracking volume $|\eta| < 2.5$, as a consequence of requiring the absence of tracks. Similar to electrons, it is usual to select photons with a p_T threshold of 20–30 GeV. The energy determination of photons will be revisited at the end of section 3.2.

3.1.4 Charged and neutral hadrons

Charged and neutral hadrons (mostly pions and kaons), resulting from quarks and gluons, deposit their energies in hadronic calorimeters (as they are heavier than electrons and photons), additionally the charged ones also leave tracks. The calorimeters in ATLAS and CMS extend up to $|\eta|$ of 4–5.

In figure 3.7 (left) a generic event display with many clusters of energy deposits in electromagnetic and hadronic calorimeters is shown. In ATLAS, energy deposits in neighbouring calorimeter cells are combined to form a three-dimensional topological cluster, as shown in figure 3.7 (right).

Hadrons are reconstructed from their energy deposit in hadronic calorimeters.

Most calorimeters are position sensitive, i.e. segmented, to measure the position of the energy deposition and the direction of the incoming particle. A calorimeter measures the energy a particle loses as it passes through. It is usually designed to stop entirely or 'absorb' most of the particles coming from a collision, forcing them to deposit all of their energy within the detector. Calorimeters typically consist of layers of 'passive' or 'absorbing' high-density material, for example, lead interleaved with layers of an 'active' medium such as solid lead-glass or liquid argon. Electromagnetic calorimeters measure the energy of electrons and photons as they interact with the electrically charged particles in matter. Hadronic calorimeters sample the energy of

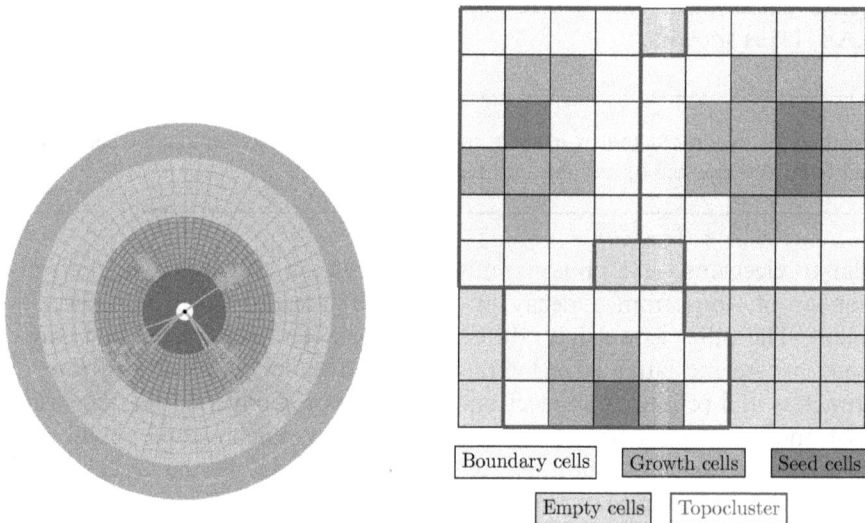

Boundary cells | Growth cells | Seed cells
Empty cells | Topocluster

Figure 3.7. A generic detector display showing energy deposits in the calorimeters (left), mostly from charged and neutral hadrons, indicated by orange squares. The charged hadrons will leave tracks as well, indicated by the green lines. These cells can be combined in spatial dimensions to form clusters (right). The red seed cells, having the most energy compared to the neighbouring cells are chosen to initiate the process. The nearby orange growth cells, which have energy over a certain predetermined noise threshold are combined. Their neighbouring yellow boundary cells are added as well, to form the cluster indicated by the blue line. The grey cells have either no energy deposited in them, or have energy less than the noise threshold. Each cluster is formed dynamically depending on the position and energy of the cells.

hadrons as they interact with atomic nuclei. Calorimeters can stop most known particles except muons and neutrinos.

A measure of quality of the calorimeters is their energy resolution, σ_E, which signifies the ability of the calorimeter to accurately determine the energy of an incoming particle (i.e. minimum energy difference required to distinguish between two energy deposits). The energy resolution is usually given by:

$$\frac{\sigma_E}{E} = \frac{a}{\sqrt{E}}$$

where E is the energy of the particle being stopped by the calorimeter, and a is a constant, which depends on the design of the calorimeter (linearity of energy deposit, leakage, etc). This is often called the stochastic term, as it is dominated by (Poisson) fluctuations related to the development and measurement of the shower. For EM calorimeters, the full resolution is given by adding a noise term, as well as a constant term:

$$\frac{\sigma_E}{E} = \frac{a}{\sqrt{E}} \oplus b \oplus \frac{c}{E}$$

The noise term accounts for the quality of the calibration, electronic noise of the readout chain, readout dead-time and non-linearity of the response. It usually affects low energy measurements. The non-uniformity from the instrumental effects and contamination from pile-up are accounted for in the constant term, which becomes important at high energies. The notation $\oplus$ indicates addition in quadrature[2].

The situation is more complicated for hadronic calorimeters, as they need to account for both electromagnetic and hadronic showers. The important measure is the relative response of e/π ratio, where the numerator and the denominator respectively denote energy deposited by electrons and pions. If this ratio is unity, then the calorimeter is termed a *compensating calorimeter*. If the response to the hadrons is lower, leading to this ratio >1, then the energy response is no longer linear with energy. The angular resolution (i.e. the minimum angle between two energy deposits to distinguish them) is another important aspect for a calorimeter. The reader is referred to dedicated resources [12, 13] for details.

3.1.5 Muons

Muons have roughly the same characteristics as the electron, but they are significantly heavier and do not interact with atomic nuclei of the calorimeter material. Therefore, they pass through the calorimeters without interacting. They do interact electromagnetically, but because they are much heavier than electrons, they only lose a small fraction of energy in collisions with electrons. For this reason, the outermost layer of the detectors, are specially designed to stop the

---

[2] Square root of the sum of squares.

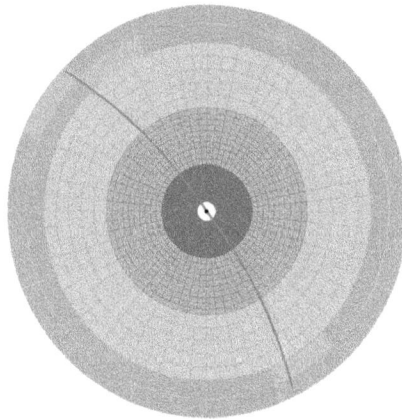

Figure 3.8. A generic detector display showing muon pair production. The red lines indicate the tracks in the tracking detector, and the orange squares indicate energy deposits in the muon chamber, which is the outermost layer in the detector. The charged muons leave tracks roughly consistent with the direction of the energy deposits. Note that the two muon tracks have opposite curvature.

muons. These are essentially tracking detectors. In figure 3.8, a generic event display with two oppositely charged muons is shown (this can correspond to a $Z \to \mu^+\mu^-$ event). The muons are seen to deposit energy in electromagnetic calorimeters, and the tracks are roughly in the same direction as the energy deposits. Similar to electrons and photons, muons are needed to be with $|\eta| < 2.5$, usually with a p_T threshold of 20–30 GeV, due to trigger threshold and gain in reconstruction efficiency. Occasionally, very energetic pions make it through the hadronic calorimeter, and stop at muon chambers. This is referred to as *punch-through* and may result in a fake muon.

Muons are reconstructed from their interaction in the muon chamber.

3.1.6 Neutrinos/missing energy

Neutrinos are the only SM particles which do not interact with any part of the detector at all, as they do not have electric charge or colour charged constituents. They only interact by weak force, which is too faint for detectors. However, since energy–momentum conservation dictates that the total p_T of all the particles after the collision is zero (as initial protons move only along the beam axis), any imbalance can be attributed to the neutrinos. Usually a quantity called missing (transverse) momentum or missing (transverse) energy (MET) is defined as the negative of sum of scalar p_T all other objects (including tracks, often denoted as the *soft term*). Technically, since energy is not a vector, the term missing momentum is more accurate, but both the terms are used interchangeably to denote neutrinos.

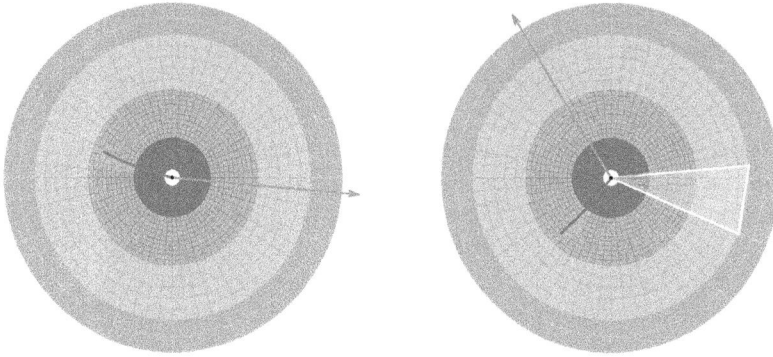

Figure 3.9. Generic detector display showing neutrino production. The production of only one electron (left), or an electron and an unbalanced jet (right), shown by the white triangle collecting a bunch of hadrons in the same direction (jets will be formally defined in the next section), is prohibited by energy–momentum conservation. The magenta lines with arrow indicate the direction of this missing (transverse) momentum, signifying the directions of the escaped neutrinos.

> Missing (transverse) momentum or energy represents the non-conservation of transverse momentum in an event, originating from neutrinos.

In figure 3.9 (left), the event is seen to have a single-electron. Energy–momentum conservation would not allow that, so we can assume there must be a neutrino balancing it, where we draw the dashed line with the arrow (this corresponds to a $W \to e\nu$ event). In figure 3.9 (right), we have an electron, and energy deposits in a bunch of calorimeter cells (which will be referred to as jets hereafter). The configuration is still not conserving energy–momentum, so we have to assume a neutrino was also produced along the direction shown by the magenta line with the arrow (this corresponds to a $W \to e\nu +$ jet event).

3.2 Jets and making them

The strong interaction of quarks and gluons (governed by QCD) is fundamentally different from the electroweak force. The quarks and gluons are often collectively referred to as *partons*. Interactions between particles in quantum field theory language is expressed in terms of excitation of a field between them, and force carrying bosons are manifestations of this field. If the distance between two quarks increases, the energy stored in the gluon field gets larger, and it takes more energy to keep the pair separated (the closest analogy is a spring). This is termed asymptotic freedom, as a consequence of which we do not see free quarks or gluons in nature. In fact, when this energy in the gluon field becomes large enough to create a quark–anti-quark pair, such a pair is produced (rather than having infinitely high energy in the gluon field). Just like charged particles can radiate photons, colour charged quarks can emit gluons. Again as photons can create electron–positron pairs, gluons can split into quark–anti-quark pairs. So a single parton usually leads to a cascade of partons, which eventually form colour-neutral hadrons (the detailed mechanism will be discussed in the next chapter). In the detectors, deposits of energy, corresponding to these (colour-neutral) charged and neutral hadrons are obtained. The energy flow of a coloured parton can be approximated by the colour-neutral hadrons created from it, and *jets* are formed specifically for this purpose.

A jet is essentially a collection of collimated bunches of hadrons, ideally each bunch coming from a single parton. Often, therefore, they are referred to as proxies for quarks and gluons, but we must keep in mind that this definition is fundamentally ambiguous, as there is no way to uniquely identify the origin of the hadrons. We will come back to this in chapter 8. Additionally, hadrons from I/FSR, MPI and pile-up would invariably become part of jets, no matter how they are formed. The term jet might have come from *Jet d'Eau*, a 140 m high water fountain located in Lake Geneva near CERN, as shown in figure 3.10. In figure 3.11, generic event displays with two- and three-jet events are shown, with each jet including charged and neutral components from the tracker and the calorimeters. The two-jet (or dijet) events have the jets in back-to-back configuration, resulting in energy–momentum balance. The three-jet (or trijet) events also have energy–momentum balanced, as they are symmetrically oriented. The jets need to be within the calorimeter expanse, and a p_T threshold of 25–30 GeV is usually required, mostly driven by calibration and trigger limitations.

Jets are not fundamental objects, rather a construction to capture hadrons originating from a single parton.

Figure 3.10. A photo of the *Jet d'Eau* in Geneva.

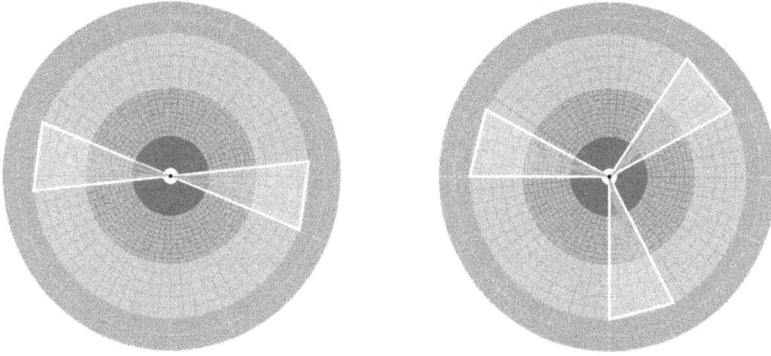

Figure 3.11. Generic detector display showing events with two (left) and three (right) reconstructed jets. The jets are formed by combining the energy deposits corresponding to hadrons roughly in the same direction by some specific algorithm and size measure. The three-jet is often referred to as a Mercedes event, indicative of the similarity to the logo of the famous car-maker.

A generic jet forming algorithm, (referred to as *jet algorithm* for brevity from now on, or as a recombination scheme) can be represented as a mapping from a set of hadrons with four-momenta $\{p_1^{\text{had}}, p_2^{\text{had}}, \ldots p_N^{\text{had}}\}$ to a set of jets with four-momenta $\{p_1^{\text{jet}}, p_2^{\text{jet}}, \ldots p_M^{\text{jet}}\}$, where usually $M < N$, and $p^{\text{jet}} = \sum_{i \in \text{jet}} p_i^{\text{had}}$. The resultant vector sum of the input four-momenta defines the *jet axis*.

There exists no unique definition of a jet, as it depends on how this mapping is performed. Once the jets are formed, the event can be fully or partially characterised in terms of the jets in that event, rather than the individual hadrons which made up the jet.

The concept of jets is not new, however, the algorithm to form jets has evolved significantly over the years. One of the earliest mentions of jets appears in 1000 GeV nuclear interactions of protons [14] in 1957. In the context of colliders, the 1977 work of George Sterman and Steven Weinberg [15] is considered the pioneering work, as well as of Feynman and Field [16]. However, rather than presenting a historical perspective, we will review the current status.

The first step is to identity the inputs to jet formation. Experimentally, clusters of energy deposit in calorimeters can be taken as inputs, but charged particle tracks, or objects formed by combining energy deposit and track information (sometimes referred to as particle flow objects) can be used. Stable final state particles, or even partons are used as inputs to jet formation when using simulation. A good jet algorithm needs to be robust against input objects used. The next step is to collect the input objects based on some distance criteria, which determines the size of the jets. The chosen jet algorithm determines which input objects go to which jets. As a distance measure, R is used, as it is equivalent to the opening angle in collider coordinates. The jet algorithms need to meet certain criteria:

- Even though there cannot be an unambiguous definition of jets, the physics interpretation of the event should not depend too sensitively on them.
- They should be independent of the design/structure of the detector.

Infrared safety

Collinear safety

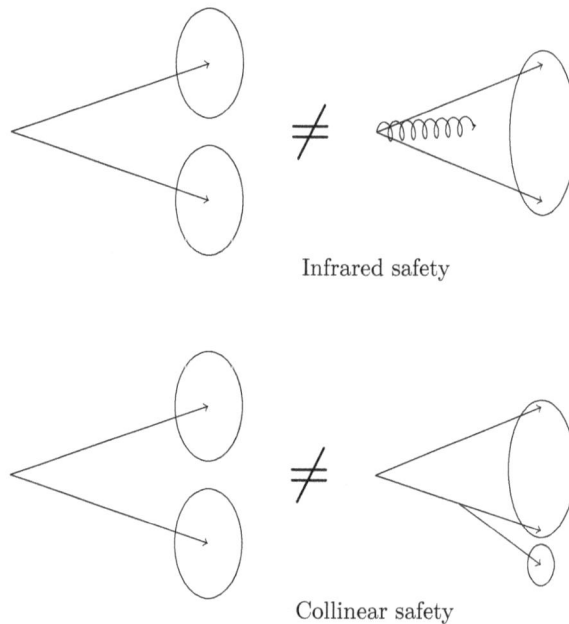

Figure 3.12. Schematic display of infrared and collinear safety. A soft gluon emission (top) between two jets should not result in one merged jet, which is termed infrared safety. A soft gluon emission at a small angle (bottom) from a jet should not result in changing the jet configuration, which is referred to as collinear safety.

- Minimal sensitivity to hadronisation, UE, PU is desired, but is not always possible in practice.
- Experimentally, they should be easy to implement and (computationally) quickly executable.
- Theoretically, they need to be *infrared* (IR) and *collinear* safe to not have divergent cross-sections in perturbative calculations. The terms infrared and collinear safety (often collectively referred to as IRC safety) deserves an explanation. Infrared safety means that the addition of a soft gluon should not change the results of the jet clustering. Collinear safety implies that splitting one parton into two collinear partons should not change the results of the jet clustering. These are illustrated in figure 3.12.

We will focus on two classes of jet algorithms, referred to as cone algorithm and sequential recombination algorithm. The first one was used in Tevatron experiments, whereas the latter one was used in HERA and is currently being used at the LHC.

- *Cone algorithms* These can be thought of as a top-down approach, where a cone of fixed radius R is constructed around the highest energy input object (termed the *seed*), and other input objects within the cone are considered part of the jet. After those inputs are combined with the seed, the jet momentum and direction (or the jet axis, which was initially just that of the seed) are re-calculated. This is then used as a new seed, and the previous step is repeated. The final jet is obtained when this iteration no longer changes the jet axis. Then the same process is repeated successively with the highest energy leftover input objects, till all the input objects are used up. JETCLU and Midpoint algorithms were used by the CDF collaboration [17, 18]. This may, however, result in overlapping jet cones, starting from different seeds. That can be resolved by defining a splitting/merging criteria depending on the overlap energy.

Even though this is rather intuitive, this definition of cone algorithm suffers from lack of IR safety. As an example, the presence of a soft gluon between two hard partons will result in all three being merged into a single cone, whereas the absence of it would result in two separate cones. A modern version of the cone algorithm, Seedless IR Safe Cone jet algorithm (or SISCone) [19], uses a geometric approach to find all distinct cones. Discussion of other cone jet algorithms, like XCone [20] is beyond the scope of this book.

- *Sequential recombination algorithms*: rather than combining inputs in co-ordinate space like cone algorithm, they are combined in momentum space by using a bottom-up approach. If the development of a jet can be thought of as a repeated $1 \rightarrow 2$ parton splitting, then that can be undone by repeatedly combining two inputs to one to get back the jet. The choice of the pair of particles to combine at any given stage and when to stop combining them are dependent upon the individual algorithm.

One of the earliest sequential recombination algorithms was by the JADE collaboration [21, 22], used in the HERA e^+e^- collider. The distance measure between two inputs i and j with energies E_i and E_j, separated by angle θ_{ij} was defined as: $d_{ij} = \frac{2E_iE_j(1 - \cos\theta_{ij})}{Q^2}$, where Q^2 is the total energy of the hadronic final state, or the c.m. energy. This is essentially the invariant mass of the two-particle system scaled by the event energy scale. If the minimum d_{ij} of any combination is less than the defined cut-off parameter y_{cut}, then these two inputs are combined, and this is repeated until all $d_{ij} > y_{cut}$. Typical values of y_{cut} were chosen to be 0.02–0.2, based on experimental factors. The remaining inputs, combined or not, are then taken as the jets. The number of jets obtained was seen to depend very sensitively on y_{cut} because smaller values will result in a higher number of jets, while higher values will give fewer jets. In fact, the same event can be classified as having 2, 3, 4 or more jets depending on y_{cut} value.

While that introduces a degree of subjectivity in this algorithm, the other problem with this distance measure is that two soft particles emitted at large angle get recombined into a single jet. The solution was to use $2\text{Min}\{E_i^2, E_j^2\}$ in place of $2E_iE_j$ in the distance measure. Since $1 - \cos\theta_{ij} \approx \theta_{ij}^2/2$, for small θ_{ij} and massless particles, the numerator reduces to the transverse momentum (squared) of the relatively less energetic input. In literature, this is known as the Durham k_t algorithm [23]. To adapt this algorithm to hadron colliders, several modifications were needed.

- Particles which go down the beam-pipe are not measured, so they should not be included in the jets. So the distance measure needs to include a beam distance parameter, d_{iB}, indicative of the distance between input object i and the beam direction (or the z-axis, according to our convention).
- The distance measure needs to be formulated in terms of longitudinally boost invariant quantities, such as rapidity, azimuthal angle and transverse momentum, not in terms of energy or polar angle, which cannot be measured in hadron colliders.
- At the hadron collider, no single fixed scale Q^2 can be determined because partons participate in the collisions. So a dimensionless angular radius parameter, R_0, is chosen, which corresponds to the angular size of the jet to be obtained.

Taking these into consideration, the distance measure is written as:

$$d_{ij} = \mathrm{Min}\left\{p_{T_i}^{2\rho}, p_{T_j}^{2\rho}\right\} \frac{\Delta R_{ij}^2}{R_0^2}$$

$$d_{iB} = p_{T_i}^{2\rho}$$

where $R_{ij}^2 = (y_i - y_j)^2 + (\phi_i - \phi_j)^2$, is the angular distance between two input objects, and the exponent ρ controls the ordering or merging depending on its chosen value, as we will see. The jet forming is then performed via the following steps:

1. First the smallest value among all possible d_{ij} and d_{iB} is determined.
2. If this is one of the d_{ij} values, inputs i and j are merged. The merging implies that the p_T are added, while the position of the merged input is the p_T weighted average position.
3. If it is one of the d_{iB} values, ith input is considered a jet, therefore not considered an input anymore.
4. This continues till all inputs are merged or classified into jets.

Typical values of R_0 are 0.4–0.6. $R_0 = 0.4$ corresponds to a half-angle of 22° and a full angle of 45°. It should be noted that this prescription by itself does not guarantee IR safe jets due to the possible presence of soft jets near the beam, which is rectified by imposing a minimum p_T threshold on the jets. Depending on the value of ρ, three different jet algorithms can be constructed, which essentially differ in the choice of the distance measure between pairs of particles.

- For $\rho = -1$, it is termed the anti-k_t algorithm [24], which is one of the most used jet algorithms at the LHC. It starts with the highest p_T input object, and that jet swallows all softer input objects within R_0. This leads to formation of roughly circular shaped jets with area of πR_0^2. In that sense, this gives cone-like jets without the associated shortcomings. This makes it the experimentally favoured algorithm, which we will see as we go along. However, the order of clustering is not related to actual parton splitting order.

- For $\rho = 1$, it is termed the k_t algorithm [25]. Since the clustering starts from the softest input objects, the shape of the jet changes with addition of harder and harder inputs, resulting in irregular amoeba-shaped jets. However, k_t algorithm represents an approximate inversion of QCD branching process, thereby preserving the actual clustering history.
- For $\rho = 0$, it is termed the Cambridge–Aachen [26, 27] algorithm. Here, inputs are clustered only based on angular separation, without considering their momenta.

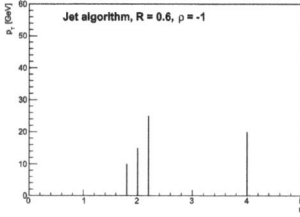

d_{ij}	1	2	3	4
1	-	0.00049	0.00071	0.03361
2	0.00049	-	0.00018	0.02778
3	0.00071	0.00018	-	0.01440
4	0.03361	0.02778	0.01440	-
d_{iB}	0.01000	0.00444	0.00160	0.0025

(a) We have 4 input objects as shown. The smaller value is indicated, which dictates 2 and 3 should be merged. The merged p_T will be 40 GeV, and the position is determined by the p_T-weighted average: $(2*15+2.2*25)/40 = 2.13$.

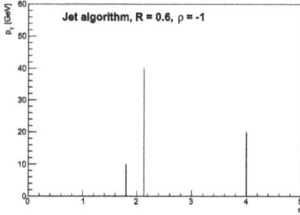

d_{ij}	1	23	4
1	-	0.00019	0.03361
23	0.00019	-	0.00607
4	0.03361	0.00607	-
d_{iB}	0.00444	0.00063	0.00250

(b) At this step, we indicate the merged input from previous step by 23. The distances indicate that inputs 1 and 23 should be merged. The merged p_T will be 50 GeV, and the position will be determined by the p_T-weighted average: $(1.8*10+2.13*40)/50 = 2.06$.

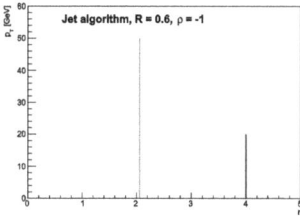

d_{ij}	123	4
123	-	0.00418
4	0.00418	-
d_{iB}	0.00040	0.00250

(c) At this step, we indicate the merged input from previous step by 123. The distances indicate the input 123 should be classified as a jet itself. Since that leaves input 4, that will be classified as a jet as well.

Figure 3.13. Step-by-step demonstration for jet reconstruction with anti-k_t algorithm. In this simple example, the x-axis depicts the position r and the y-axis denotes the p_T of the inputs.

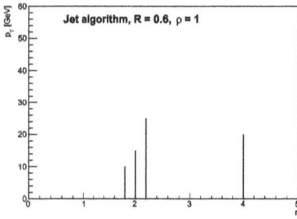

d_{ij}	1	2	3	4
1	-	11	44	1344
2	11	-	25	2500
3	44	25	-	3600
4	1344	2500	3600	-
d_{iB}	100	225	625	400

(a) We have 4 input objects as shown. The smaller value is indicated, which dictates 1 and 2 should be merged. The merged p_T will be 25 GeV, and the position is determined by the p_T-weighted average: $(1.8 * 10 + 2 * 15)/25 = 1.92$.

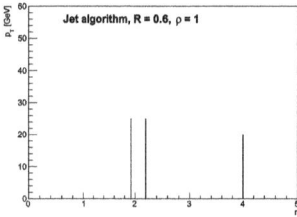

d_{ij}	12	3	4
12	-	136	4807
3	136	-	3600
4	4807	3600	-
d_{iB}	625	625	400

(b) At this step, we indicate the merged input from previous step by 12. The distances indicate that inputs 12 and 3 should be merged. The merged p_T will be 50 GeV, and the position will be determined by the p_T-weighted average: $(1.92 * 25 + 2.2 * 25)/50 = 2.06$.

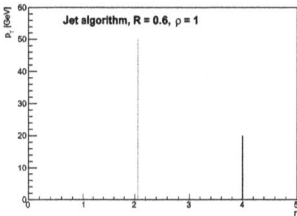

d_{ij}	123	4
123	-	4182
4	4182	-
d_{iB}	2500	400

(c) At this step, we indicate the merged input from previous step by 123. The distances indicate the input 4 should be classified as a jet itself. Since that leaves input 123, that will be classified as a jet as well.

Figure 3.14. Step-by-step demonstration for jet reconstruction with k_t algorithm. In this simple example, the x-axis depicts the position r and the y-axis denotes the p_T of the inputs.

The jet formation is schematically shown in figures 3.13 and 3.14 for anti-k_t and k_t algorithms. Although in this overly simplistic case the end result is the same in both cases, the intermediate step is different. Then figure 3.15 shows the shape of these three types of jets. The anti-k_t jets are rather circular, which is why experimentally they are preferred. The k_t and Cambridge–Aachen jets end up amoeba-shaped. For $\rho = 1$ and an unnormalised distance measure (i.e. without using the R_0 parameter), the algorithm is known as an exclusive k_t algorithm, which is used in some specific cases. In this case, a specific number of jets or a minimum p_T threshold for a final jet

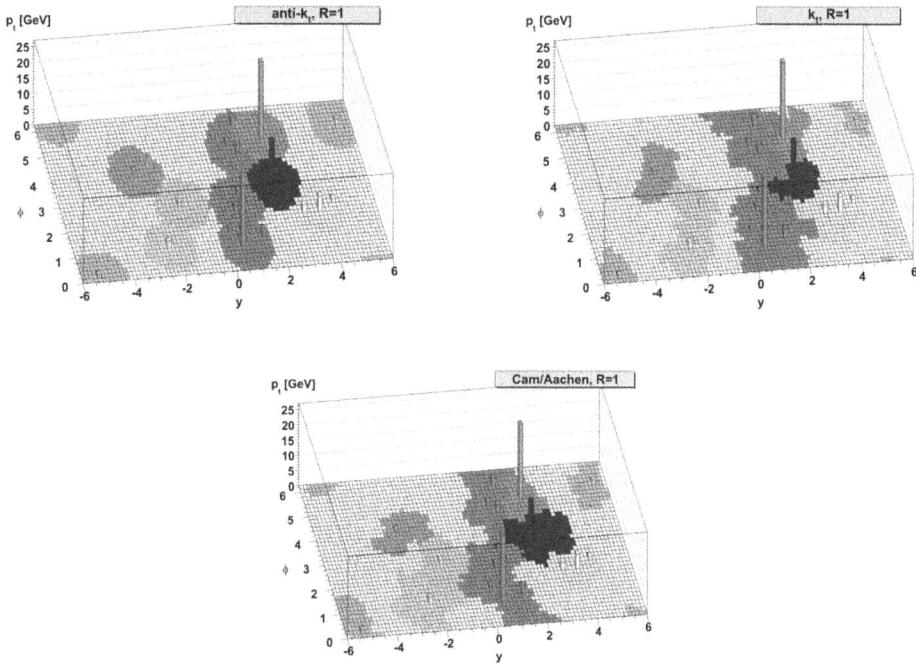

Figure 3.15. Jets obtained using anti-k_t (top left), k_t (top right), and Cambridge–Aachen (bottom) algorithms with $R = 1$. The colours have no physical meaning, they only served to distinguish energy deposits clustered together according to some algorithm. Figure taken from [24], © Matteo Cacciari, Gavin P Salam and Gregory Soyez.

is demanded, and the algorithm stops when that number is reached. The choice of jet algorithm depends on physics goal, as will be seen in chapter 9.

In practice a program named FASTJET [28], containing the implementation of sequential recombination algorithms (as well as SISCone) is used to construct the jets.

Apart from the constituents mentioned above, *particle flow objects* are also used at inputs for jet forming, for example in the CMS experiment [29]. The idea is to use (charged particle) tracks as they have better angular resolution. The tracks are matched with energy clusters in the calorimeter, and calorimeter signal associated with tracks are fully (whole cluster) or partially (cells from cluster) removed. Then the inputs to jet forming are the charged (matched and unmatched tracks) and the neutral (original or modified clusters) particle flow objects. Once constructed, jets are treated as any other physics objects present in an event. While the jet axis determines the direction of the jet, and the four-vector combinations during the jet clustering give its p_T, a few other properties of jets are worth mentioning.

Jet mass: the constituents of a jet usually have zero or very small mass, however, the jet itself often ends up having a non-negligible mass, as the jet mass can be calculated as:

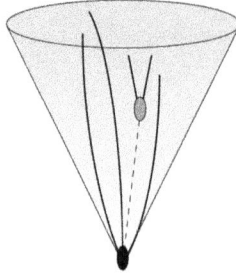

Figure 3.16. Illustration of jet containing a displaced vertex, corresponding to the decay of slightly long-lived *B*-hadron away from the PV.

$$m = \sqrt{\left(\sum_{i\in\text{jet}} E_i\right)^2 - \left(\sum_{i\in\text{jet}} \vec{p}_i\right)^2}$$

where E_i and $\vec{p}_i$ respectively denote the energy and momentum of the ith constituent. This is especially important when the jet is created from a heavy object, as will be seen in chapter 8.

B-tagging: identifying jets containing *B* mesons (or alternatively originating from a *b*-quark) is important for distinguishing many physics processes. A *B*-hadron has certain unique properties, it has a much longer lifetime (1.6×10^{-12} s) compared to the most commonly produced hadron, the pion (8×10^{-17} s). This implies a *B*-hadron will travel roughly 0.5 mm (obtained by multiplying the lifetime by speed of light), as compared to 25 nm for pion. The bottom quark is also heavier than all other quarks except the top quark, and it produces more charged particles during its decay compared to other hadrons. It also has a 10% branching fraction to leptons. So, by reconstructing the secondary vertex corresponding to the *B*-hadron decay, as shown in figure 3.16, and utilising the other properties, a jet containing a *B*-hadron is identified, a process which is called *b-tagging*. A jet can also be *c-tagged* depending on if it originates from a *c*-quark, which is a more difficult process.

Ghost-matching [30]: in certain situations, it becomes necessary to identify which tracks are inside a jet, where the jet has been formed using calorimeter energy deposits. The obvious way will be to perform a simple ΔR matching, by demanding the angular distance of the track from the jet axis be less than the jet radius. However, this approach would not properly work for jets with irregular cross-sectional shapes. Ghost association provides a correct matching of tracks to calorimeter jets. In this technique, tracks are treated as infinitesimally soft, low p_T particles by scaling their p_T by a very small number, such as 10^{-100}. These tracks are then added to the list of inputs to the jet algorithm. The low scale implies that the tracks do not affect the reconstruction of jets. However, after the jets are formed, it is possible to identify which tracks were clustered into which jets.

Jet area [30]: jets consist of point-like particles, which themselves have no intrinsic area. But a measure of area can be obtained by adding ghost particles and identifying the area in η–ϕ space where those ghost particles were clustered in that given jet.

It must be noted that photons need to be selected without significant hadronic energy around them (the presence of hadronic energy would indicate that the photon originated in soft processes). For theoretical calculations of isolated photon cross-section, demanding no soft gluons near photons constrains the phase space of soft gluons, which spoils the cancellation of infrared divergences. To mitigate that, jet algorithms are used to identify photons, with a widely used approach known in the literature as *Frixione isolation* [31].

3.3 Trigger

As discussed previously, the limitations in data transmission and storage capacity means that every collision event cannot be saved. This necessitates the use of a trigger and data acquisition system, which tries to rapidly identify *interesting* events and record only those. This is perfectly reasonable, as cross-sections for interesting physics processes are very small. The LHC has event rates greater than 40 MHz and trigger rates are required to be below 1 kHz. This corresponds to a factor of 10^4 reduction, saving approximately a few hundred events per second. The ratio of the trigger rate to the event rate is referred to as the selectivity of the trigger.

> Triggers are designed to record only the potentially interesting collision events, discarding a large fraction of events.

Triggering is usually performed in successive stages, usually referred to as trigger levels. The idea is that each level selects the data that becomes an input for the following, which has more time available and more information to take a better decision. The Level 1 trigger operates under the tightest time constraint, so it can for example only use a subset of information from the calorimeter and muon detectors. The Level 2 trigger analyses in greater detail specific regions of interest (RoI) identified by the Level 1 system for each event. The Level 3 trigger (also known as the high level trigger or the HLT) performs a detailed analysis of the full event data. A schematic diagram illustrating these steps from ATLAS is shown in figure 3.17 (left).

The HLT selects events with one or two leptons with some minimum p_T, or a jet with some minimum p_T. The minimum bias events are recorded with a minimally biased trigger, usually requiring just one charged particle in the detector. A zero-bias trigger can also be used, but that will be dominated by electronics noise, particles from cosmic rays (but often that is useful to get a measure of the contamination). The triggers must have a high efficiency for benchmark physics processes, as discarded events are lost forever[3]. The other performance indicators of a trigger

---

[3] The computing bottleneck is often not storage, but processing the data. Since the LHC goes through periods of shutdown, experiments save more data than they can process during the run, and process them when LHC is not running. This is known as Data Parking at CMS and the Delayed Data Stream at ATLAS.

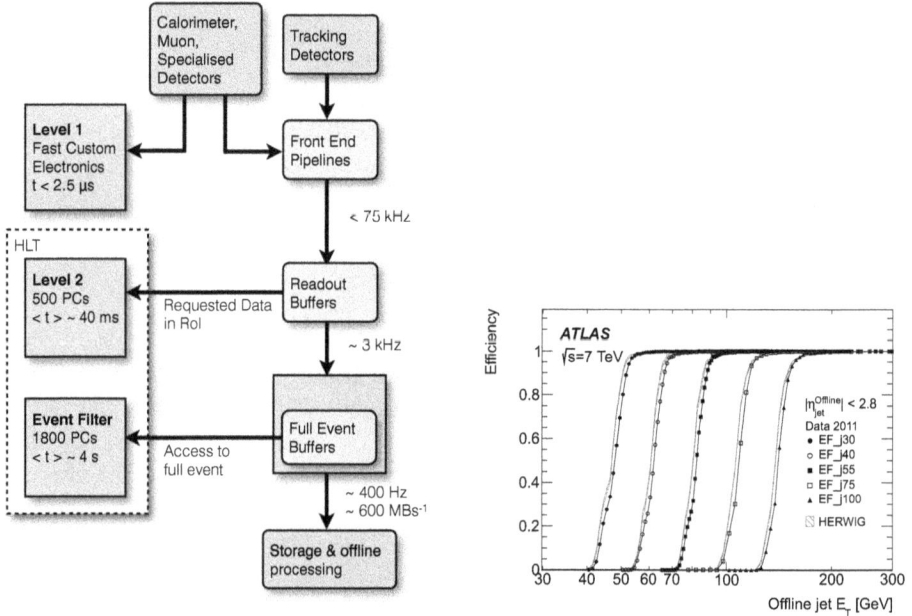

Figure 3.17. Schematic illustration of successive triggering steps (left), and an example trigger turn-on curve plot (right), The Level 1 trigger runs hardware algorithms over data with reduced spatial granularity from the calorimeter and muon subsystems to identify geometrical regions of interest (RoI) in the detector, containing candidate physical objects which should be examined more closely in subsequent trigger levels. The Level 2 (L2) algorithms have access to the data at the full detector granularity but only from those detector elements that lie within an RoI. Following the L2 processing, all events with RoIs that satisfy a set of predefined selection criteria are passed to the event builder which reads out the detector at full granularity. These fully built events are then processed by the event filter (EF). The turn-on curve is showing the efficiency as a function of jet E_T, for different E_T thresholds at EF level. Both figures are from ATLAS [32] (ATLAS Experiment © 2018 CERN).

system is its latency, which is the time required to form the trigger decision. Longer latency implies good events will be rejected due the trigger being busy, which is a source of inefficiency in data acquisition.

During a run, usually specific bandwidth is applied to different triggers, so that all types of events are recorded. Generally the bulk of the selected events are those useful for the physics analysis, but trigger also selects events for instrumental and physics background studies, detector and trigger efficiency measurement from data, and for calibrations. Sometimes *pre-scales* are used to reduce trigger rate for a specific process, which means only a certain fraction of events (selected randomly) satisfying the criteria is recorded. Since trigger rate changes with luminosity, dynamic pre-scales are sometimes used, which reduce the pre-scales as luminosity falls.

3.4 Preparing the data

We have seen in previous sections how different objects are reconstructed using information from the detector system. The term *online* refers to this stage, as the reconstruction is performed in real time. When these objects are later used in analysis (or even reconstructed later with more refined algorithms), that is referred to as *offline*. However, before these objects can be used to derive physics conclusions, a few further steps are necessary. These are calibration, efficiency corrections, applying isolation, overlap removal, and correction for pile-up, not necessarily in that order. Of course the details of all of these is very dependent on the particular detector (and its shortcomings and the physics to be done), but here we give an overview focussing on the commonly used techniques.

3.4.1 Trigger efficiency

Every event is selected by triggering on physics objects, as described earlier in the chapter. That implies that there can be an associated inefficiency if the trigger does not pick up every event with the object of interest. Triggering means (events with) low p_T objects can be missed. Therefore, trigger efficiency can be defined as: $N_{accepted}/N_{produced}$, where the numbers in the numerator and denominator corresponds to events accepted by triggers containing the object of interest, and total events with the object of interest.

The challenge is to estimate the denominator. One way is to measure the efficiency relative to a looser trigger, which is expected to be more inclusive in event selection. Another way is to trigger on a different object to measure the selection efficiency of the object of interest. Triggering is usually performed using a p_T threshold of the object, however, due to the potential difference between object reconstruction at HLT and full object reconstruction (and also due to resolution smearing), the trigger efficiency does not behave like a step function at the p_T threshold. Generally, a gradual rise of efficiency to plateau corresponding to unity is observed as in figure 3.17 (right). These curves are referred to as trigger *turn-on curves*.

Another aspect to keep in mind is the so-called trigger bias. Since application of triggers only selects certain classes of events and reject others, it is possible to miss some interesting events entirely. The preferred way is to apply the corresponding selections to the simulation as well, and re-running the analysis using an alternate trigger. An oft-repeated saying is: 'The Trigger does not determine which Physics Model is right, only which Physics Model is left'.

3.4.2 Pile-up correction

Pile-up is becoming more and more active with increasing instantaneous luminosity. Pile-up affects reconstruction of all the objects, and calculation of MET by polluting the event with extra energy and tracks. Here we mention some commonly used methods to reduce the effects of pile-up.

Although pile-up contributes extra tracks, usually those will not originate from the primary vertex (i.e. from the hard scatter, HS). So tracks coming from pile-up are comparatively easy to identify and discard. However, if a pile-up vertex is located very near to the PV, the vertices can end up being merged due to the finite resolution of the tracker, and then the tracks from pile-up will contaminate the PV tracks.

For jets, pile-up can either add extra energy to the jets, or create entirely new jets. The concept of jet area is useful to subtract the effect of pile-up [33]. The pile-up contamination is roughly uniform as a function of the jet area, so the corrected p_T of the jet can be written as: $p_T^{corr} = p_T - \rho A_{jet}$, where ρ is the momentum contamination due to pile-up per unit area. Since anti-k_t jets have a circular area, it is easier to subtract pile-up from them. Charge Hadron Subtraction (CHS) [34] is another method to remove charged particles associated with pile-up from jet reconstruction. The association is done by comparing the fit quality of a track with PV or pile-up vertex.

To identify jets coming mainly from pile-up, tracking information is used to calculate a variable called the jet-vertex-fraction (JVF). It is defined as the fraction of the total momentum of tracks in the jet which is associated with the primary vertex [35]. Imposing a lower limit on this variable rejects the majority of pile-up jets. The Pile-up Per Particle Identification (or Puppi) [36] technique popularised by CMS is another way to mitigate the effect of pile-up. Each particle i is assigned a weight α_i based on the distribution of j particles around it:

$$\alpha_i = \log \sum_{j \in \text{event}} \frac{p_T^j}{R_{ij}} \Theta(R_{\min} < R_{ij} < R_0)$$

where the first factor is the sum p_T weighted by distance, and Θ is a step function. The step function ensures only particles around a certain distance of R_{ij} are considered, as otherwise it goes to zero. If the distribution of the particles corresponds to an isotropic distribution, then the weight is closer to zero (pile-up-like), and for a small number of high p_T objects around it, the weight is closer to unity (HS-like). It is assumed that a particle in the midst of other pile-up particles has a high probability of originating in pile-up. The four-momenta of the particle is then scaled by the weight, making possible pile-up particles contribute very little to calculated observables. There are dedicated techniques to subtract pile-up for large-radius jets, which will be discussed in chapter 9.

3.4.3 Calibration

The detectors cannot correctly measure the actual energy and momenta of the particles (or jets constructed from them). Two distinct effects can be caused by the detector—the p_T^{reco} of the reconstructed object may be systematically shifted from the actual value p_T^{ref}, and additionally there may be random fluctuations, i.e. the same original p_T^{ref} would not result in identical p_T^{reco} values. Usually these random fluctuations result in Gaussian distribution, so the p_T^{reco} can be represented by a Gaussian function with centre at $\mu \times p_T^{ref}$ and of width σ. Then $\mu = p_T^{reco}/p_T^{ref}$ is referred to as energy scale (or response) and σ is referred to as the resolution. Another aspect is detector resolution, as detectors can only differentiate between objects with some minimal separation in coordinates or energy. Representative

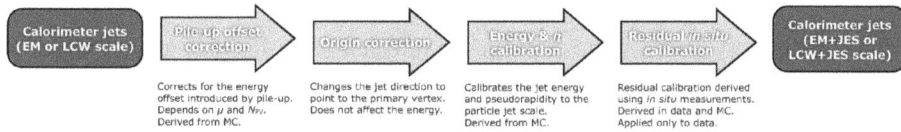

Figure 3.18. Successive steps for jet calibration, as employed in ATLAS [37] (ATLAS Experiment © 2018 CERN). First, the origin correction recalculates the four-momentum of jets to point to the hard scatter primary vertex rather than the centre of the detector, while keeping the jet energy constant. Next, the pile-up correction removes the excess energy due to in-time and out-of-time pile-up. It consists of two components; an area-based p_T density subtraction, applied at the per-event level, and a residual correction derived from the MC simulation. The absolute JES calibration corrects the jet four-momentum to the particle-level energy scale, as derived using generator level simulation. Further improvements to the reconstructed energy and related uncertainties are achieved through the use of calorimeter, muon spectrometer, and track-based variables in the global sequential calibration. Finally, a residual *in situ* calibration is applied to correct jets in data using well-measured reference objects, including photons, Z bosons, and calibrated jets.

numbers for ATLAS and CMS detectors were presented in table 3.1 earlier in the chapter.

Calibration refers to the calculation of the response and resolution, so that they can be applied to the relevant objects. There are different calibration techniques: they can be based on simulation (a topic which will be discussed in the next chapter) or on data, the latter is referred to as *in situ* calibrations. Calibration can also be performed based on the response of the detector in specific test setups. An example of the last strategy is the calibration of calorimeter energy clusters. Test beam measurements during the commissioning of the detector are used to obtain the energy response (i.e. the mapping between measured and actual energy). This is usually validated by some *in situ* techniques, by comparing with a sample of $Z \rightarrow e^+e^-$ and $J/\psi \rightarrow e^+e^-$ events, where the resonance mass peak is well defined. Single-electron response is also derived from $W \rightarrow e\nu$ events using E/p, the ratio of reconstructed energy to its track momentum of the electron. For photons, $\pi^0 \rightarrow \gamma\gamma$ events are used.

Calibration of the jets is usually a multi-step procedure, as shown in figure 3.18 from ATLAS. The first calibration stage involves calibrating the clusters to account for hadron response. This is done by comparing energy measured from calorimeter with momentum measured in tracker for charged hadrons. Then a sequence of calibration steps are applied to the reconstructed jets, most of which are dependent on the particular detector. However, some general features to be addressed can be outlined:

- Calorimeter non-compensation: correction for difference responses in EM and hadronic calorimeter.
- Dead material: energy loss in the inactive material of the calorimeter.
- Noise: energy deposits below noise thresholds.
- Leakage/punch-through: showers reaching the outer edge of the calorimeter.
- Pile-up: residual effects after initial noise suppression, often done using jet area, as discussed before.
- Non-linear response: while EM calorimeters usually have a constant response as function of energy, it is in general not true for hadronic calorimeters.

For jets, the energy scale and resolution are referred to as JES (jet energy scale) and JER (jet energy resolution). When using simulation, p_T of the particle-level jet matched (either by ΔR or by ghost-matching) can be used as reference. For data-driven *in situ* calibration, p_T balance in Z+jets, γ+jets or multijet events are used. As the p_T of the Z or the photon can be measured very accurately from their decays, so the actual energy or the p_T of the jet can be ascertained by enforcing energy–momentum conservation.

While the actual energy scale and resolution numbers depend on the specific experimental and reconstruction set-up, and the p_T and η of the objects, we can give some rough estimates to give a sense of the numbers. For jets in ATLAS, the energy scale calibrations tend to be in the order of 15%–60%, with about 1%–2% uncertainty. The energy resolution corrections are about 10%–15%, with about 10% relative uncertainty. The uncertainties are determined by comparing different calibration methods, or using different simulated samples.

A simulation should describe the data perfectly after applying the calibrations to both, any difference between them needs to be addressed. A scale factor (SF) is computed based on residual data and simulation difference is applied on the simulation to correct it. If the difference is not too large, an extra systematic uncertainty (discussed in chapter 6) is assessed on the simulation results.

3.4.4 Isolation

Reconstructed prompt leptons and photons are not expected to have other particles produced very close to them (as they are not correlated with other activity in the event), whereas that will not be the case for non-prompt or misidentified leptons and photons. So for identification of leptons and photons, an *isolation* criteria is applied. Isolation can be estimated in terms of calorimeter clusters, or tracks, although they are very correlated. The sum of energy or transverse momentum in a cone centred around the object with a specific radius (usually around 0.1 or 0.2) is demanded to have a value smaller than a threshold to classify the object as isolated. Sometimes the sum of energy or p_T in this cone divided by the energy or p_T of the object is used as a measure. Calorimeter-based isolation is more sensitive to the surrounding particle activity because it measures the energy of both neutral and charged particles. Track-based isolation, on the other hand, can only protect against the charged particle component, but it is more robust against pile-up, as tracks can be associated with the primary vertex.

3.4.5 Reconstruction and identification efficiency

Not all particles created in a collision can be detected. This can happen due to many reasons, if the particle landed in an area of the detector which is not *active* (possibly containing electrical connections, or mechanical support, or simply not working), or if it had too little energy so that it gets swamped by electronic noise which we cut out. An efficiency can be defined for reconstructing the detector objects, which is simply the fraction of originally produced objects (or fraction of originally produced energy or p_T) reconstructed by the detector. Of course there is no obvious way to know the denominator of this fraction, so indirect methods are used.

At any stage, if an object produced in the collision is *lost*, that introduces an inefficiency, and an efficiency correction needs to be applied in order to estimate the actual number of that object (or events containing that object). So the efficiencies are calculated for every *selection* stage for an object, usually corresponding to (but not limited to) reconstruction, identification (as tight/medium/loose if that is the case), isolation, and trigger (if the triggering was performed on that object). The net efficiency is a product of all the individual efficiencies.

The efficiency determination (for any of these stages) of electrons and muons in data is usually performed using a widely used technique called the *tag-and-probe method*. A pure sample of Z or $J/\psi \rightarrow e^+e^-$ or $\mu^+\mu^-$ events is chosen. In this pure sample, in every event there should be a dilepton pair satisfying the invariant mass requirement, and not finding a pair will indicate a loss of efficiency. The efficiency needs to be determined for a specific selection. A *tag* lepton is chosen satisfying a stricter selection criteria than the criteria whose efficiency is being determined. The motivation is that the tag lepton must unequivocally be a real lepton. Then a *probe* lepton is chosen, with the dilepton invariant mass satisfying the Z boson or the J/ψ mass, to make sure that probe lepton is actually also a real lepton coming from the decay under consideration. The probe lepton must also satisfy the selection criteria whose efficiency is being determined. There will be cases where the probe lepton will fail that requirement, so the pair will not be complete. The efficiency is then given by the probability of the probe being selected: $\varepsilon = N_{\text{selected probe}}/N_{\text{all}}$, where N_{all} denotes all leptons in the sample which were considered as probes initially. An example of determination of the isolation efficiency shown in figure 3.19. Often the efficiencies are reported as a function of the transverse momentum and pseudorapidity of the lepton. The advantage of this method is that it can be completely data-driven, without requiring simulated events.

While for jets, calibration is the most important part, when jets are b- (or c-) tagged, there is an associated (in)efficiency of the tagging algorithm. Several methods have been developed to measure the b jet tagging efficiency, the c jet tagging efficiency and the mistag rate in data. The b jet tagging efficiency is measured in an inclusive sample of jets with muons inside and in samples of $t\bar{t}$ events with one or two leptons in the final state. The c jet tagging efficiency is measured in an inclusive sample of jets associated with D^{*+} mesons (as they decay into charm quarks) as well as in a sample of $W + c$ events. Analyses typically use a fixed efficiency working point. The mistag rate (probability of mistakenly tagging a jet containing a c hadron or a light-flavour parton jet as a b jet) is measured in an inclusive jet sample. The results are presented as data-to-simulation scale factors, derived from the ratio of the efficiency or mistag rate measured in data to that obtained in a simulated event.

Track reconstruction has its associated inefficiencies, which can occur due to mismodelling of the amount of detector material, or misidentifying or mis-fitting the hits. The efficiency is determined using simulation, by matching generator level charged particles to reconstructed tracks satisfying a set of criteria, and then the efficiency is parametrised in two-dimensional bins of p_T and η of the tracks.

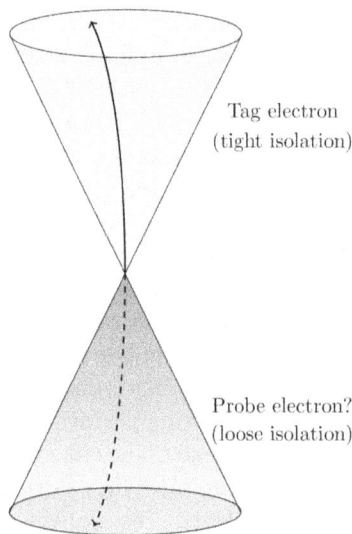

Figure 3.19. A schematic illustration of tag-and-probe procedure for determining isolation efficiency. The tag electron is represented by a light cone, indicating that it satisfies some tight isolation requirement. Then the probe electron, satisfying a looser isolation criteria, corresponding to the analysis selection needs to be found. This is depicted in a darker shade.

Usually these efficiencies are determined both in data and simulation. The ratio of the measured efficiency in data, with that measured with the same method in simulation is addressed by scale factors: $SF = \varepsilon_{data}/\varepsilon_{simulation}$. These scale factors, just like the ones discussed for calibrations, represent per-object weights and are usually measured as a function of the object kinematics as well. If the selection of an event depends on multiple objects being selected, the weight to be applied to the event from scale factors is the product of individual scale factors in the simplest approximation. The scale factors are always applied on simulated events, never on data.

3.4.6 Fakes and overlap removal

The object reconstruction algorithms usually run independent of each other in experiments. For example, jet reconstruction would not know what energy clusters have been used to form electrons. That implies that the same detector signature can result in different reconstructed objects. This results in misidentification of objects, and potential double counting.

> Fakes denote the misidentification of one object as another.

The misidentification of hadrons or jets as electrons was mentioned earlier. The converse is also true, some leptons are misidentified as jets. These are commonly called fakes. It must be noted that for lepton identification, the jets-faking-leptons is the only relevant quantity, not how many leptons are faking jets. Dedicated methods for estimating these fake fractions will be discussed in chapter 5.

To address the potential double counting, *overlap removal* procedures are used. Jets deposit a significant fraction of their energy in the electromagnetic calorimeter, for example because they can contain neutral pions that decay as $\pi^0 \to \gamma\gamma$, as well as bottom/charm mesons with semi-leptonic decays. Therefore, the jet reconstruction algorithms have to use energy deposits from both electromagnetic and hadronic calorimeters, so that the jet four-momentum can be estimated as precisely as possible.

An electron leaves most of its energy in EM calorimeter, but since the jet reconstruction algorithm is 'democratic' in forming jets from all energy deposits, the energy deposit from the electron also results in the formation of a jet by itself. So an isolated electron in most cases has an extremely close-by jet. Overlap removal algorithms typically address this problem by removing any jet close (i.e. within ΔR of 0.2) to an electron. After that, if there is still a jet near the electron, then the electron is removed, as the assumption then is that the electron is one produced by a semi-leptonic meson decay or it is misreconstructed from a jet. Similar steps are also followed for electron close to muons, or for muons close to jets.

Exercises

1. What was the main design goal for the ATLAS and CMS detectors?
2. What is the most prevalent final state particle in LHC collisions, and why? What is the next most common final state?
3. In a collision event, you find a Z boson from two oppositely charged electrons. But the electron tracks do not seem to extrapolate back to the centre of the detector. What can be the cause?
4. There are cases when the charge of an electron can be misidentified. How would that relate to p_T of the electron? How you can use the tag-and-probe method to calculate the charge misidentification rate?
5. In an event where no neutrino is produced, the value of MET turns out be non-zero. Discuss why this may not be a mistake.
6. The extra jets coming from pile-up are removed by JVF cut in ATLAS, as mentioned. What can be a limitation of this method? In fact, this limitation, along with noise in a certain part of EM calorimeter resulted in a striking data-simulation mismatch in η distribution of jets in ATLAS. This was, interestingly enough, termed 'Batman ears' inside the collaboration, as the distribution resembled that shape, and a variable named *BadBatman* was used to remove the problematic events. Who says physicists are not creative?
7. Do we need IRC safe observables theoretically? What about experimental quantities?
8. Typically jets with radius 0.4–0.6 are used in experiments, but how small can we go and why?
9. How do the size of jets change as they are more forward?

10. Jets are typically constructed to include hadrons. What other particles can be inside jets? If an electron is produced very close to a real jet, will it be used in jet reconstruction? What about a muon? If yes for either case, what possible problems can it lead to?

References

[1] Aad G *et al* 2008 The ATLAS experiment at the CERN Large Hadron Collider *JINST* **3** S08003

[2] Chatrchyan S *et al* 2008 The CMS experiment at the CERN LHC *JINST* **3** S08004

[3] Green D (ed) 2010 *At the Leading Edge* (Singapore: World Scientific)

[4] Chatrchyan S *et al* 2014 Description and performance of track and primary-vertex reconstruction with the CMS tracker *JINST* **9** P10009

[5] Aaboud M *et al* 2016 Charged-particle distributions at low transverse momentum in $\sqrt{s} = 13$ TeV pp interactions measured with the ATLAS detector at the LHC *Eur. Phys. J. C* **76** 502

[6] ATLAS Collaboration 2015 Track reconstruction performance of the ATLAS inner detector at $\sqrt{s} = 13$ TeV *Technical Report* ATL-PHYS-PUB-2015-018, CERN, Geneva. http://cdsweb.cern.ch/record/2037683

[7] Froidevaux D and Sphicas P 2006 General-purpose detectors for the large hadron collider *Annu. Rev. Nucl. Part. Sci.* **56** 375–440

[8] Asquith L *et al* 2018 *Jet Substructure at the Large Hadron Collider: Experimental Review*

[9] Ragusa F and Rolandi L 2007 Tracking at lhc *New J. Phys.* **9** 336

[10] Billoir P 1984 Track fitting with multiple scattering: a new method *Nucl. Instrum. Meth. A* **225** 352–66

[11] Aaboud M *et al* 2017 Evidence for light-by-light scattering in heavy-ion collisions with the ATLAS detector at the LHC *Nat. Phys.* **13** 852–8

[12] Fabjan C W and Gianotti F 2003 Calorimetry for particle physics *Rev. Mod. Phys.* **75** 1243–86 (CERN-EP-2003-075, 96 p)

[13] Cavallari F 2011 Performance of calorimeters at the lhc *J. Phys.: Conf. Ser.* **293** 012001

[14] Edwards B, Losty J, Perkins D H, Pinkau K and Reynolds J 1958 Analysis of nuclear interactions of energies between 1000 and 100 000 bev *Philos. Mag.: J. Theor. Exp. Appl. Phys.* **3** 237–66

[15] Sterman G and Weinberg S 1977 Jets from quantum chromodynamics *Phys. Rev. Lett.* **39** 1436–9

[16] Field R D and Feynman R P 1977 Quark elastic scattering as a source of high-transverse-momentum mesons *Phys. Rev. D* **15** 2590–616

[17] Abe F *et al* 1992 The topology of three jet events in $\bar{p}p$ collisions at $\sqrt{s} = 1.8$ TeV *Phys. Rev. D* **45** 1448–58

[18] Blazey G C *et al* 2000 Run II jet physics *Proc. of QCD and Weak Boson Physics in Run II (Batavia, USA, March 4–6, June 3–4, November 4–6, 1999)* pp 47–77

[19] Soyez G 2008 The SISCone and anti-k(t) jet algorithms *Proc. of 16th Int. Workshop on Deep Inelastic Scattering and Related Subjects (DIS 2008) (London, UK, April 7–11, 2008)* p 178

[20] Salam G P and Soyez G 2007 A practical seedless infrared-safe cone jet algorithm *JHEP* **05** 086

[21] Bartel W *et al* 1981 Experimental study of jets in electron–positron annihilation *Phys. Lett. B* **101** 129–34

[22] Bartel W *et al* 1983 Particle distribution in three jet events produced by e+ e− annihilation *Z. Phys.* C **21** 37

[23] Catani S, Dokshitzer Y L, Seymour M H and Webber B R 1993 Longitudinally invariant K_t clustering algorithms for hadron hadron collisions *Nucl. Phys.* B **406** 187–224

[24] Cacciari M, Salam G P and Soyez G 2008 The anti-k_t jet clustering algorithm *JHEP* **04** 063

[25] Ellis S D and Soper D E 1993 Successive combination jet algorithm for hadron collisions *Phys. Rev.* D **48** 3160–6

[26] Dokshitzer Y L, Leder G D, Moretti S and Webber B R 1997 Better jet clustering algorithms *JHEP* **08** 001

[27] Wobisch M and Wengler T 1998 Hadronization corrections to jet cross-sections in deep inelastic scattering *Proc. of the Workshop of Monte Carlo Generators for HERA Physics (Hamburg, Germany, 1998–99)* pp 270–9

[28] Cacciari M, Salam G P and Soyez G 2012 FastJet user manual *Eur. Phys. J.* C **72** 1896

[29] Sirunyan A M *et al* 2017 Particle-flow reconstruction and global event description with the CMS detector *JINST* **12** P10003

[30] Cacciari M, Salam G P and Soyez G 2008 The catchment area of jets *JHEP* **04** 005

[31] Frixione S 1998 Isolated photons in perturbative QCD *Phys. Lett.* B **429** 369–74

[32] Aad G *et al* 2016 The performance of the jet trigger for the ATLAS detector during 2011 data taking *Eur. Phys. J.* C **76** 526

[33] Cacciari M and Salam G P 2008 Pileup subtraction using jet areas *Phys. Lett.* B **659** 119–26

[34] ATLAS Collaboration 2014 Pileup removal algorithms *Technical Report* CMS-PAS-JME-14-001, CERN, Geneva. https://cds.cern.ch/record/1751454?ln=en

[35] Aad G *et al* 2016 Performance of pile-up mitigation techniques for jets in *pp* collisions at $\sqrt{s} = 8$ TeV using the ATLAS detector *Eur. Phys. J.* C **76** 581

[36] Bertolini D, Harris P, Low M and Tran N 2014 Pileup per particle identification *JHEP* **10** 059

[37] Aaboud M *et al* 2017 Jet energy scale measurements and their systematic uncertainties in proton-proton collisions at $\sqrt{s} = 13$ TeV with the ATLAS detector *Phys. Rev.* D **96** 072002

IOP Publishing

Experimental Particle Physics
Understanding the measurements and searches at the Large Hadron Collider
Deepak Kar

Chapter 4

Theoretical view of collisions and simulating them

… But it often happens that the physics simulations provided by the MC generators carry the authority of data itself. They look like data and feel like data, and if one is not careful they are accepted as if they were data.

—James Daniel Bjorken [1]

In the previous chapters, we discussed how protons are made to collide at the LHC, and what we observe in the detectors. Another important aspect is predicting or modelling the outcome of the collisions, otherwise it will be impossible for us recognise a deviation from Standard Model (SM) expectations, as the predictions are based on the SM. The theoretical modelling is implemented in simulation programmes based on the Monte Carlo technique. The resulting simulated events not only allow us to compare data to SM predictions, but are critical for any data analysis. They are used to predict cross-sections and topologies (as in the configuration of the reconstructed objects in the event) of various processes, which helps in determining the feasibility of an analysis, and devising the analysis strategy. The simulated events are also used to calculate detector response and acceptance/ efficiency, estimate the background contribution and correct the data for detector effects, as we will see in the next chapters.

This chapter is intended to provide an overview of the collisions from a theoretical standpoint. We introduce Feynman diagrams, an easy way to represent the individual processes taking place in section 4.1. The discussion includes dominant production and decay modes of the most commonly encountered particles. Then the basic principles of the simulation programmes (including the ones modelling the resulting particles interaction with detector material) will be discussed in section 4.2.

doi:10.1088/2053-2563/ab1be6ch4

4.1 Theoretical overview from an experimentalist's perspective

4.1.1 Feynman diagrams

To calculate the probabilities for relativistic scattering processes, like those occurring in hadron colliders, we need to find out the Lorentz-invariant scattering amplitude which connects an initial state containing some particles with well defined momenta to a final state. A pictorial way (connecting Lagrangians to cross-sections), where each graph represents a contribution to scattering matrix amplitude, is often used, and these are called Feynman diagrams. We refer the reader to quantum field theory texts, where formal calculation of scattering amplitudes from Feynman diagrams is explained. There are precise Feynman rules for transcribing diagrams into scattering amplitudes. Here we use the diagrams to display the possible contributions to a given process up to a certain order in the coupling constant.

The strength of an interaction in field theory is determined by a factor which is called the coupling constant.

We start with a simple (but incomplete) diagram depicting a quantum electrodynamical (QED) process, as in figure 4.1. The diagrams represent possible space–time configurations, with the horizontal axis representing time. The direction of time is usually taken from left to right (it is also common to find Feynman diagrams using the convention that time flows from the bottom of the diagram to the top). There are three elements in this diagram, wiggly and straight directional lines, and the points where they meet. Each line represents a particle, and each point represents an interaction vertex (they are simply symbols, and do not however correspond to the PV constructed from tracks, for example). The sum of the momentum going into each vertex (i.e. from the left) is equal to the momentum going out of it (to the right).

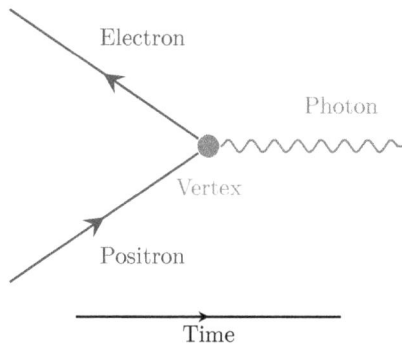

Figure 4.1. Components of a Feynman diagram. The solid blue lines represent fermions, with the forward arrow (toward the flow of time) depicting a particle and the backward arrow (against the flow of time) an anti-particle. The green wiggly lines represent force-carrier bosons. The red dot denotes an interaction vertex.

The straight lines here represent electrons and positrons, denoted by e. The particles are depicted as moving toward the direction of time, while the anti-particles are depicted as moving opposite to the direction of time. The wiggly lines representing a (virtual) photon, denoted by γ, which are the force carriers for the electromagnetic interactions. Virtual particles (i.e. off-mass shell particles with all the quantum numbers identifying the particle by its name, but not the mass), are internal lines in the Feynman diagrams. In contrast, the on-shell particles are produced with their characteristic mass and decay following their natural lifetime.

Each interaction vertex adds an order of the coupling constant in the calculation. The lines with arrows are fermions, where the arrow indicates whether it is a particle (pointed along the direction of time) or an anti-particle (moving back in time). The wiggly lines are bosons. The lines are usually slanted for representational purpose.

Let us now take a look at the diagrams corresponding to a simple process of two electrons interacting by the exchange of a photon, as shown in figure 4.2. The figure 4.2 (left) diagram depicts an electron coming into contact with a positron, annihilating each other and transforming into a photon. The resulting photon then splits into another electron and positron. When the diagram is rotated, as in figure 4.2 (right), it displays a different interaction with the same set of incoming and outgoing particles as before. Here the electron or the positron is emitting a photon and changing its direction (required for conservation of energy), and the emitted photon collides with either a positron or an electron and changes its direction. Changing the position of the internal vertices does not affect the Feynman diagram (the incoming and outgoing particles are the same, just the time ordering is different), it represents another contribution to the overall process. For a particular process, we have no way of knowing exactly what interaction occurred, which is why in principle we need to consider all possible diagrams with the same incoming and outgoing particles.

These diagrams represent the leading-order (LO) contribution to the process in the relevant coupling constant. However, to calculate the full amplitude, so-called higher-order diagrams need to be included as well.

Diagrams with extra emissions are also possible, as shown in the figure 4.3 (left). Here, the electron–positron annihilation leads to the production of a quark–antiquark pair, with one of them emitting a gluon (the helical line). It is called a multi-leg diagram. Diagrams can have one or more closed loops (in fact an infinite number),

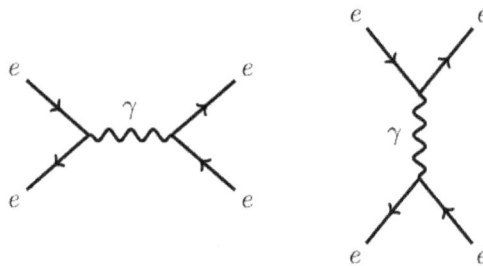

Figure 4.2. Feynman diagram for electron–positron annihilation. Both diagrams represent contributions to the same overall process.

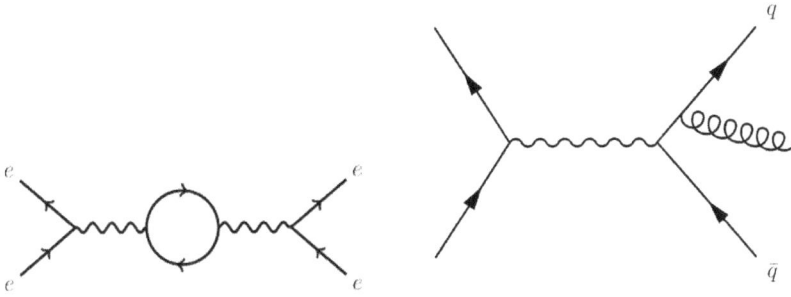

Figure 4.3. Feynman diagrams depicting multi-loop (left) and multi-leg processes (right).

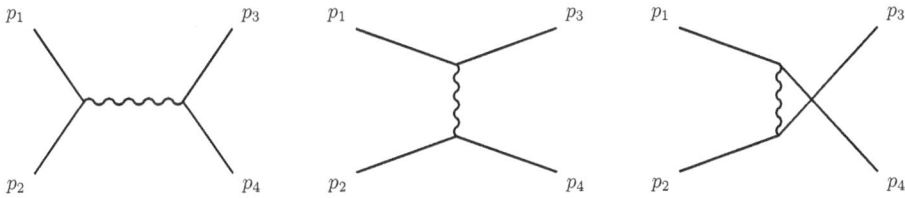

Figure 4.4. Feynman diagrams depicting s (left), t (middle) and u (right) channel processes.

as shown in figure 4.3 (right). The loops represent production (and subsequent annihilation) of intermediate virtual particles with arbitrary value of momentum. Each closed loop effectively means two more orders of the coupling constant as a loop results in two extra vertices. This represents one or more terms in perturbative series, therefore they are called next-to-leading-order or NLO. The diagrams not containing loops are termed tree level diagrams.

The leading-order contribution to a process gives a ballpark estimate of the cross-section of the process, and the shape of the distributions. The higher-order contributions lead to more accurate prediction of normalisation of the distributions and certain kinematic features, and they also provide numerical stability to the calculation, as the answer is less dependent on scale choices. The loops are sensitive to physics at higher energy scales.

4.1.2 Mandelstam variables

Let us consider a 2-to-2 scattering, with p_1 and p_2 being the four-momenta of the incoming particles and p_3 and p_4 being the four-momenta of the outgoing particles. The four-momenta conservation implies: $p_1 + p_2 = p_3 + p_4$. There are three possible ways by which the interaction can take place, as shown in figure 4.4. In the first case, the two incoming particles merge into a virtual intermediate particle that finally splits into the two final particles. In the second case, particle 1 emits a virtual particle and in doing so it turns into particle 3. The virtual particle is absorbed by particle 2, that as a consequence of this interaction turns into particle 4. In the third case, the role of particles 3 and 4 are interchanged from the second case.

The energy-squared of the intermediate virtual particle determines how the calculated scattering amplitude behaves, and can be represented in these three cases by:

$$s = (p_1 + p_2)^2 = (p_3 + p_4)^2$$
$$t = (p_1 - p_3)^2 = (p_4 - p_2)^2$$
$$u = (p_1 - p_4)^2 = (p_3 - p_2)^2$$

These variables, s, t, and u are termed Mandelstam (named after the proposer, who was a South African physicist) variables, and the corresponding processes are termed s, t, and u channel processes. The variable s can be shown to represent the square of the centre-of-mass energy. Since the energy transfer is maximum in this case, the chances of observing a new heavy particle is the highest in the s-channel. It can also be shown:

$$s + t + u = \sum_{i=1,2,3,4} p_i^2 = \sum_{i=1,2,3,4} m_i^2,$$

where p_i and m_i respectively denote the momenta and mass of the incoming and outgoing particles.

4.1.3 QCD and perturbative expansion

To predict the scattering cross-section for collider processes, we should ideally start by drawing all possible Feynman diagrams for that process. However, in practice, usually we restrict ourselves to LO or NLO (and multi-leg). Then for each diagram, we can calculate the matrix element, or the amplitude corresponding to that diagram by integrating. Then the individual amplitudes, corresponding to the diagrams considered initially, are summed over. This may lead to different amplitudes contributing positively or negatively to the total amplitude. Finally, the squared total amplitude represents the probability density for that process, which integrated over the phase space of the process gives the cross-section. For interactions involving quarks and gluons governed by quantum chromodynamics (QCD), the individual amplitudes can be expressed in terms of a perturbation series with strong coupling constant, α_S as the expansion parameter. The total amplitude receives contributions from both quark and gluon loop diagrams beyond the leading-order diagrams. If Q represents the momentum transfer in that process, the effects of these loops are absorbed in α_S, by making it a function of Q^2 (expressed in terms of squared momentum transfer to be sign-independent). Thus it is referred to as running of the coupling constant.

For high Q^2 processes energetic partons behave almost like free particles inside hadrons. As a consequence of this asymptotic freedom $\alpha_S(Q^2)$ is large at low Q^2 (which corresponds to large distances) but decreases to zero at large Q^2. The convergence of perturbative series is not automatic, but at large Q^2, the first few terms approximate the value of the amplitude reasonably well, as α_S remains within a small value ($\ll 1$). It also becomes very computationally expensive to calculate to higher orders. The running of the coupling constant is consistent with the experimental observations [2, 3]. It is customary to represent the value of α_S at $Q^2 = M_Z$, where M_Z represents the mass of the Z boson. The value turns out to be ≈ 0.12, which is large compared to electromagnetic coupling constant of 1/137, confirming that *strong* interaction is indeed strong.

For low Q^2 processes (where partons are confined, $Q^2 \approx 1$ GeV^2), this perturbative expansion does not work, as $\alpha_S \rightarrow 1$. A QCD scale parameter Λ_{QCD} is defined, corresponding to the value Q^2 where the coupling constant becomes large, leading to the breakdown of the perturbative QCD. The amplitudes for these processes cannot be calculated, rather they need to be modelled phenomenologically by looking at data distributions sensitive to these processes,

While Q^2 sets the hard scattering scale for an interaction, there are several other scales involved. A *renormalisation scale* μ_R, with the dimension of mass is introduced to subtract the (high-energy or ultra-violet, UV) divergences which result from summing up the higher-order Feynman diagrams, and it is the scale where the coupling constant is evaluated. A *factorisation scale* μ_F is introduced to separate (factorise) non-perturbative dynamics (i.e. from PDF) from perturbative hard cross-section. An exact calculation in all orders of perturbation series would have eliminated any dependence of the final result on the scale choices.

4.1.4 Processes at hadron colliders

The LO Feynman diagrams representing the production of the different SM particles in $p\bar{p}$ collisions are shown in the following sections.

> Anything that is not forbidden (allowed by conservation laws) will occur. But some may be more frequent than others.

Hadrons and jets: The allowed QCD vertices are ggg, $gggg$ and $gq\bar{q}$, as shown in figure 4.5. Therefore, some example processes for the production of quarks and gluons can be the ones shown in figure 4.6.

Top quarks: The production of top–anti-top quark pair (at the LHC) happens via the interaction of two gluons (90%) or of a quark–anti-quark pair (10%), depicted in

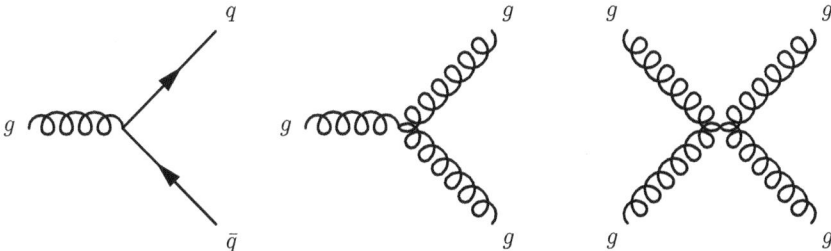

Figure 4.5. Feynman diagrams depicting allowed QCD vertices.

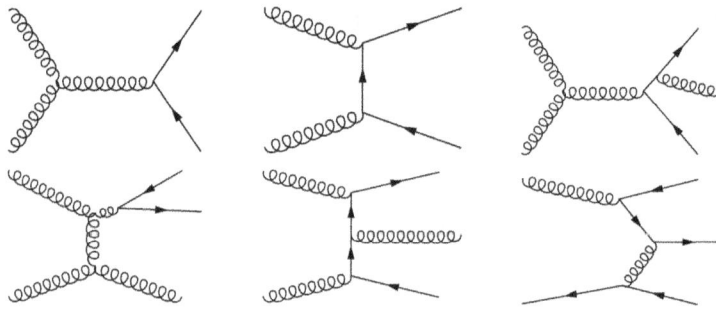

Figure 4.6. Feynman diagrams depicting some exemplar processes in hadron collisions resulting in quarks and gluons. The incoming particles can be gluon pair, quark–anti-quark pair or a mixture, as long as each vertex is allowed.

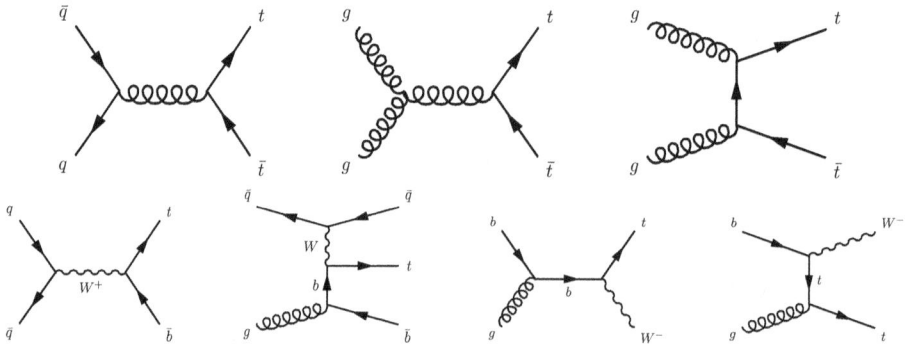

Figure 4.7. Feynman diagrams depicting top quark pair (top) and single top quark (bottom) production.

figure 4.7 (top row). The former process is usually termed gluon–gluon fusion (ggF). Production of a single top quark happens in three ways. The first is the s-channel production (4%) of the top quark along with a bottom quark from a quark–anti-quark pair via a virtual W boson. The t-channel interaction (71%) results in a top quark along with a quark, produced from an (anti-)quark and gluon, via a virtual W boson and bottom quark exchange. Finally, the associated production with a W boson (25%) happens from a gluon and bottom quark interaction, mediated by (as in, with a virtual) a top or a bottom quark. These are shown in figure 4.7 (bottom row). However, at NLO, t-channel and s-channel production mix with each other, and the Wt production interferes with the $t\bar{t}$ production, which makes the distinct classification of these processes ambiguous. The $t\bar{t}$ decay modes are shown in figure 4.8.

Z/W bosons: Drell–Yan process (named after Sidney Drell and Tung-Mow Yan [4], who first performed theoretical calculations for this process) is defined as where a $q\bar{q}$ pair forms a Z boson or a (virtual) photon, which decays to a lepton and anti-lepton pair, without any jets. However pp collisions can give rise to a Z boson in many different ways. It can be produced in association with a quark (which gives a jet), or the Z boson can be radiated off a quark, along with the production of a gluon. The possibilities are shown in figure 4.9. W boson production occurs

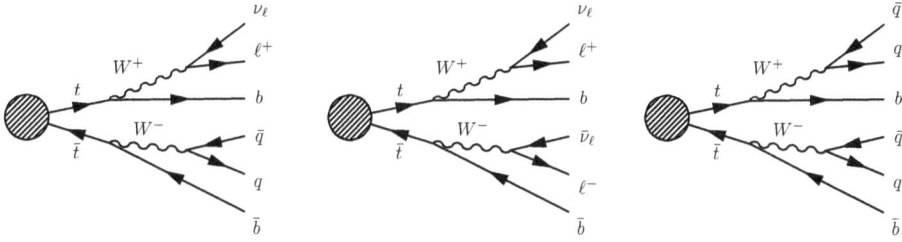

Figure 4.8. Feynman diagrams depicting the three possible top quark pair decay modes, where one top quark decays leptonically and the other decays hadronically (semi-leptonic, left), where both decays leptonically (leptonic, middle), and where both decays hadronically (all-hadronic, right). The blob on the left denotes the top quark pair production.

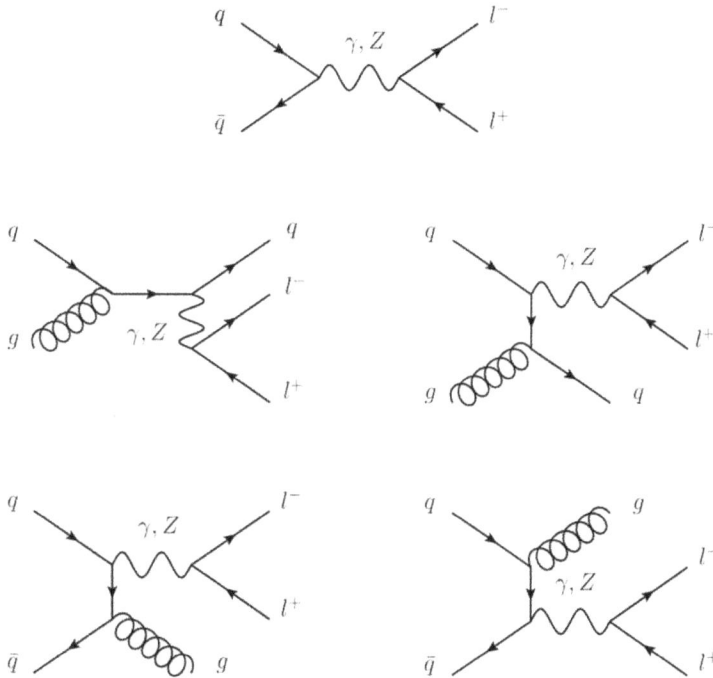

Figure 4.9. Feynman diagrams depicting the Z boson (and photon) production and decay in Drell–Yan mechanism (top), and in association with a quark or a gluon (middle and bottom row). The same diagrams can represent W boson production, but then the decay will be either via a quark–anti-quark pair, or via a lepton and corresponding anti-neutrino.

similarly. Production of each extra jet is suppressed by one order of α_S. They are also produced by much rarer electroweak processes. In these processes, a quark or anti-quark interacts with another quark or anti-quark by exchange of a virtual electroweak gauge boson, namely a W or Z boson or a photon. When this interaction produces a single Higgs boson or electroweak gauge boson, then it is called Vector Boson Fusion (VBF), while emission of two electroweak bosons is termed Vector

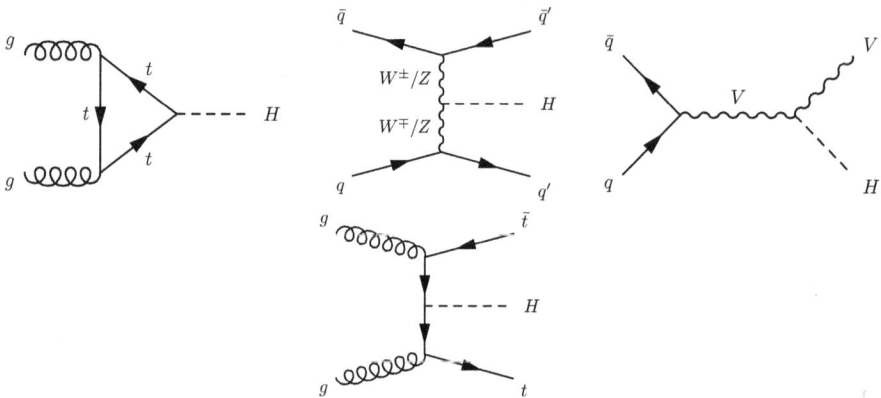

Figure 4.10. Feynman diagrams depicting Higgs boson production modes, gluon–gluon fusion, vector boson fusion, associated production with vector bosons, and associated production with top–anti-top quark pair (from left to right). For the associated production with vector bosons, V is used to denote W or Z bosons.

Boson Scattering (VBS). The two quarks from which the vector bosons radiated produces two forward jets, which is a tell-tale signature of these processes.

Higgs boson: The overwhelmingly dominant production mode (87%) for the Higgs boson is via *ggF*, through a virtual top quark loop [5] (as massless gluons cannot interact directly with the Higgs field). The next most occurring mode (7%) via VBF, where two W or Z bosons, each radiated off a quark in a proton, fuses together to create the Higgs, and two (anti-)quarks remain. In the Higgs-strahlung or associated production mode with W/Z bosons (5%), the Higgs is essentially radiated off the W/Z boson. Finally, it can also be produced along with two quarks (1%), where the $t\bar{t}H$ is of most interest. The Higgs boson production and decay modes are shown in figure 4.10 and in figure 4.11.

The major decay modes with branching ratios [6] of these particles (and also of τ leptons) at $\sqrt{s} = 13$ TeV are summarised in table 4.1. Although the numbers for $t\bar{t}$ can be obtained from W branching fractions, they are mentioned separately for convenience. The decay modes with sub-percent contributions are not shown, except for the Higgs boson.

A discussion of τ leptons is in order. Their decay modes are shown in figure 4.12. In the leptonic mode, the τ lepton decays to an electron or muon and a pair of neutrinos. The difficulty here is that the electron or muon from the τ lepton decay is nearly impossible to distinguish from an electron or muon from another process, so τ leptons are hard to be identified in these processes. As a result, leptonically decaying τ leptons may already be included in analyses which look for leptonic signatures from electrons or muons. They are usually identified in using their hadronic decay mode, from isolated charged hadron candidates (either one alone, or three nearby), which happens ≈65% of the time. This results in a narrow jet of an odd number of charged particles (usually pions) and any number of neutral pions. The neutral and charged hadrons stemming from the τ lepton decay make up the visible part of the

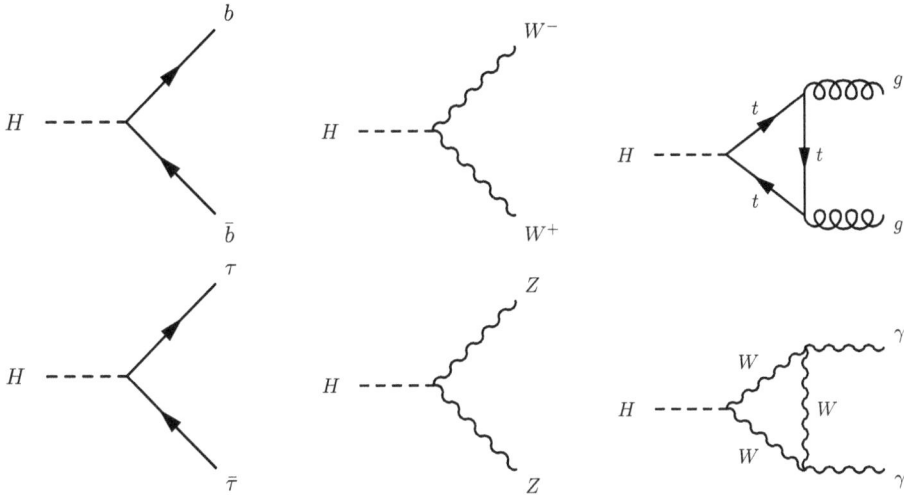

Figure 4.11. Feynman diagrams depicting major decay modes of the Higgs boson.

tau lepton (the rest being the neutrino). They are categorised by isolated hadron candidates (1-prong or 3-prong, indicative of the number of charged particles they yield). However, the main difficulty lies in discriminating them from light quark and gluon jets, and from electrons faking jets. Variables such as the number of isolated and collimated tracks, the area of the energy deposit, the reconstructed visible mass, are used. The details of the reconstruction algorithms are beyond the scope of this book, but they often involve multivariate analysis and/or boosted decision tree (BDT). The performance of these algorithms is measured with Z boson or top quark decays to τ leptons, and energy calibration and necessary correction factors are determined. Their unique position as the heaviest leptons ensures that they have a crucial role in Higgs and BSM phenomenology as in the strongest direct probe of Higgs couplings to leptonic sector.

Before concluding this section, a brief discussion of decays of quarks other than top quark are in order, which also sheds light on how hadrons decay. The strong decays result in resonances. They can also decay via weak interaction, characterised by much longer lifetime compared to strong decays.

The weak decays follow a pattern, a quark of charge $+2/3$ (u, c, t) is transformed to a quark of charge $-1/3$ (d, s, b) and vice versa, by exchange of virtual W bosons (that is why the charge changes by one unit). Generally a quark decays to the next most massive quark possible, leading to a successive decay chain like: $t \rightarrow b \rightarrow c \rightarrow s \rightarrow u \leftrightarrow d$, which always involves a change of quark flavour. The virtual W boson can decay into a lepton and a neutrino, which will lead to a lepton being present in a meson decay. This explains why a lepton can be present in jets originating from B-hadrons (b-tagged jets). The hadrons have other rare decay modes as well.

The weak decays are represented in terms of mixed states (d', s', b') of down-type quarks. Cabibbo–Kobayashi–Maskawa (CKM) matrix [7, 8] relates them as:

Table 4.1. Major decay modes with branching ratios for different standard model particles. When both the top quarks in $t\bar{t}$ process decay leptonically, we can have same flavour leptons or different flavour leptons coming from the two. These are marked as the same or mixed in the parenthesis. The decay modes with sub-percent contributions are not shown, except for the Higgs boson.

Particle	Decay modes	BF (in %)
Z	$\ell^+\ell^-$	10 (equally divided into three flavours)
	$\nu\nu$	20 (equally divided into three flavours)
	$q\bar{q}$	70
W	$\ell^\pm\nu$	33 (equally divided into three flavours)
	$q\bar{q}$	70
Top	Wb	100
$t\bar{t}$	$\ell\nu b, q\bar{q}b$	14.8 (equally divided into three flavours)
	$\ell\nu b, \ell\nu b$	2.5 (equally divided into three flavours, mixed)
	$\ell\nu b, \ell\nu b$	1.25 (equally divided into three flavours, same)
	$q\bar{q}b, q\bar{q}b$	44.4
Higgs	$b\bar{b}$	60
	W^+W^-	21
	gg	9
	$\tau^+\tau^-$	5
	$c\bar{c}$	2.5
	ZZ	2.5
	$\gamma\gamma$	0.2
	$Z\gamma$	0.15
τ	$e\nu_e\nu_\tau$	17.8
	$\mu\nu_\mu\nu_\tau$	17.4
	$\pi^-\nu_\tau$	10.8
	$\pi^-\pi^0\nu_\tau$	25.5
	$\pi^-\pi^0\pi^0\nu_\tau$	10.3
	$\pi^-\pi^0\pi^0\pi^0\nu_\tau$	1.1

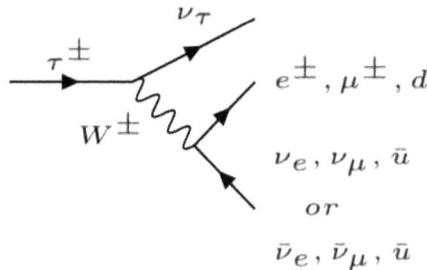

Figure 4.12. Feynman diagram depicting τ lepton decays.

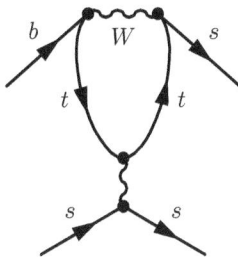

Figure 4.13. An example of a penguin diagram.

$$\begin{bmatrix} d' \\ s' \\ b' \end{bmatrix} = \begin{bmatrix} V_{ud} & V_{us} & V_{ub} \\ V_{cd} & V_{cs} & V_{cb} \\ V_{td} & V_{ts} & V_{tb} \end{bmatrix} \cdot \begin{bmatrix} d \\ s \\ b \end{bmatrix} \tag{4.1}$$

The V_{ij} elements of the CKM matrix represent strength of the amplitude of the flavour changing quark transition, $i \rightarrow j + W$ (i.e. transition of quark i to quark j, along with a W boson), which is also referred to as weak charged current process (as opposed to weak neutral current processes involving a Z). The experimentally determined values of the elements show smaller but non-zero off-diagonal contributions. It needs to be unitary ($V_{CKM} V_{CKM}^{\dagger} = 1$), which makes the CKM matrix expressible in terms of four independent parameters, usually taken to be the three mixing angles, and a complex phase signifying CP violation.

While in first-order there are no allowed transitions between quarks of the same charge, higher-order diagrams predict such transitions (which are highly suppressed). They are termed Flavour Changing Neutral Current (FCNC). The corresponding diagrams are often referred to as penguin diagrams [9] (apparently as a result of a bet requiring use of the word in a paper), as shown in figure 4.13.

4.2 Simulation programmes

4.2.1 Overview

The simulation programmes (commonly termed *event generators*) are designed to simulate the final states of high-energy collisions in full detail. Some of them stop at parton level, while some others go down to the level of individual stable particles. Each generated event is represented by a list of final-state particles and their momenta (often with information of particles produced at intermediate stages saved). Usually a large number of such simulated events are generated, such that the probability to produce an event with a given configuration (in terms of the particles produced and their four-momenta) is proportional (approximately) to the probability that the corresponding event is produced in actual collisions.

As discussed earlier in this chapter, the hard scattering can be represented in terms of Feynman diagrams, and the scattering amplitudes can be calculated by integrating the amplitudes coming from Feynman diagrams in all orders. The probability of a particular event happening is then just the amplitude squared. The integration phase space is multi-dimensional, and many attributes have to be assigned to particles (colour charge, four-momentum, spin, etc) in the generation process, consistent with conservation principles. Also an N-body process receives contribution from a large number of diagrams, including interference effects. So it becomes computationally challenging to perform numerical integrations. Additionally, the analytical output of the hard process does not correspond to the stable final-state particles, as some of the intermediate steps are not calculable by perturbative expansions.

In order address this *curse of dimensionality*, the Monte Carlo method is used to generate the events. The Monte Carlo (MC) method [10, 11] in general is an integration tool, to estimate the area enclosed by a function by checking what fraction of random numbers land inside the area. As an example, we can take the example of determining π by determining the area of a circle. In figure 4.14, there is a circle inside a square, whose area we know. Now, if we throw a dart enough times uniformly over the square, the area of the circle will be given by the fraction of times the dart falls inside the circle times the area of the square. The accuracy increases with the number of tries. The method is named after the gambling resort of Monte Carlo in France due to the use of random numbers.

The integration phase space can be imagined as the area enclosed by the function, and random numbers are generated in this multi-dimensional phase space. Rather than using purely random numbers (which are hard to generate), pseudo-random numbers are used, which are generated using deterministic algorithms, initialised with a seed number. The benefit of that is reproducibility, as the same seed produces the same random number sequence. Also, they offer better control to generate numbers uniformly spanning the range

While not based on MC methods, the first use of computer simulation to describe quark fragmentation, and therefore to model jets was by Feynman and Field [12]. So it is perhaps not out of place to recall their statement: 'The predictions of the model are reasonable enough physically that we expect it may be close enough to reality to be useful in designing future experiments and to serve as a reasonable approximation to compare to data. We do not think of the model as a sound physical theory ⋯'.

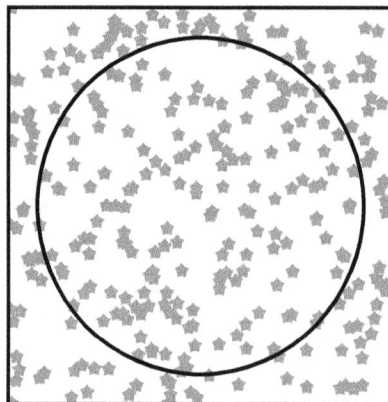

Figure 4.14. Finding the area of a circle by the Monte Carlo method.

The takeaway message is that MC simulations are our best guess of what can happen at the LHC, but we cannot take them as absolute truth. Additionally, we can only generate specific (known) physics processes (with limited accuracy), so simulation is not always expected to describe the data.

Monte Carlo event generator programmes produce simulated collision events, according to Standard Model principles. The event generation is usually factorised into different steps.

4.2.2 Hard process and parton distribution function

The hard scattering characterises the event and sets the event energy scale Q^2, which is estimated using MC methods. The (inclusive) event cross-section is given by this part. While the LO corresponds to tree-level Feynman diagrams, contributions of diagrams with one or more extra loops increases the precision of the cross-sections to NLO or to next-to-next-to-leading-order (NNLO). It must be noted that cross-sections calculated for individual diagrams at orders higher than LO can be negative, but the total sum cannot be negative. We also have to consider extra emissions (multi-leg), which results in more outgoing partons from the hard process. This part is usually referred to as fixed-order (FO) or matrix element (ME) component.

The energy fraction carried by incoming partons participating in a particular collision is not known *a priori* because of uncertainty principle, therefore the parton distribution function (PDF) is introduced [13, 14]. They state the probability of a parton carrying a certain fraction of the proton energy. The longitudinal momentum fraction $x = p_{\text{parton}}/p_{\text{proton}}$ (often referred to as Bjorken-x) of the proton momentum carried by the parton is given by $f_i(x; \mu_F)$, where i is the parton flavour and μ_F is the factorisation scale appropriate to the interaction. Then the total cross-section for processes in hadronic collisions is calculated by convolving the partonic cross-section with the corresponding PDFs, which absorbs the non-perturbative divergences by using the factorisation scale. PDFs extracted from a process can be used for different processes.

The PDF evaluation is usually factorised into two parts. QCD does not predict the parton content of the proton, so that the non-perturbative part is parametrised and the parameters are fitted from experimental data (deep inelastic scattering (DIS)[1] as well as hadron collider data) at a low initial scale. Then using the Dokshitzer–Gribov–Lipatov–Altarelli–Parisi (DGLAP) equations [17–19], the evolution to higher Q^2 (i.e. to the scale where the measurement is being performed) is carried out. This part follows perturbative QCD and thus it can be evaluated in required order (LO, NLO or NNLO). The DGLAP splitting function is a universal

---

[1] Experiments where electrons lose energy and get deflected by scattering with a hadron, leading to break-up of the hadron [15, 16].

probability distribution for the radiation of a collinear gluon in any process producing a quark.

For simplicity, most PDFs assume all quarks other than the top to have zero masses (termed *five flavour scheme, 5FNS*). To improve accuracy, sometimes b quarks are also assigned mass (termed *four flavour scheme, 4FNS*). In the former, a single (massless) b quark can be produced in PDF evolution, while in the latter, b quarks can only be produced from gluon splitting. Often the top quark and the bottom quark are referred to as heavy quarks.

The ME cross-section depends on the renormalisation and factorisation scale values. The factorisation scale in this context separates the ME and the PDF. Often they are set equal to the energy scale Q set by the hard scattering (simply because there is no other scale in the problem). It is common to vary the renormalisation and factorisation scale over some interval, perhaps $Q/2 < \mu < 2Q$ to assess uncertainties for this choice of scales, but there is no objective argument for either the central value or the range of variation.

4.2.3 Parton shower

Most processes at hadron colliders are effectively $2 \to N$ processes, where not all N of the final-state particles are created directly from the hard scatter (or calculable from ME). This complex final state is achieved by starting from a simple $2 \to 2$ process that approximately defines the directions and energies of the hardest partons, and then adding a succession of simple parton branchings to build up the full event structure. This is known as the *parton shower* (PS) approach [20, 21], which determines the structure of energy flow in the event, but the total cross-section is not modified. It does not depend on the physics process that generated the initial partons.

In case the full event is generated at LO with PS generator, often a *k-factor* is calculated as a ratio of (theoretical) NLO or NNLO cross-section over the LO cross-section, and the event is reweighted by this k-factor, without regenerating the event with higher-order accuracy. However, this does not take into account the difference of kinematics due to higher-order effects.

The branching processes $q \to qg$, $g \to gg$ and $g \to q\bar{q}$ are governed by the DGLAP equation with the probability that a branching occurs at a given time given by Sudakov form factors [22]. The Sudakov form factor essentially determines which emissions would occur, and which emissions will be *vetoed* (alternatively termed as the no emission probability). In order for these perturbative approaches to be valid, each branching must be at small angles, and the emitted parton must have a small fraction of the energy of the original parton. This is referred to as collinear and soft safety, enforced by a cutoff scale (usually around 1 GeV).

Parton branchings can happen both before and after the QCD interaction vertex giving rise to initial and final-state parton showers, respectively. The final-state shower evolution is considered *time-like*, as consecutive and iterative emissions provide the full shower structure, shown in figure 4.15 (left). To decide which of the allowed emissions occur first, an ordering is introduced [23]. Priority can be given to

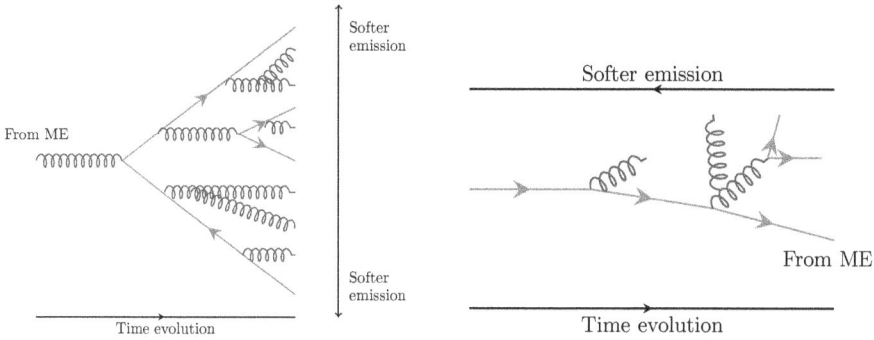

Figure 4.15. Simplified schematic illustration of final state (left) and initial state (right) shower development. The final-state shower (FSR) progresses with time, with allowed collinear parton branchings. The latter splittings result in softer or more small-angle emissions. The initial state shower (ISR) is developed backward in time, with progressively softer or small-angle emissions. The drawings are not to scale.

emissions with largest mass, p_{T} or emission angle (usually weighted by energy). The initial state shower is considered *space-like*, which stands for backward evolution [24]. It is convenient to start at the hard interaction and then the branchings are traced going backward in time to the incoming parton, as shown in figure 4.15 (right). If the emissions can be thought of as two external particles connected by a single vertex, then the squared sum of their four-momenta is termed the virtuality, which is analogous to the internal mass of the virtual particle. The virtuality comes out to be greater than zero for final-state shower, and less than zero for initial state shower (basically implying the mass is imaginary), consistent with their time-like and space-like nature.

The parton shower starts at energy scale near Λ_{QCD}, evolving down from hard scale. Then each branching results in further decrease in the energy scale, with new branches starting at lower energy scales. The evolution continues until the energy scale of partons reach Λ_{QCD}, when they undergo transitions to colour neutral hadrons.

One of the assumptions of the parton shower model is IRC safety, as introduced earlier. This requires that infra-red divergence caused by higher-order perturbative contributions due to virtual gluons are exactly cancelled by radiation of undetected real gluons, a technique known as *resummation* [25]. However, when two scales are involved in the calculation, i.e. the hard scale Q and a cutoff scale (for shower emission) Q_0, it can be shown that for $Q_0 \gg Q$ (or equivalently for $x \to 1$) large logarithmic terms in the form $\log Q^0/Q$ arising in the expansion in α_S may invalidate the perturbative expansion even for $\alpha_S \ll 1$ (or make the real and virtual terms highly unbalanced). The parton shower model effectively performs the resummation to leading logarithmic (LL) order accuracy. Parton shower models accurate to next-to-leading-logarithmic accuracy (NLL) necessitates the need for NLO (i.e. $1 \to 3$ splitting, such as $q \to qgg$, $q \to qq'\bar{q}'$, $g \to ggg$, $g \to q\bar{q}g$) calculation of parton branching.

There are alternative shower models as well. We can consider the example of the so-called *dipole shower* (sometimes the general idea is referred to as *antenna shower* as well, with slightly different implementation), which takes into account the colour of the partons [26]. Analogous to an electric dipole formed by two opposite charges, a quark–anti-quark pair with complementary (as in red and anti-red) colour charges is treated as a dipole. Then a gluon emission is generated according to the dipole radiation pattern (similar to photon radiation from an electric dipole), rather than just as an emission from a quark. The resulting gluons can form newer dipoles, which can split again, and so on until a threshold scale of dipole mass or relative transverse momentum is reached.

4.2.4 Multiple parton interactions

The composite nature of protons indicates that several parton pairs can interact when two protons collide. This is termed as multiple parton interactions (MPI) [27–29]. Often the term double parton interaction (DPI) is used to specify only one extra interaction. The presence of many interactions per event (which has been proved conclusively in experiments) also follows from the need to regulate the divergence of total QCD $2 \to 2$ cross-section dominated by t-channel gluon exchange (diverges as $1/p_T^4$ for $p_T \to 0$). The regularisation of the divergence in the cross-section for low p_T is done by the enforcing cutoff p_T^0 for MPI, replacing $1/p_T^4$ by $1/(p_T^2 + (p_T^0)^2)^2$. The MPI depends on the hadronic matter distribution inside the proton. Different shapes like a simple Gaussian, double-Gaussian or even an x-dependent Gaussian (x is the momentum fraction of the partons, as before) are used.

Each multiple interaction is associated with its set of initial- and final-state radiation showers. However, this raises an interesting question. Should all MPI be generated before the (initial and final-state) showering begins, or should they be independent? In an approach termed *interleaving*, all three effectively compete, and the process having the highest p_T value initiates the evolution.

4.2.5 Hadronisation

Finally, after the termination of the shower determined by the shower cutoff scale(s), the process of fragmentation and hadronisation converts the partons to observed hadrons, which are ultimately detected in the detector. This part technically occurs outside of the proton radius, and is non-perturbative. Two of the most used phenomenological models are the cluster and Lund string models, shown schematically in figure 4.16.

- Lund [30, 31]: Each quark–anti-quark pair, q and $\bar{q}$, are imagined to be connected by a massless relativistic string, with no transverse degrees of freedom. As they move apart from their common production vertex, the potential energy stored in the string increases, and the string may break by the production of a new quark–anti-quark pair $q'\bar{q}'$, so that the system splits into two colour-singlet systems $q\bar{q}'$ and $q'\bar{q}$. If the invariant mass of either of these string pieces is large enough, further breaks may occur. Gluons add kinks to

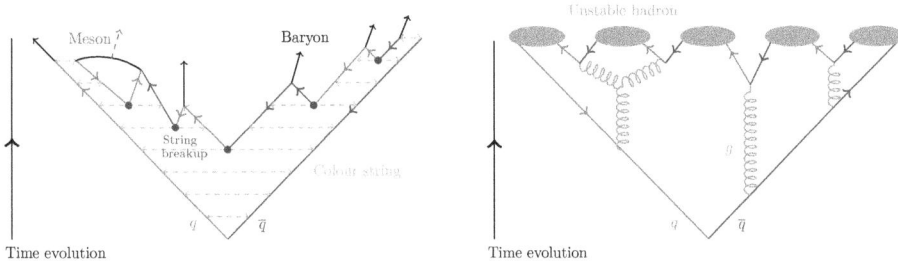

Figure 4.16. Simplified schematic illustration of Lund string (left) and cluster (right) hadronisation steps. In the Lund string model, string break-up, denoted by purple dots creates quark–anti-quark pairs, and eventually they combine to form colour neutral hadrons. In cluster model, gluons split into quark–anti-quark pairs, and intermediate colour neutral clusters are formed combining them, which then decay into hadrons. The drawings are not to scale.

the string. In the Lund string model, the string break-up starts in the middle and spreads out until there is no more energy to form hadrons in the string. The model cannot predict which hadrons will be formed, since that is determined by phenomenological parameters.

- Cluster [32–34]: at the end of the parton shower, all gluons are made to split into quark–anti-quark pairs. Neighbouring pairs form colour neutral clusters which decay into two hadrons. The decay is assumed to be isotropic in the rest frame of the cluster. The hadron type is determined by the available phase space.

At the end, the hadrons may decay further depending on their lifetimes. Roughly it has been observed than ten times more mesons are produced in comparison to baryons at the LHC [35, 36].

An additional component is colour connection and reconnection. Partons from hard scatter, MPI and beam remnants can interact by exchanging colour, represented by *connection* of colour strings between them. These strings may overlap in space, and interactions between them may lead to merging of two strings, which is referred to as colour reconnection (CR) [37, 38]. CR results in reduction of string length, which in turn leads to the reduction of hadronic multiplicity.

The full event generation is schematically depicted in figure 4.17.

4.2.6 Electroweak corrections

The strong interaction processes dominate at the LHC, however, electroweak processes often have non-negligible contributions, often as much as 10%. For the measurements or searches which are not limited by statistical uncertainty, proper estimating QED effects become important. Again, this is a vast topic by itself, so only the salient features will be described in brief. Since the $\mathcal{O}(\alpha_{EW}) \approx \mathcal{O}(\alpha_s^2)$, it can be safely assumed NLO EW effects are of similar magnitude to NNLO QCD contributions[2]. However, systematic enhancement in certain phase space regions

---

[2] Where the symbol $\mathcal{O}$ denotes orders of magnitude.

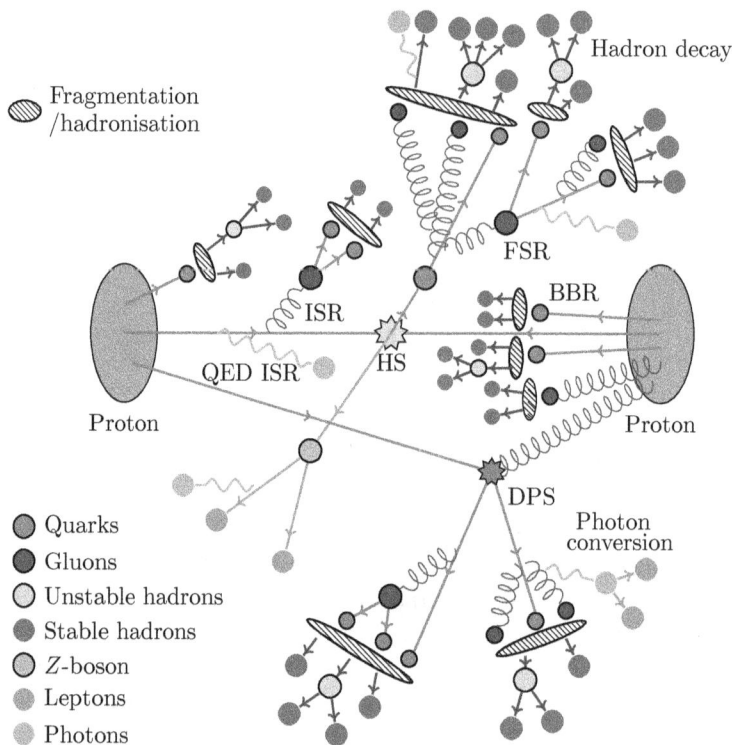

Figure 4.17. Simplified schematic diagram of a complete event generation, illustrated by a Z+jets process. The hard scatter (HS), denoted by the grey star, results in a Z boson and a quark. The former decays to an oppositely charged same flavour lepton pair, with one of them emitting a photon radiation. The latter creates a parton shower via final-state radiation (FSR), leading to the formation of unstable, then stable hadrons. The QED initial state radiation (ISR) results in a photon, while the QCD initial state radiation leads to its own parton shower, resulting in hadrons. The beam–beam remnants (BBR) similarly result in hadrons. The secondary scatter (DPS), denoted by the teal star, produces two quarks, each giving rise to its own parton shower, finally yielding hadrons. A radiated photon from a gluon converts to an electron–positron pair. The drawing is not to scale.

make their contributions non-negligible, such as photon emission or virtual exchange of soft or collinear massive weak gauge bosons. The former effect, especially for FSR photons is addressed by using *dressed* leptons, where four-momenta of all (non-hadronic-origin) photons inside a cone of certain radius (usually $R = 0.1$) of the lepton are summed up with the lepton four-momentum.

4.2.7 Matching/merging

Parton shower generators are accurate only in leading logarithmic order, while there are programmes which calculate the fixed-order ME using multi-leg or multi-loop calculations. Not only are the NLO calculations more accurate than LL, they are also more robust with respect to scale choices. Kinematic features that cannot occur at LO, begin to arise at NLO. For example, the W and Z bosons can only be produced with zero transverse momentum at the LO, by not taking into account the intrinsic (non-perturbative) transverse motion of the quarks and gluons inside the colliding hadrons, nor the possibility of generating large transverse momentum by recoil against additional energetic partons produced in the hard scattering. So it is advantageous to combine them in order to generate a more precise ME result, and then use the parton shower for the full evolution of the event [39, 40]. These two approaches are complimentary in the sense that fixed-order ME calculations describe the individual hard objects better, while the PS is responsible for an accurate simulation of the evolution of the individual jet.

They cannot be run on top of each other, as that will produce double counting of jets. The parton shower already approximates the loops and legs, which will also be coming from higher-order matrix element calculations. For example, in Z boson production, the first additional jet can either come from parton shower by way of QCD radiation, or in an NLO calculation, it can come from the ME, as shown in figure 4.18.

To avoid this overlap and also to make sure that the entire phase space is covered, the most intuitive way is to define an energy scale, below which PS will be used, and above which ME will be used. This effectively corresponds to the calculation of the higher multiplicities with this cutoff scale but multiply with no emission probability of shower. Many approaches are used, and these are typically referred to as *matching* or *merging* algorithms. Historically the terms meant slightly different procedures[3], but now both are used almost interchangeably. The scale that defines the border between ME and PS is, therefore, referred to as matching or merging scale. As the main aim is to not double count parton emissions, the matching/merging scales are defined using p_T or energy of jets.

It is beyond the scope of this text to do justice to individual methods, rather a very simplified overview of some commonly used approaches are presented. The first three involve (multi-leg) LO ME, while the next ones consider NLO ME.

- MLM [41] (named after Michaelangelo Luigi Mangano): jets from PS are matched (by ΔR) with original ME partons, and the event is only kept if each PS jet matches to one parton in the ME event. The motivation is that a parton shower should not produce emissions not consistent with original partons. This approach has a drawback in that it throws out a large fraction of events, thereby making the event generation process inefficient.
- CKKW [42, 43] (Catani–Krauss–Kuhn–Webber): the idea is to do the PS starting from the highest possible scale, but to veto shower emissions which are above matching scale, set by the ME. In order to avoid dependence on the chosen matching scale, a weight for the event is calculated based on the Sudakov form factors and the running coupling from PS, and the event is reweighted. This does not lead to the discarding of any events unlike MLM scheme.

---

[3] Merging referred to division of the phase space and restricting contribution of ME and PS to hard and soft regions separated by a merging scale, while matching referred more to combination of ME and PS by applying corrections. Currently merging mostly refers to combining exclusive ME multiplicities.

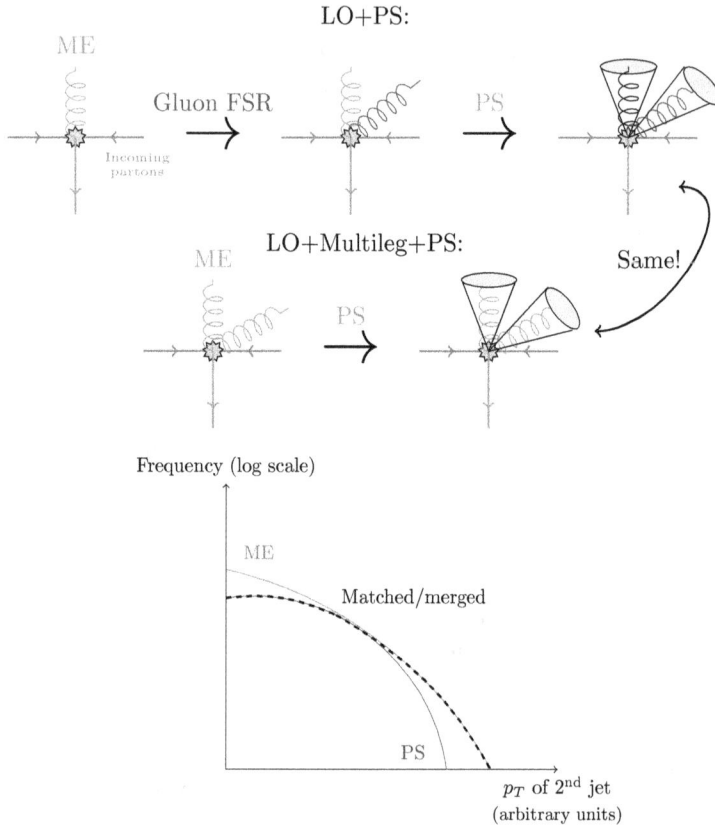

Figure 4.18. The problem of double counting, when combining matrix element (ME) with parton shower (PS) results, illustrated by Z+jets process (left), and the desired p_T spectra of the additional jet (right, inspired from Johan Alwall). For leading-order (LO) matrix element and subsequent parton shower, a hard gluon final-state radiation (FSR) can result in a jet, while for next-to leading-order (NLO) matrix element and subsequent parton shower, a hard gluon from the matrix element will give rise to a jet through parton shower. Now if a parton shower creates the same jet from hard gluon final-state radiation, then effectively the same jet will be created twice. When performing matching or merging, the p_T of this contentious jet should initially follow the shape given by the matrix element, but then should smoothly transition to the behaviour given by the parton shower.

- CKKW-L [44] (CKKW–Lonneblad) this is very similar to CKKW, but differs in the detail of calculation of weights (based on parton shower histories), and how the shower veto is applied.
- NLO merging: To preserve the accuracy of NLO ME, subtraction of the 'unwanted' part of the PS is needed [45–48]. Two approaches are most common. In the MCatNLO [49] method, this is achieved by assigning negative weights to certain PS configurations, which contain radiations that are already accounted for at the NLO calculation. Then summing the weights for the full sample gives the correct normalisation.

In the Powheg (POsitive Weight Hardest Emission Generator) [50] approach, the double counting is avoided by defining a threshold cutoff equal to the momentum of the hardest emission. In the PS any emissions that have energy greater than this threshold are vetoed. The correct normalisation is attained by (multiplicative) event weights. An additional damping parameter D (within 0–1) for real emission cross-section is used to suppress the resummation of higher-order effects, having the form: $D = h^2/(p_T^2 + h^2)$, where p_T sets the hard scale, and h is a tunable parameter corresponding to the resummation scale, named h_{damp} (the h comes from the fact that it was first used in calculations for Higgs boson production).

- There exist dedicated schemes for multi-jet merging, which can combine several NLO orders, taking care that the real emission parts of NLO calculations do not overlap. Some of the schemes are FxFx (MCatNLO for different jet multiplicities combined with MLM like matching) [51], MENLOPS [52], UNLOPS (focus on unitarity of the shower, i.e. preserving all emission probabilities to add up to unity) [53].

4.2.8 Tuning

These MC models usually have many free parameters controlling different aspects of the phenomenological models, and these need to be set based on data. Technically, the model parameters can be varied independently of each other, but the physical requirement of a sensible description of a set of data leads to correlations and anti-correlations between the parameters. Hence, it is necessary to produce *tunes*, not of one parameter at a time, but simultaneously for a group of them.

First, we will try to describe some of the parameters usually tuned and their effect in table 4.2. This list is by no means exhaustive. Each programme also has some specific modelling features, resulting in parameters specific to that programme as well[4].

Tuning is the process where the values of parameters which cannot be determined from first principles in the event generator programmes, are set using data distributions sensitive to that specific physics aspect.

Although in an ideal situation, it would have been desirable to tune all the parameters corresponding to these different modelling stages together, it becomes impossible to do that because of computational limitations. Therefore, the tuning is usually performed in stages, by finding specific classes of measurements sensitive to different modelling aspects (the measurements themselves will be discussed in detail in chapter 6). However, this factorisation ignores the interplay between different

---

[4] Since the PYTHIA 8 programme has the most extensive PS model, this table is strongly influenced by PYTHIA 8 terminology. However, the specific PYTHIA 8 (or PYTHIA 6, or HERWIG++) parameter names have not been mentioned to keep the discussion sufficiently general. For PYTHIA 8, the reader is referred to the excellent online manual [54], and for HERWIG++, (Herwig7) a very concise list of parameters exists [54].

Table 4.2. Tuning parameters. It has become standard practice use the mass of the Z boson, M_Z, which is simply a convenient reference scale large enough to be in the perturbative domain. The table is adapted and expanded from [29].

Process	Parameter	Effect
Hard process	α_S value at scale $Q^2 = M_Z^2$	More/less activity from hard process for higher/lower value
MPI	α_S value at scale $Q^2 = M_Z^2$	More/less activity from MPI for higher/lower value
MPI	p_T cutoff scale	Larger/smaller values result in less/more MPI
MPI	Energy extrapolation exponent	Determines the p_T cutoff scale as a function of $\sqrt{s}$
MPI	Impact parameter/inverse hadron radius dependence	Smaller/larger values indicate more/less overlap between colliding hadrons, hence more/less MPI
BBR/CR	Strength of colour reconnection	Higher/lower value indicates more/less chances of two systems merging
BBR	Length of colour strings	Longer string length gives more multiplicity per interaction, resulting in more high p_T particles
ISR	Primordial k_T	Parton intrinsic transverse momentum due to uncertainty principle
ISR	α_S value at scale $Q^2 = M_Z^2$	More/less activity from ISR for higher/lower value
ISR	α_S order	Order at which ISR strong coupling constant runs
ISR	p_T cutoff scale	Larger/smaller values result in less/more ISR
ISR	Setting of maximum shower evolution scale	Depends on the matching to hard process, multiplicative/damping factors can be applied
FSR	α_S value at scale $Q^2 = M_Z^2$	More/less activity from FSR for higher/lower value
FSR	α_S order	Order at which FSR strong coupling constant runs
FSR	p_T cutoff scale	Larger/smaller values result in less/more FSR
FSR	Setting of maximum shower evolution scale	Depends on the matching to hard process, multiplicative/damping factors can be applied
Frag/had	Model dependent string or cluster parameters, separate for heavy-quark fragmentation, strangeness suppression	
Matching/ merging	Cutoff scale	More/less extra jets from ME/PS for higher/lower value

processes. For example, in any interleaved shower model, MPI and I/FSR parameters are not totally independent. Fragmentation/hadronisation parameters usually depend on MPI and CR modelling. Typically, the tuning is performed in the reverse order of the event generation steps:

- Fragmentation and hadronisation (and decays): one can justifiably assume that this part will be independent of the details of particle production. Since electron–positron collisions at LEP experiments (ALEPH, DELPHI, L3, OPAL) offered a much cleaner final state (no QCD radiation from ISR), most of the fragmentation and hadronisation parameters are tuned using identified particle spectra and multiplicities, and b-fragmentation results. Then the same modelling is used in LHC tunes, assuming the previously stated universality.

Final-state shower: the same considerations as the previous step apply here as well. Event shape observables, fragmentation function of track jets and jet production rates help to constrain FSR parameters. Jet substructure measurements from hadron colliders are also being used.

- Initial state shower: this obviously needs to be tuned using hadronic collision data. The most used measurement is Z boson p_T (as it is sensitive to recoil against jet originating in ISR), and measurements directly sensitive to extra jets like dijet angular (de-)correlation (azimuthal angle difference of the two highest p_T jets), or gap fraction (energy flow between the leading and subleading jets), 3–2 jet ratio.

Intrinsic or primordial k_T: this is essentially a smearing (to approximate the effect of resummation of large logarithmic terms at high x) added to model the shape of Z boson p_T distribution at very low values. This is essentially independent of other ISR parameters.

Multiple parton interaction and colour reconnection: again, this is very specific to hadronic collisions. The most commonly used observables are charged track multiplicities and energy flow in inclusive events (i.e. minimum-bias events) or events with an identified hard scatter (underlying event measurements). An important aspect is to parametrise the $\sqrt{s}$ dependence of MPI, which is done by looking at data from different experiments and centre-of-mass energies. It has been seen [55] that the correlation of the average charged particle p_T with the multiplicity is very sensitive to colour reconnection. These parameters are sensitive to the PDF choice as well.

- Diffraction: usually they control the inelastic cross-section in a certain detector volume (not in the total cross-section), so such measurements are used to tune that.
- Matching/merging parameters: while they should be theoretically well motivated, some tuning is often necessary, mostly based on jet multiplicity or p_T spectrum.

There is obviously a great deal of subjectivity involved in this tuning workflow:
- What parameters to include in tuning?
- What should be the optimal order of tuning them?
- What are the physical ranges of those parameters?
- How to judge their relative importance?
- Which distributions to tune to?
- How are the different distributions correlated?
- Which part of the distributions are most important?
- How good an agreement with data is enough?

We should especially be careful such that tuning is based on distributions (or regions of distributions) which are actually sensitive to the non-perturbative parameters. Tuning cannot, or should not, compensate (often referred to as *overtuning*) for fundamental limitations in the model, be it missing higher-order contributions, or not having enough freedom in the model (to describe some feature of the data).

Table 4.3. Particle numbering convention.

Particle	PDG ID	Particle	PDG ID	Particle	PDG ID
u quark	1	e^-	11	Gluon	21
d quark	2	ν_e	12	Photon	22
s quark	3	μ^-	13	Z^0	23
c quark	4	ν_μ	14	W^+	24
b quark	5	τ^-	15	Higgs	25
t quark	6	ν_τ	16	Proton	2212

4.2.9 Practicalities and software programmes

While the previous sections gave a generic overview of event simulation, the discussion will be incomplete without a mention of the tools and approaches commonly used.

Particle IDs: In simulated events, it is convenient to designate every particle with a unique number. The particle data group (PDG) convention [56] is used, and table 4.3 shows the number for some common SM particles. The corresponding anti-particles are denoted by the same number but with negative sign. The commonly predicted BSM and SUSY particles also are assigned PDG IDs.

PDF: PDF sets (LO, NLO and beyond) are produced by a few different groups, CTEQ (Coordinated Theoretical-Experimental Project on QCD) [57], HERAPDF (only using HERA data) [58], MSTW/MRST/MMHT (Martin, Roberts, Stirling, Thorne and Watt) [59], and NNPDF (Neural Net PDF) [60–62] are the prominent ones. They are usually made available via the LHAPDF (Les Houches Accord PDF) [63–65] interface, ready to be used in user's code[5].

MC event generators: There are many available event generators [66–68], some are general purpose, and some are geared more towards specific processes. Some of the common and *currently used* generators are listed:

- PYTHIA series (named after the mythical oracle of the temple of Apollo at Delphi): originated from JETSET [69, 70], the original PYTHIA 6 [71] was in Fortran, the current version is C++ PYTHIA 8 [72, 73]. It employs the Lund string hadronisation model, and p_T ordered shower (in the latest versions). This has been one of the most widely used general purpose event generators, and over the years the maximum amount of effort has been spent on development of PYTHIA.

---

[5] As it turns out, the Fortran version was developed in the University of Florida when the author was a graduate student there, and the C++ version was developed at Glasgow University, when the author was a post-doctoral researcher there. So LHAPDF was developed literally in the next office, although the author was not involved at all!

- HERWIG series (Hadron Emission Reactions With Interfering Gluons): The initial Fortran version was named HERWIG [74, 75], which had to be interfaced to another programme named JIMMY [76] for MPI. It had evolved to a C++ version named HERWIG++ [77, 78], and the current version is HERWIG7 [79]. It uses cluster hadronisation, and angular ordering parton shower.
- SHERPA (Simulation of High-Energy Reactions of PArticles) [80–82]: provides automated merging of LO or NLO (NNLO for some processes) matrix element results (computed by AMEGIC model [83]) with its own parton shower model. The MPI and hadronisation has been based on PYTHIA model, but new developments in soft modelling are taking place, with SHRiMPS model based on Khoze–Martin–Ryskin model [84]. The first CKKW implementation appeared in Sherpa.
- MADGRAPH (named after Madison, Wisconsin, where the programme originated from) [51]: is a multi-leg matrix element generator, which calculates the Feynman diagrams corresponding to SM or BSM processes (via HELAS [85] (HELicity Amplitude Subroutines) routines). The code has now been integrated with AMC@NLO, which calculated ME at NLO. This needs interfacing with a PS shower generator, and as the name suggests, the AMC@NLO merging scheme was part of the development of this code.
- POWHEG (box) [50, 86–88]: is a generic framework which implements NLO calculation for several processes. This needs interfacing with a PS shower generator, and as the name suggests, the POWHEG merging scheme was part of the development of this code.

There are also generators focused on modelling a specific aspect of the event (so-called after-burners, as they are run on top of other set-ups, the first three on the list), or specific processes. It is beyond the scope of this book to cover them in adequate detail, but for completeness, we present a (incomplete) list.

- PHOTOS [89] for precision electroweak photon emission.
- EVTGEN [90] for B-hadron decay.
- TAULA [91] for τ lepton decay.
- WHIZARD [92, 93] is another multi-purpose parton level generator, but it has been mostly used for VBF and VBS processes.
- ALPGEN [94] is a parton level ME generator, divides the inclusive processes based on exclusive parton multiplicity.
- ACERMC [95] is a parton level ME generator, with built in libraries for many SM processes considered background to searches.
- VINCIA [96–98] and Dire [99] (DIpole REsummation) implement parton shower to NLL accuracy.
- GENEVA [100, 101] is a parton level ME generator based on resummed NNLO + NNLL calculations.
- DEDUCTOR [102, 103] is another parton shower program, with a strong emphasis on the handling of quantum interference effects, notably in colour and spin.

- EPOS [104, 105] was developed for cosmic ray spectrum, but gives a good description of soft hadronic interaction phenomena.
- HIJING [106] or HEJ [107, 108] (Heavy-Ion Jet INteraction Generator, High-Energy Jets) are used for simulation of jets in heavy-ion collisions.

There are also so-called fixed-order ME calculation codes, such as NLOJET++ [109], BLACKHAT [110], JETPHOX [111–113], the output of which need to be passed through MC generators for full simulation of events (or can be compared to experimental results after applying non-perturbative corrections, which approximates the effect of going from parton to hadron level).

For models with hundreds of free parameters (such as Supersymmetry), *spectrum generator* programmes, like SPHENO [114, 115], SOFTSUSY [116], FLEXIBLESUSY [117, 118] are used to calculate possible mass spectra from the model to be fed into programmes which would calculate Feynman diagrams corresponding to those mass points. Finally, SARAH [119, 120] is effectively a spectrum generator, which outputs source codes to be compiled with SPheno or FlexibleSusy.

FEYNRULES [119] is a MATHEMATICA [121] package, which can calculate Feynman rules associated with input Lagrangian corresponding to a new model in UFO (universal FEYNRULES output) format. The UFO is then used to implement the new physics model into MC generators. SLHA (SUSY Les Houches Accord) [122] files are the output from SUSY (or other BSM models) spectrum generators, used as input to event generators. The reader is referred to the corresponding manuals of each generators for a full discussion of their model and available features.

Event format: In order to have consistent output format across all generators, certain conventions are adapted. Parton shower generators (are made to) save events in HEPMC [123] format. Generators producing parton level output save events in LHEF (Les Houches Event File) [124] format, agreed upon at the bi-annual series of workshops held in the French village of that name. Often experiments use their own internal format, based on these.

Event generation in experiments: The LHC experiments usually end up running the MC event generators inside their experimental framework. The event generated corresponding to a process is usually referred to as a sample. Samples need to be generated for all SM processes, and for all BSM signatures being investigated. Since many searches span a range of parameter space, a sufficient number of samples need to be produced covering the parameter space. For SM processes, to span as much of the phase space as possible, *sliced* samples (or subsamples) are generated, by dividing the phase space in different slices in terms of jet p_T, mass or H_T (scalar sum of visible hard objects in the event). This ensures that a sufficient number of events are generated with high p_T objects, as that part of the phase space may be of interest in an analysis. For example, inclusive jet production has a steeply falling cross-section with leading jet p_T, we would not get any events at all with a leading jet p_T of say 1 TeV without specifically generating events with high minimum p_T threshold. Of course these events will have a very small event weight to signify that we would only find a small number of such events in data. The slicing can be done in different ways. For some processes, simply a p_T or mass cut at generation can be applied. For more complicated slicing strategies, a *filter* is applied, which only saves events satisfying some specific selection. In order to get the correct inclusive distribution, we then add up all the slices, by scaling them with event weights, cross-section for that slice, and filter efficiency, if a filtering strategy is applied. The filter efficiency is the ratio of the accepted events divided by total number of event generation attempts. Sometimes the subsamples for a particular process do not involve slicing, rather than generating them in individual decay modes. A common example is $t\bar{t}$ samples, where subsamples corresponding to semi-leptonic, dileptonic and hadronic decay modes are generated. Depending on the analysis final state, two or more of them can be combined, again by scaling them by individual cross-sections. Sometimes subsamples are generated in terms of additional partons produced along with the hard scattering process.

Weights in MC generators either account for unrealistically sampling the phase space, or to avoid double counting when matching/merging between NLO and PS. The weights can be unity, fractional positive numbers, large positive or negative numbers. Always the sum of the weights should be used rather than total number of events.

Proper use of event generators in analysis: The generated events usually contain a full history depicting how the final state was arrived at, starting from the ME calculation, via parton shower and hadronisation, and adding MPI. It is critical to understand that this is not a physical picture, and using this information in an analysis is fraught with danger. Quantum mechanically, a single such history just denotes one of the many ways a final state could have been achieved, rather than a definitive description of the evolution. Equally importantly, the order of particle emissions in parton shower is not physically meaningful but rather an artefact of the model used by the generator, primarily useful for debugging by generator authors. So devising an analysis strategy based on these histories can be very generator dependent, which is far from ideal. Also, the internal evolution is documented differently in different generators. For example, SHERPA does not write matrix element particles (W/Z or Higgs bosons) into the event record in most processes, while PYTHIA does. So it is better to treat these bosons that are written in the event record as intermediate objects, rather than physical objects. Many generators contain multiple copies of single internal particles, as a bookkeeping tool for various stages of event processing. The general advice is to only use the particles from generated events which are meant to be used, designated by *status code* of unity (and those with two, if decays of physically meaningful particles like hadrons are being studied).

Tunes: Tuning of a (parton shower) generator is usually performed either by the generator authors, or by experimental collaborations. The latter is often driven by the need to improve the modelling of a particular process, so specialised tunes (as opposed to a tune in principle applicable to all processes, which is more desirable) are often used.

The first tuning (in hadron collider context) was performed at CDF by Rick Field. His approach can be referred to as 'by-eye' tuning [125]. A set of correlated parameters, corresponding to a specific aspect of the model is varied at a time and the resulting distributions (by running the MC generator with the new set of parameters) are visually compared with the same distributions obtained from data, till a satisfactory agreement is achieved. This procedure is repeated for all the parameters, covering different aspects of the model that need to be tuned. This approach, and carefully chosen experimental distributions corresponding to each step, allows the expert to concentrate on just a few parameters and distributions at a time, reducing the full parameter space to manageable-sized pieces. Still it is usual for the changes made in subsequent steps to change the agreement obtained in previous ones by a non-negligible amount, requiring additional iterations from the beginning to properly tune the entire generator framework. However, with an increasing number of parameters and distributions, the process can become rather inefficient.

A different approach to tuning, first adapted in LEP [126], subsequently developed more by the PROFESSOR [127, 128] collaboration is now used extensively. Full generator runs are performed only for a limited set of parameter points, and then interpolating between these to obtain approximations to what the true generator result would have been for any intermediate parameter point. Then the set of predictions can be fitted to data, identifying the configuration with the best fit. This fit can be repeated for different combinations of observables, and putting different weights on certain distributions, and the parameter set which gives the best overall description for the most important observables is picked as the tune.

Historically the PYTHIA series has received most attention in terms of tuning. Tevatron era PYTHIA 6 tunes (last stable version: Tune DW [129]) were seen to be inadequate for the LHC. As a result tunes were provided by the authors, Perugia 2012 [130] by Peter Skands being the last stable version (named after the Italian town where the first MPI@LHC workshop was held in 2008), and experimental collaborations. ATLAS made the AMBT and AUET series of tunes [131], focussing on minimum-bias and underlying event measurements separately, while CMS had the Z series of tunes [132]. After the migration to PYTHIA 8, there were a series of author tunes (4C [133], 4Cx [134]), the current version being Monash [135] (again by Peter Skands, named after the Australian university where he is based). ATLAS came up with A14 [136], A2 [137], A3 [138], ATTBAR [139], and AZNLO [140] tunes, the first of which, a general purpose tune combing shower and MPI, and the next ones are specialised for minimum-bias, $t\bar{t}$ and W/Z processes respectively. CMS has CUET and CDPS series of tunes [141], focussing on UE and DPI respectively.

HERWIG++ had a series of author tunes, culminating in UE-EE-5 [142], focussing on describing UE at different centre-of-mass energies. The current HERWIG7 due to update of the model requires its own set of tunes, provided by the authors. SHERPA so far has a default tune for a specific release, performed by the authors.

Data preservation: The Durham High-Energy Physics Database, HEPDATA [143] is an open access repository, comprising of data points from plots and tables from decades of experimental publications, often including bin-to-bin correlations of uncertainties. CMS [144] and ATLAS [145] experiments have made part of their dataset public, with detailed instructions on how to use it. There has already been a publication using the CMS open data [146].

4.2.10 Detector simulation

The output of MC based simulation programmes discussed above corresponds to an idealised situation without the presence of any detector. This output needs to be input to *detector simulation* programmes, which simulate the passage and interaction of the produced particles through the different layers of the detector (again using the MC principle). While the event generation part is independent of the experiments, the detector simulation is obviously different for each experiment. It also depends sensitively on the detector and run conditions.

Detector simulation completes the event generation process, as real data will always be obtained using a detector.

The detector components are modelled in terms of geometrical objects with specific material, through which particles are *transported*, one at a time. The interactions of the particle with the detector material and the electric and magnetic fields is treated in steps, with the step sizes determined by the energy of the particle, and geometrical boundaries. At each step, the particle deposits energy in the part of the detector it is traversing by appropriate mechanisms, and creates a shower of new particles (which then are transported independently) consistent with the electric or magnetic field present in the detector. This continues until the particle reaches the end of the detector, or its energy goes down below a certain threshold.

Electromagnetic (happening both in tracker and electromagnetic calorimeter) and hadronic interactions need to be modelled differently. Table 4.4 shows how different particles lose energy in electromagnetic interaction with atoms of the detector material.

Among these, bremsstrahlung (braking radiation in German) and ionisation creates an increasing number of new (secondary) particles, corresponding to discrete energy losses by the primary particle. A threshold cut is applied, below which these effects are stopped, and the primary particle is made to lose energy conterminously. Particles are transported roughly down to an energy of 1 keV. The cumulative effect of elastic small-angle Coulomb scatterings is considered as a net deflection of the path of the particle. The neutral pion produces photon pairs copiously in the whole detector.

The next part is the hadronic interaction with the material, which results in the production of secondary hadrons of lower energies, which in turn can produce other hadrons, and so on. This cascade is called a hadronic shower, which spans many

Table 4.4. Possible electromagnetic interactions by different particles, which need to be considered to model the behaviour of electromagnetic calorimeters.

Process	Particles			
	Electron	Photon	Muon	Charged hadrons
Bremsstrahlung (photon emission)	✓	—	✓	✓
Multiple Coulomb scattering (deflection in path)	✓		✓	✓
Rayleigh scattering (deflection in path)	✓	—	✓	✓
Ionisation (production of free electron)	✓	—	✓	✓
Compton scattering (production of free electron from bound electron)	—	✓	—	—
Photoelectric effect (production of free electron)	—	✓	—	—
Conversion (lepton pair production)	—	✓	—	—
Pair production (electron pair production)	—	—	✓	—
Pair annihilation (electron–positron pair giving two photons)	✓	—	—	—

orders of energy, from TeV to keV. The whole process involves hadrons interacting with nucleus of the material, soft scattering, re-scattering, and hadronization (non-perturbative processes). These are implemented in different hadronic models, depending on particle type, their energies, and target material. So usually a combination of these models are used. The models contain adjustable parameters, which are tuned and validated using thin-target (microscopic, single-interaction) measurements, and from response of hadronic calorimeters in test beam set-ups.

Although neutrons do not register any response in the detector material, due to continuous elastic interactions with the nuclei over a period of time, they can damage the material. This does not need to be simulated unless the radiation damage needs to be studied.

While such a detailed simulation of detector is ideal (implemented in Geant4 software [147–149], stands for GEometry ANd Tracking), it needs a large amount of computing time and resources. In order to speed things up, certain approximations are made. Rather than creating the full shower for every event, a library of *frozen showers* is used depending on the initiating particle. The so-called fast simulation programmes (Delphes [150], Atlfast [151, 152]) make further approximations of the geometry and parametrisations of hadronic shower (which is usually the slowest component), and reduce the number of steps. Often the shortcomings of fast simulation are addressed by dedicated jet calibrations. Additional proton–proton collisions are overlaid to simulate the effect of pile-up at this stage, dependent on the pile-up seen in data.

In experimental software framework, this step, where interactions of particles with different layers of the detector are simulated, is said to produce *hits*. The next step is *digitisation*, where the hits are converted to electronic signals from the detector. Finally, the same triggering and reconstruction algorithms that are applied

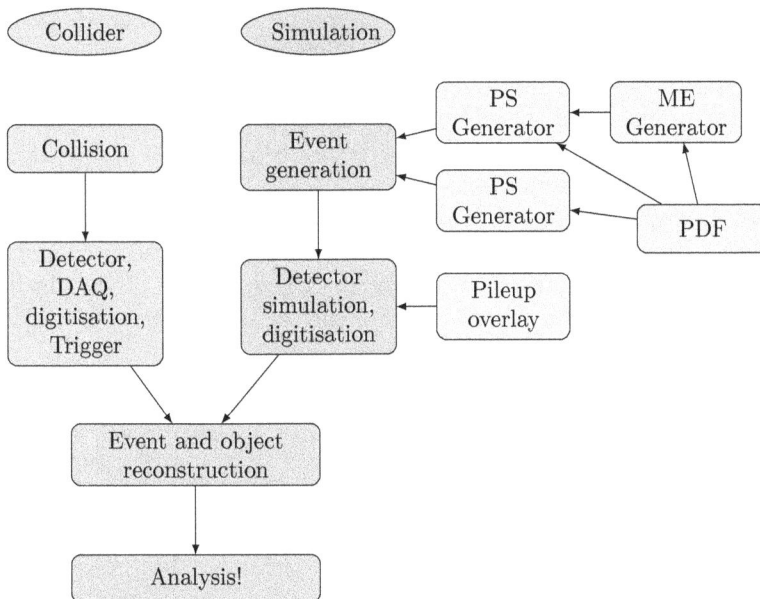

Figure 4.19. Diagram showing the steps for simulating events, compared with the processing steps for the data. The event and object reconstruction set-ups are identical.

on data to identify different physics objects are applied in MC as well, taking the digitised hits as inputs.

We end this chapter with a schematic diagram of the full event generation process, compared with the corresponding steps for real data in figure 4.19.

Exercises

1. Let us look at some striking differences in particle production between Tevatron and LHC.
   - Why is dijet production more gluon dominated at the LHC compared to Tevatron, where it was quark dominated?
   - The $t\bar{t}$ production at Tevatron was dominantly via quark–anti-quark annihilation, while at the LHC, it is predominantly via gluon–gluon fusion. Why?
   - For single top quark production, at Tevatron, s-channel production is the second most important channel ($\approx 30\%$) but it is negligible at the LHC, where Wt production has the second largest cross-section ($\approx 25\%$). Why?
2. Following the example in figure 4.14, estimate the value of π by MC method. Check how the accuracy increases with more 'darts' thrown.
3. We claim a parton gives rise to a jet. Then why do the jets not have fractional electric charge, or do they?

4. As we mentioned in the text, α_S is typically evaluated at M_Z scale. At the LHC, we are often dealing with TeV scale jets. How would you reconcile these two?

5. Let us assume you are adding different jet slices and obtain the leading jet p_T distribution. What are the potential problems you can think of, if the distribution is not smooth?

6. An example of next-to-leading-order Feynman diagram for single top quark production in association with a W boson can be represented in the left figure. However, an intermediate $t\bar{t}$ state, as in the right figure can also give the same diagram. How can we deal with this overlap when calculating single top quark production in NLO?

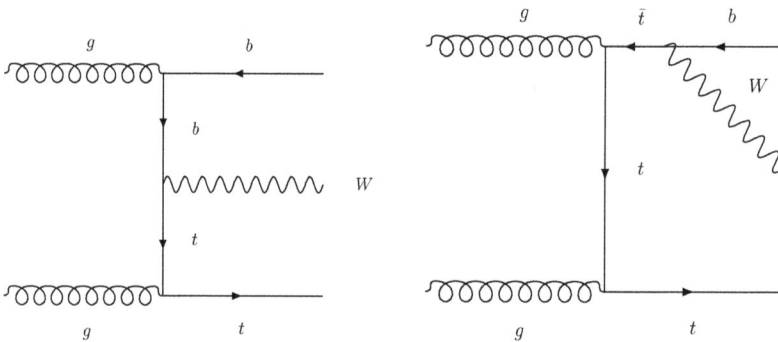

Question 6

7. Let us look two cases where we need to be careful with using event generator records.
   • Let us assume you want to look at some properties of Z boson events at generator level. You build your analysis requiring particles with PDG ID 23 (of Z boson). This gives sensible results with one generator, let us say PYTHIA 8. Then with another generator, let us say SHERPA, you generate Z boson events again, but your analysis finds none. Can you think of a reason why?
   • Let us assume you are performing an analysis with 4 b-jets in the final state. You can generate a sample with four b quarks produced in ME. Why will that be a problem?

8. Often to estimate the effect of UE, people switch off MPI in the generator. Why is this approach not robust enough?

9. A long-standing debate in the community is about using PDFs with NLO accuracy with LO generators. An example where choice of PDFs makes a significant difference is W or Z boson rapidity distributions. Using any LO MC generator programme, generate this distribution, once using a LO PDF, then using an NLO PDF. Comment on the difference, and which one do you think will be closer to data? Why?

10. What observables are expected to be most discrepant between full simulation and fast simulation?

References

[1] Bjorken J 1992 Particle physics - Where do we go from here? *SLAC Beam Line* **22N4** 8

[2] Amati D, Bassetto A, Ciafaloni M, Marchesini G and Veneziano G 1980 A treatment of hard processes sensitive to the infrared structure of QCD *Nucl. Phys.* B **173** 429–55

[3] Deur A, Brodsky S J and de Teramond G F 2016 The QCD Running Coupling *Prog. Part. Nucl. Phys.* **90** 1–74

[4] Drell S D and Yan T-M 1970 Massive lepton-pair production in hadron-hadron collisions at high energies *Phys. Rev. Lett.* **25** 316–20

[5] Georgi H M, Glashow S L, Machacek M E and Nanopoulos D V 1978 Higgs bosons from two gluon annihilation in proton proton collisions *Phys. Rev. Lett.* **40** 692

[6] Tanabashi M *et al* 2018 Review of particle physics *Phys. Rev.* D **98** 030001

[7] Cabibbo N 1963 Unitary symmetry and leptonic decays *Phys. Rev. Lett.* **10** 531–3

[8] Kobayashi M and Maskawa T 1973 CP violation in the renormalizable theory of weak interaction *Prog. Theor. Phys.* **49** 652–7

[9] Shifman M A 1995 Foreword to ITEP lectures in particle physics *ITEP Lectures in Particle Physics and Field Theory* vol 1 ed M Shifman (Singapore: World Scientific) pp v–xi

[10] Hammersley J M 1960 Monte Carlo methods for solving multivariable problems *Ann. N. Y. Acad. Sci.* **86** 844–74

[11] Weinzierl S 2000 *Introduction to Monte Carlo Methods*

[12] Field R D and Feynman R P 1978 A parametrization of the properties of quark jets *Nucl. Phys.* B **136** 1–76

[13] Dittmar M *et al* 2009 *Parton Distributions*

[14] Stewart I W, Tackmann F J and Waalewijn W J 2010 Factorization at the lhc: From parton distribution functions to initial state jets *Phys. Rev.* D **81** 094035

[15] Taylor R E 1991 Deep inelastic scattering: The early years *Rev. Mod. Phys.* **63** 573–95

[16] Engelen J and Kooijman P 1998 Deep inelastic scattering at HERA: A review of experimental results in the light of quantum chromodynamics *Prog. Part. Nucl. Phys.* **41** 1–47

[17] Dokshitzer Y L 1977 Calculation of the structure functions for deep inelastic scattering and e+ e− annihilation by perturbation theory in quantum chromodynamics *Sov. Phys. JETP* **46** 641–53

Dokshitzer Y L 1977 Calculation of the structure functions for deep inelastic scattering and e+ e− annihilation by perturbation theory in quantum chromodynamics *Zh. Eksp. Teor. Fiz.* **73** 1216

[18] Gribov V N and Lipatov L N 1972 Deep inelastic e p scattering in perturbation theory *Sov. J. Nucl. Phys.* **15** 438–50

Gribov V N and Lipatov L N 1972 Deep inelastic e p scattering in perturbation theory *Yad. Fiz.* **15** 781

[19] Altarelli G and Parisi G 1977 Asymptotic freedom in parton language *Nucl. Phys.* B **126** 298–318

[20] Nagy Z and Soper D E 2018 What is a parton shower? *Phys. Rev.* D **98** 014034

[21] Höche S 2015 Introduction to parton-shower event generators *Proc. of Theoretical Advanced Study Institute in Elementary Particle Physics: Journeys Through the Precision Frontier: Amplitudes for Colliders (TASI 2014) (Boulder, Colorado, June 2–27, 2014)* pp 235–95

[22] Collins J C 1989 Sudakov form-factors *Adv. Ser. Direct. High Energy Phys.* **5** 573–614

[23] Nagy Z and Soper D E 2014 Ordering variable for parton showers *JHEP* **06** 178

[24] Gottschalk T D 1986 Backwards evolved initial state parton showers *Nucl. Phys.* B **277** 700–38

[25] Catani S 1997 Soft gluon resummation: A Short review *Proc. of 32nd Rencontres de Moriond QCD and High Energy Hadronic Interactions (Les Arcs, France, March 22–29, 1997)* pp 331–6

[26] Platzer S and Gieseke S 2011 Coherent parton showers with local recoils *JHEP* **01** 024

[27] Sjöstrand T and van Zijl M 1987 A multiple-interaction model for the event structure in hadron collisions *Phys. Rev.* D **36** 2019–41

[28] Collins J C, Soper D E and Sterman G 1988 Soft gluons and factorization *Nucl. Phys.* B **308** 833–56

[29] Gaunt J R and Bartalini P (ed) 2018 *Multiple Parton Interactions at the LHC* (Singapore: World Scientific Publishing)

[30] Andersson B, Gustafson G, Ingelman G and Sjostrand T 1983 Parton fragmentation and string dynamics *Phys. Rep.* **97** 31–145

[31] Andersson B 1997 The lund model *Camb. Monogr. Part. Phys. Nucl. Phys. Cosmol.* **7** 1–471

[32] Marchesini G and Webber B R 1984 Simulation of QCD jets including soft gluon interference *Nucl. Phys.* B **238** 1–29

[33] Webber B R 1984 A QCD model for jet fragmentation including soft gluon interference *Nucl. Phys.* B **238** 492–528

[34] Marchesini G and Webber B R 1988 Monte Carlo simulation of general hard processes with coherent QCD radiation *Nucl. Phys.* B **310** 461–526

[35] Chliapnikov P V 1999 Hyperfine splitting in light-flavour hadron production at LEP *Phys. Lett.* B **462** 341–53

[36] Grigoryan L 2008 *Correlation Function in Field: Feynman Hadronization Model*

[37] Gustafson G, Pettersson U and Zerwas P M 1988 Jet final states in ww pair production and colour screening in the qcd vacuum *Phys. Lett.* B **209** 90–4

[38] Sjostrand T and Khoze V A 1994 On color rearrangement in hadronic W+ W- events *Z. Phys.* C **62** 281–310

[39] Mrenna S and Richardson P 2004 Matching matrix elements and parton showers with HERWIG and PYTHIA *JHEP* **05** 040

[40] Alwall J *et al* 2008 Comparative study of various algorithms for the merging of parton showers and matrix elements in hadronic collisions *Eur. Phys. J.* C **53** 473–500

[41] Mangano M L, Moretti M and Pittau R 2002 Multijet matrix elements and shower evolution in hadronic collisions: $Wb\bar{b} + n$ jets as a case study *Nucl. Phys.* B **632** 343–62

[42] Catani S, Krauss F, Kuhn R and Webber B R 2001 QCD matrix elements + parton showers *JHEP* **11** 063

[43] Krauss K 2002 Matrix elements and parton showers in hadronic interactions *JHEP* **08** 015

[44] Lonnblad L 2002 Correcting the color dipole cascade model with fixed order matrix elements *JHEP* **05** 046

[45] Lavesson N and Lonnblad L 2008 Extending CKKW-merging to one-loop matrix elements *JHEP* **12** 070

[46] Gehrmann T, Hoche S, Krauss F, Schonherr M and Siegert F 2013 NLO QCD matrix elements + parton showers in $e+$ $e-$ *rightarrow* hadrons *JHEP* **01** 144

[47] Plätzer S 2013 Controlling inclusive cross sections in parton shower + matrix element merging *JHEP* **08** 114

[48] Bellm J, Gieseke S and Plätzer S 2018 Merging NLO Multi-jet Calculations with Improved Unitarization *Eur. Phys. J.* C **78** 244

[49] Frixione S and Webber B R 2002 Matching NLO QCD computations and parton shower simulations *JHEP* **06** 029

[50] Frixione S, Nason P and Oleari C 2007 Matching NLO QCD computations with Parton Shower simulations: the POWHEG method *JHEP* **11** 070

[51] Alwall J, Frederix R, Frixione S, Hirschi V, Maltoni F, Mattelaer O, Shao H S, Stelzer T, Torrielli P and Zaro M 2014 The automated computation of tree-level and next-to-leading order differential cross sections, and their matching to parton shower simulations *JHEP* **07** 079

[52] Hamilton K and Nason P 2010 Improving NLO-parton shower matched simulations with higher order matrix elements *JHEP* **06** 039

[53] Lönnblad L and Prestel S 2013 Merging multi-leg NLO matrix elements with parton showers *JHEP* **03** 166

[54] *Root* http://root.cern.ch/ (accessed: 2018-12-30)

[55] Aaltonen T *et al* 2010 Studying the underlying event in Drell-Yan and high transverse momentum jet production at the Tevatron *Phys. Rev.* D **82** 034001

[56] *Monte Carlo particle numbering scheme* http://pdg.lbl.gov/2018/reviews/rpp2018-rev-monte-carlo-numbering.pdf (accessed: 2018-12-30)

[57] Dulat S, Hou T-J, Gao J, Guzzi M, Huston J, Nadolsky P, Pumplin J, Schmidt C, Stump D and Yuan C P 2016 New parton distribution functions from a global analysis of quantum chromodynamics *Phys. Rev.* D **93** 033006

[58] Zhang Z 2015 HERA inclusive neutral and charged current cross sections and a new PDF fit, HERAPDF 2.0 *Acta Phys. Polon. Supp.* **8** 957

[59] Thorne R S, Martin A D, Stirling W J and Watt G 2009 Status of MRST/MSTW PDF sets *Proc. of 17th Int. Workshop on Deep-Inelastic Scattering and Related Subjects (DIS 2009) (Madrid, Spain, April 26–30, 2009)* pp 22

[60] Forte S, Garrido L, Latorre J I and Piccione A 2002 Neural network parametrization of deep inelastic structure functions *JHEP* **05** 062

[61] Ball R D, Del Debbio L, Forte S, Guffanti A, Latorre J I, Rojo J and Ubiali M 2010 A first unbiased global NLO determination of parton distributions and their uncertainties *Nucl. Phys.* B **838** 136–206

[62] Ball R D *et al* 2017 Parton distributions from high-precision collider data *Eur. Phys. J.* C **77** 663

[63] Whalley M R, Bourilkov D and Group R C 2005 The Les Houches accord PDFs (LHAPDF) and LHAGLUE *HERA and the LHC: A Workshop on the Implications of HERA for LHC Physics. Proc. Part B* pp 575–81

[64] Bourilkov D, Group R C and Whalley M R 2006 LHAPDF: PDF use from the Tevatron to the LHC *TeV4LHC Workshop—4th meeting (Batavia, Illinois, October 20–22, 2005)*

[65] Buckley A, Ferrando J, Lloyd S, Nordström K, Page B, Rüfenacht M, Schönherr M and Watt G 2015 LHAPDF6: parton density access in the LHC precision era *Eur. Phys. J.* C **75** 132

[66] Buckley A *et al* 2011 General-purpose event generators for LHC physics *Phys. Rep.* **504** 145–233

[67] Sjöstrand T 2016 Status and developments of event generators *PoS* **LHCP 2016** 007

[68] Alioli S *et al* 2019 Monte Carlo Event Generators for High Energy Particle Physics Event Simulation

[69] Sjöstrand T 1983 The Lund Monte Carlo for e+ e− jet physics *Comput. Phys. Commun.* **28** 229–54

[70] Sjöstrand T 1986 The Lund Monte Carlo for jet fragmentation and e+e− physics - jetset version 6.2 *Comput. Phys. Commun.* **39** 347–407

[71] Sjostrand T, Mrenna S and Skands P Z 2006 PYTHIA 6.4 physics and manual *JHEP* **05** 026

[72] Sjostrand T, Mrenna S and Skands P Z 2008 A brief introduction to PYTHIA 8.1 *Comput. Phys. Commun.* **178** 852–67

[73] Sjöstrand T, Ask S, Christiansen J R, Corke R, Desai N, Ilten P, Mrenna S, Prestel S, Rasmussen C O and Skands P Z 2015 An introduction to PYTHIA 8.2 *Comput. Phys. Commun.* **191** 159–77

[74] Marchesini G, Webber B R, Abbiendi G, Knowles I G, Seymour M H and Stanco L 1992 HERWIG: A Monte Carlo event generator for simulating hadron emission reactions with interfering gluons. Version 5.1 - April 1991 *Comput. Phys. Commun.* **67** 465–508

[75] Marchesini G, Webber B R, Abbiendi G, Knowles I G, Seymour M H and Stanco L 1996 *Herwig Version 5.9*

[76] Butterworth J M and Forshaw J R 1993 Photoproduction of multi-jet events at HERA: a Monte Carlo simulation *J. Phys. G: Nucl. Part. Phys.* **19** 1657

[77] Bahr M *et al* 2008 Herwig++ physics and manual *Eur. Phys. J.* C **58** 639–707

[78] Gieseke S, Stephens P and Webber B 2003 New formalism for QCD parton showers *JHEP* **12** 045

[79] Bellm J *et al* 2016 Herwig 7.0/Herwig++ 3.0 release note *Eur. Phys. J.* C **76** 196

[80] Gleisberg T, Hoeche S, Krauss F, Schalicke A, Schumann S and Winter J-C 2004 SHERPA 1. alpha: A proof of concept version *JHEP* **02** 056

[81] Schumann S and Krauss F 2008 A Parton shower algorithm based on Catani-Seymour dipole factorisation *JHEP* **03** 038

[82] Winter J-C and Krauss F 2008 Initial-state showering based on colour dipoles connected to incoming parton lines *JHEP* **07** 040

[83] Krauss F, Kuhn R and Soff G 2002 AMEGIC++ 1.0: A matrix element generator in C++ *JHEP* **02** 044

[84] Ryskin M G, Martin A D and Khoze V A 2009 Soft processes at the LHC. I. Multi-component model *Eur. Phys. J.* C **60** 249–64

[85] Murayama H, Watanabe I, Hagiwara K and Ko enerugi Butsurigaku Kenkyujo (Japan) 1992 *HELAS: HELicity Amplitude Subroutines for Feynman Diagram Evaluations* (Japan: National Laboratory for High Physics Ibaraki-ken)

[86] Alioli S, Nason P, Oleari C and Re E 2010 A general framework for implementing NLO calculations in shower Monte Carlo programs: the POWHEG BOX *JHEP* **06** 043

[87] Oleari C 2010 The POWHEG-BOX *Nucl. Phys. Proc. Suppl.* **205–6** 36–41

[88] Nason P 2004 A new method for combining NLO QCD with shower Monte Carlo algorithms *JHEP* **11** 040

[89] Golonka P and Was Z 2006 PHOTOS Monte Carlo: a precision tool for QED corrections in Z and W decays *Eur. Phys. J.* C **45** 97–107

[90] Ryd A, Lange D, Kuznetsova N, Versille S, Rotondo M, Kirkby D P, Wuerthwein F K and Ishikawa A 2005 *EvtGen: A Monte Carlo Generator for B-Physics*

[91] Was Z 2001 TAUOLA the library for tau lepton decay, and KKMC / KORALB / KORALZ /... status report *Nucl. Phys. Proc. Suppl.* **98** 96–102
      Was Z 2000 TAUOLA the library for tau lepton decay, and KKMC / KORALB / KORALZ /... status report *Nucl. Phys. Proc. Suppl.* 96

[92] Kilian W, Ohl T and Reuter J 1742 WHIZARD: Simulating multi-particle processes at LHC and ILC *Eur. Phys. J.* C **71** 1742

[93] Moretti M, Ohl T and Reuter J 2002 *O'Mega: An Optimizing Matrix Element Generator* pp 1981–2009

[94] Mangano M L, Moretti M, Piccinini F, Pittau R and Polosa A D 2003 ALPGEN, a generator for hard multiparton processes in hadronic collisions *JHEP* **07** 001

[95] Kersevan B P and Richter-Was E 2013 The Monte Carlo event generator AcerMC versions 2.0 to 3.8 with interfaces to PYTHIA 6.4, HERWIG 6.5 and ARIADNE 4.1 *Comput. Phys. Commun.* **184** 919–85

[96] Giele W T, Kosower D A and Skands P Z 2008 A simple shower and matching algorithm *Phys. Rev.* D **78** 014026

[97] Giele W T, Kosower D A and Skands P Z 2011 Higher-order corrections to timelike jets *Phys. Rev.* D **84** 054003

[98] Ritzmann M, Kosower D A and Skands P 2013 Antenna showers with hadronic initial states *Phys. Lett.* B **718** 1345–50

[99] Höche S and Prestel S 2015 The midpoint between dipole and parton showers *Eur. Phys. J.* C **75** 461

[100] Alioli S, Bauer C W, Berggren C J, Hornig A, Tackmann F J, Vermilion C K, Walsh J R and Zuberi S 2013 Combining higher-order resummation with multiple NLO calculations and parton showers in GENEVA *JHEP* **09** 120

[101] Alioli S, Bauer C W, Berggren C, Tackmann F J, Walsh J R and Zuberi S 2014 Matching fully differential NNLO calculations and parton showers *JHEP* **06** 089

[102] Nagy Z and Soper D E 2007 Parton showers with quantum interference *JHEP* **09** 114

[103] Nagy Z and Soper D E 2014 A parton shower based on factorization of the quantum density matrix *JHEP* **06** 097

[104] Werner K, Karpenko I, Pierog T, Bleicher M and Mikhailov K 2010 Event-by-event simulation of the three-dimensional hydrodynamic evolution from flux tube initial conditions in ultrarelativistic heavy ion collisions *Phys. Rev.* C **82** 044904

[105] Pierog T, Karpenko I, Katzy J M, Yatsenko E and Werner K 2015 EPOS LHC: Test of collective hadronization with data measured at the CERN Large Hadron Collider *Phys. Rev.* C **92** 034906

[106] Gyulassy M and Wang X-N 1994 HIJING 1.0: A Monte Carlo program for parton and particle production in high-energy hadronic and nuclear collisions *Comput. Phys. Commun.* **83** 307

[107] Andersen J R, Lonnblad L and Smillie J M 2011 A parton shower for high energy jets *JHEP* **07** 110

[108] Andersen J R and Smillie J M 2011 Multiple jets at the LHC with high energy jets *JHEP* **06** 010

[109] Nagy Z 2003 Next-to-leading order calculation of three-jet observables in hadron-hadron collision *Phys. Rev.* D **68** 094002

[110] Berger C F, Bern Z, Dixon L J, Febres Cordero F, Forde D, Ita H, Kosower D A and Maitre D 2008 An automated implementation of on-shell methods for one-loop amplitudes *Phys. Rev.* D **78** 036003

[111] Catani S, Fontannaz M, Guillet J P and Pilon E 2002 Cross-section of isolated prompt photons in hadron-hadron collisions *JHEP* **05** 028

[112] Aurenche P, Fontannaz M, Guillet J-P, Pilon E and Werlen M 2006 A New critical study of photon production in hadronic collisions *Phys. Rev.* D **73** 094007

[113] Belghobsi Z, Fontannaz M, Guillet J P, Heinrich G, Pilon E and Werlen M 2009 Photon - jet correlations and constraints on fragmentation functions *Phys. Rev.* D **79** 114024

[114] Porod W 2003 SPheno, a program for calculating supersymmetric spectra, SUSY particle decays and SUSY particle production at e+ e− colliders *Comput. Phys. Commun.* **153** 275–315

[115] Porod W and Staub F 2012 SPheno 3.1: extensions including flavour, CP-phases and models beyond the MSSM *Comput. Phys. Commun.* **183** 2458–69

[116] Allanach B C 2002 SOFTSUSY: a program for calculating supersymmetric spectra *Comput. Phys. Commun.* **143** 305–31

[117] Athron P, Park J-h, Stöckinger D and Voigt A 2015 FlexibleSUSY—A spectrum generator generator for supersymmetric models *Comput. Phys. Commun.* **190** 139–72

[118] Athron P, Bach M, Harries D, Kwasnitza T, Park J-h, Stöckinger D, Voigt A and Ziebell J 2018 FlexibleSUSY 2.0: Extensions to investigate the phenomenology of SUSY and non-SUSY models *Comput. Phys. Commun.* **230** 145–217

[119] Adam A, Neil D C, Cline D, Duhr C and Fuks B 2014 FeynRules 2.0 - A complete toolbox for tree-level phenomenology *Comput. Phys. Commun.* **185** 2250–300

[120] Staub F 2015 Exploring new models in all detail with SARAH *Adv. High Energy Phys.* **2015** 840780

[121] Wolfram Research, Inc. 2018 *Mathematica, Version 11.3* Champaign, IL

[122] Skands P Z *et al* 2004 SUSY Les Houches accord: Interfacing SUSY spectrum calculators, decay packages, and event generators *JHEP* **07** 036

[123] Dobbs M and Hansen J B 2001 The HepMC C++ Monte Carlo event record for high energy physics *Comput. Phys. Commun.* **134** 41–6

[124] Alwall J *et al* 2007 A standard format for Les Houches event files *Comput. Phys. Commun.* **176** 300–4

[125] Field R 2011 Min-bias and the underlying event at the LHC *Acta Phys. Polon.* B **42** 2631–56

[126] Abreu P *et al* 1996 Tuning and test of fragmentation models based on identified particles and precision event shape data *Z. Phys.* C **73** 11–60 (CERN-PPE-96-120)

[127] Buckley A, Hoeth H, Lacker H, Schulz H and von Seggern J E 2010 Systematic event generator tuning for the LHC *Eur. Phys. J.* C **65** 331–57

[128] Buckley A 2008 Tools for event generator tuning and validation *Proc. of HERA and the LHC Workshop Series on the Implications of HERA for LHC physics: 2006–2008* pp 768–73

[129] Field R 2018 *Cdf Run 2 Monte-carlo Tunes*

[130] Skands P Z 2010 Tuning Monte Carlo generators: the Perugia tunes *Phys. Rev.* D **82** 074018

[131] ATLAS Collaboration 2011 ATLAS tunes of PYTHIA 6 and Pythia 8 for MC11 *Technical Report* ATL-PHYS-PUB-2011-009, CERN, Geneva https://cds.cern.ch/record/1363300

[132] Field R 2010 Early LHC underlying event data—findings and surprises *Proc. of 22nd Conf. on Hadron Collider Physics, HCP 2010 (Toronto, Canada, August 23–27, 2010)*

[133] Corke R and Sjostrand T 2011 Interleaved parton showers and tuning prospects *JHEP* **03** 032

[134] Corke R and Sjostrand T 2011 Multiparton interactions with an x-dependent proton size *JHEP* **05** 009

[135] Skands P, Carrazza S and Rojo J 2014 Tuning PYTHIA 8.1: the monash 2013 tune *Eur. Phys. J.* C **74** 3024

[136] ATLAS Collaboration 2014 ATLAS Run 1 Pythia8 tunes *Technical Report* ATL-PHYS-PUB-2014-021, CERN, Geneva http://cdsweb.cern.ch/record/1966419

[137] ATLAS Collaboration 2012 Summary of ATLAS Pythia 8 tunes *Technical Report* ATL-PHYS-PUB-2012-003, CERN, Geneva https://cds.cern.ch/record/1474107

[138] ATLAS Collaboration 2016 A study of the Pythia 8 description of ATLAS minimum bias measurements with the Donnachie-Landshoff diffractive model *Technical Report* ATL-PHYS-PUB-2016-017, CERN, Geneva https://cds.cern.ch/record/2206965

[139] ATLAS Collaboration 2015 A study of the sensitivity to the Pythia8 parton shower parameters of $t\bar{t}$ production measurements in pp collisions at $\sqrt{s} = 7$ TeV with the ATLAS experiment at the LHC *Technical Report* ATL-PHYS-PUB-2015-007, CERN, Geneva http://cdsweb.cern.ch/record/2004362

[140] Aad G *et al* 2014 Measurement of the Z/γ^* boson transverse momentum distribution in pp collisions at $\sqrt{s} = 7$ TeV with the ATLAS detector *JHEP* **09** 145

[141] Khachatryan V *et al* 2016 Event generator tunes obtained from underlying event and multiparton scattering measurements *Eur. Phys. J.* C **76** 155

[142] Seymour M H and Siodmok A 2013 Constraining MPI models using σ_{eff} and recent tevatron and LHC underlying event data *JHEP* **10** 113

[143] Maguire E, Heinrich L and Watt G 2017 HEPData: a repository for high energy physics data *J. Phys. Conf. Ser.* **898** 102006

[144] *CMS open data* http://opendata.cern.ch/docs/about-cms (accessed: 2018-12-30)

[145] *ATLAS open data* http://opendata.atlas.cern (accessed: 2018-12-30)

[146] Tripathee A, Xue W, Larkoski A, Marzani S and Thaler J 2017 Jet substructure studies with CMS open data *Phys. Rev.* D **96** 074003

[147] Agostinelli S *et al* 2003 Geant4a simulation toolkit *Nucl. Instrum. Meth. Phys. Res. Sect.* A **506** 250–303

[148] Allison J *et al* 2006 Geant4 developments and applications *IEEE Trans. Nucl. Sci.* **53** 270–8

[149] Allison J *et al* 2016 Recent developments in geant4 *Nucl. Instrum. Meth. Phys. Res. Sect.* A **835** 186–225

[150] de Favereau J, Delaere C, Demin P, Giammanco A, Lemaître V, Mertens A and Selvaggi M 2014 DELPHES 3, A modular framework for fast simulation of a generic collider experiment *JHEP* **02** 057

[151] ATLAS Collaboration 2010 The simulation principle and performance of the ATLAS fast calorimeter simulation FastCaloSim *Technical Report* ATL-PHYS-PUB-2010-013, CERN, Geneva https://cds.cern.ch/record/1300517

[152] Lukas W 2012 Fast simulation for atlas: Atlfast-ii and isf *J. Phys.: Conf. Ser.* **396** 022031

Chapter 5

Analysis

Pretend that you're Hercule Poirot: Examine all clues, and deduce the truth by order and method.

—Popular latex compiler error message

The previous chapters presented descriptions of the different physics objects constructed from the snapshot of collisions in the detectors, and how Monte Carlo (MC) based programmes simulate the collisions. An event in our terms is essentially a collection of all available information (including, but not limited to the four vectors) of all reconstructed objects from that collision. The next step is to extract physics information by looking at billions of such collision events from real data, using guidance from MC. The term *analysis* broadly describes this process, which includes *measurements* and *searches,* described in section 5.1. An integral part of performing an analysis is to construct suitable *observables*, which should be sensitive to the physics being probed. This is discussed in section 5.2. The main steps of an analysis, including object and event selection, background estimation and unfolding is described in section 5.3. Statistical methods are integral components of any data analysis, so we review some basic concepts in section 5.4, which will be needed in subsequent chapters. ROOT [1] is the most commonly used software to perform high energy physics analyses, so a brief translation of the analysis steps in ROOT terms is included in section 5.5 for convenience. However, there are other analysis frameworks as well, RIVET (Robust Independent Validation of Experiment and Theory) [2] and MADANALYSIS [3] are worth mentioning, but they are designed to process simulated events, not real data.

5.1 Measurements and searches

The analyses can be divided into measurements and searches, although measurements can also be used to set constraints on new physics models, an aspect that will be pointed out in the next chapter. The former involves measuring specific standard

doi:10.1088/2053-2563/ab1be6ch5

model (SM) parameters (such as the top quark or the *W*-boson mass, or the strong coupling constant), or comparing distributions of observables to SM predictions, in order to verify the SM model predictions (often at a new collision energy) and/or to improve the MC simulation programmes. Searches look for deviations from SM predictions. We start by reviewing the main features of searches and measurements.

> Analysis is the procedure by which we extract physics information from data (with a lot of help from simulation). Measurements are designed to verify Standard Model predictions, while the searches are designed to look for beyond the Standard Model signatures.

5.1.1 Measurements

An important aspect of measurements is that they must be made in a well-defined *fiducial region*. Fiducial region refers to an area of detector acceptance, usually (but not limited to) given by specific ranges in transverse momentum and pseudorapidity of the analysis objects. Performing measurements in the fiducial region avoids extrapolation to unmeasured regions based on simulation models, which inherently assumes the validity of the model. The model independent aspect of the measurements is important because ultimately measurements are compared with (current and future) model predictions, so any bias introduced by a model dependent analysis strategy cannot be disentangled later. An example is the CMS measurement [4] of charged particle distributions in LHC Run 1 in so-called non-single diffractive region (NSD). The analysis strategy to select such events was determined by PYTHIA generator, introducing an inherent bias.

> Measurement should always be performed in well-defined fiducial regions, in a model independent manner.

The analyses results can correspond to different *levels*. The *detector or reconstructed level* refers to the results calculated using reconstructed objects from the detector, as described in chapter 3. These results are usually experiment-specific, as detector performances vary between the different experiments. For searches, this is usually enough, as a deviation from SM prediction (provided by MC simulation, which are passed through detector simulation) still would indicate the presence of new physics. However, for the measurements to be useful, theorists and MC model developers need to be able to use them to improve their calculations or models, and that can only be done if the result has no dependence on the particular detector. So for the measurements, the detector level results are corrected for detector effects by a procedure called *unfolding*, and the results are presented at *particle or generator level*. In order to avoid model dependent corrections as elucidated above, the fiducial selections need to be as close as possible in detector and particle level. Some LHC experiments also refer to particle level as *truth level*. For measurements, especially ones involving top quarks, results are often presented at *parton level*, although that does not correspond to anything measured directly in the detector. Normally the results at parton level can be corrected back to particle level by using non-perturbative correction factor derived from MC simulation[1].

Data is always detector level, but simulated events are initially at particle level, passing them through detector simulation yields detector level simulation.

However, there are subtleties defining physics objects in simulation. When measurements are done assuming top quarks are stable physical objects, that effectively takes into account QCD ISR, but ignores QCD FSR. The same problem appears in constructing W or Z-bosons using so-called *born-level* leptons, i.e. leptons immediately originating from their decay, which ignores the effect of QED FSR. Leptons are always measured in the detector after they undergo QED FSR, so born-level objects (or assumed intermediate states like W, Z, Higgs, top) are unrealistic. Rather, measurements performed in terms of observable final state particles (e.g. leptons, photons) or objects constructed from such particles (e.g. jets, missing energy) are more realistic. Any input to measurements which depends on a specific generator prediction or on a particular signal model (be it the SM) suffers from the inherent problem that would not be valid if that model breaks down in future. Then to compare to parton level theory, the corrections should be applied to theory calculations, preserving the universality of the data measurement. Since born-level leptons do not reach the detector, rather they lose energy by QED radiation, the leptons reconstructed in the detector usually have less energy than what they started with (referred to as bare leptons). So stable final leptons from MC simulation correspond to bare leptons. In order to reach closer to born leptons, energies of all photons within a cone of $\Delta R = 0.1$ is added to the lepton, creating what is referred to as *dressed leptons*.

Dressed leptons (at particle level) include the effect of QED FSR in a model independent way.

A concept of *truth-matching* is often used in analyses, which essentially means the particle level object corresponding to the detector level object is identified, in order to check some aspects of the analysis (like efficiency calculation for a particular selection). This is usually done in terms of angular proximity, as for example a particle level electron is expected to correspond to an electron in roughly the same area of the detector after the detector simulation. While this is useful for cross-checks in setting up the analysis, this cannot be part of the actual analysis, as no truth-matching is possible in data. Finally, differential measurements, which are performed in different sub-samples of the event, divided by p_T range for example, provide more detailed information than just one measurement over the whole range.

---

[1] For an LO MC programme, this will simply be: result $_{\text{with PS+MPI+Had}}$/result $_{\text{with only PS}}$, where PS, MPI and Had refer to parton shower, multiple parton interactions and hadronisation respectively.

5.1.2 Searches

As we have seen in section 1.2.2, the SM is unable to explain several experimental observations. So the SM can be thought of as valid up to some energy scale, where the SM is replaced by a new, larger theory. Two terms need to be introduced in this context: *hierarchy* and *naturalness*, which in some sense have driven the theory community over past decades. The value of the Higgs field is about 246 GeV, which gives the W and Z bosons their mass. However, there is no fundamental reason why that is so small compared to Planck scale, which is one way of stating the hierarchy problem. If there is physics beyond the SM, extra contribution to the Higgs field coming from this extended sector of the SM must be balanced by the contribution coming from the SM. *Naturalness* is defined to indicate the amount of mismatch (and resultant fine tuning) we are willing accept, or how high the scale of this new theory can be. So the new physics scale depends on the value of this naturalness parameter.

The searches are performed to look for deviations from SM predictions, which may indicate presence of signals of new physics. Most searches are performed targeting a new physics model (with a specific value or a range of values for the masses, coupling between particles, etc), which predicts specific final state signatures. The same (or similar) final state configuration coming from other processes are considered background. So for a search, the contributions of all relevant SM processes (usually estimated from simulation) will be our background.

> Signal is what we are looking for. Background is anything else that can yield the same final state. Analyses are designed to enhance the signal over background!

Resonance searches are those where a hypothetical new particle is predicted to decay into other particles. The usual way is to calculate the *invariant mass* of the expected decay particles for all *selected* events in data. Background events would not result in a specific value of this invariant mass, rather in a continuous spectrum. In contrast, if this hypothetical particle exists, the invariant mass in those events where it was produced will correspond to its mass. So we look for a localised excess of events corresponding to the mass of the new particle in the invariant mass distribution, as shown in figure 5.1. This widely used approach is referred to as *bump-hunting*. Examples are resonances searches looking for a peak in continuum dijet, multi-lepton or $t\bar{t}$ invariant mass distributions. They cover a wide range of new physics possibilities.

> Resonance refers to the situation where a short-lived particle is reconstructed from its easily identifiable decay products, and a clear mass peak is visible.

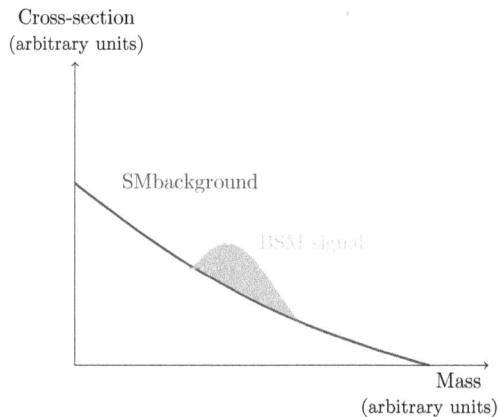

Figure 5.1. A sketch showing a bump in mass spectrum, corresponding to a beyond the standard model signal over continuously falling standard model background.

If new particles do not interact with the detector, then an excess of missing energy will be their signature, which is the typical way to look for dark matter signatures. Finally, difference from SM prediction in terms of production or decay rate, spin, coupling can potentially indicate new physics.

We can broadly divide the searches into two categories. A class of new physics models have been postulated to address specific theoretical shortcomings, such as the hierarchy problem. They yield specific collider signatures, which we search for. Then there are more experiment-driven searches, where phenomenological models have essentially been formulated to probe specific final states. However, this categorisation is neither exhaustive nor rigid, and there is a fair amount of overlap, such as supersymmetric models can predict long-lived particles or dark matter candidates, and so on. We also point the reader to dedicated reviews with the current status of the searches [5, 6].

Theoretical model driven searches
- Supersymmetry (SUSY) is a symmetry principle, which states that every SM particle has a super-partner with a spin that differs by half a unit, and it aims to solve the hierarchy problem [7–9]. Since it is a broad framework, it leads to a vast number of models. As these SUSY partners must be heavier than the corresponding SM particles (as they have not been observed), this leads to what is called SUSY breaking. The models are usually classified according to how this breaking occurs, with MSSM (minimal supersymmetric standard model) being the simplest implementation, where an SM-like Higgs doublet is also added. The super-partners for SM leptons, quarks (top, bottom), gluons and the Higgs boson are referred to as sleptons and squarks (stop, sbottom), gluinos, and the higgsino. Most models lead to an invisible lightest super-symmetric particle (LSP), so usually SUSY signatures involve missing energy (and many jets), and a dark matter candidate. The mass range of SUSY particles depends strongly on the naturalness parameter, with certain values

of it pushing them beyond the reach of the LHC. They also depend on the $\tan \beta$ parameter, which is the ratio of the vacuum expectation values[2] corresponding to these Higgs doublets. Strong production of squarks and gluions have the largest cross-section, yielding many jets leptons, and missing energy, followed by the direct electroweak gaugino and slepton production, resulting in leptons and missing energy.

- Models with extra dimensions (ED) [10, 11] is not a new idea, with the original motivation being the unification of forces. The models with large extra dimensions (LED) [12, 13] are relevant for collider searches, which solves the hierarchy problem by postulating that gravity is distributed through a higher-dimensional space and the SM particles are confined to a subspace (termed brane). There are many signatures for LED models, including copious production of microscopic or quantum black holes [14, 15] (QBH). Two-particle QBH decays to a final state consisting of a lepton and a quark-jet violate lepton and baryon number conservations, producing a distinctive collider signature of high multiplicity events. The discriminating variable is usually S_T, which is the p_T sum of all the objects with a certain p_T threshold.

  Alternative models with extra dimensions include Randall–Sundrum [16, 17] models, with Kaluza–Klein [15] (KK) excitations yielding new particles at the Planck scale. They lead to many final states such as dibosons, diquarks, diHiggs, diphotons.

- Compositeness [18] models suggest quarks and leptons are made of fundamental constituents that are bound by a new strong interaction with a characteristic energy scale termed compositeness scale. In this case the hierarchy problem is solved due to the finite size of the Higgs boson, which is, therefore, not sensitive to corrections from scales above. They predict excited states of quarks and leptons, produced via so-called contact interactions (a generic term describing new interactions with a scale above the energy scale probed), which can distort the inclusive jet or dijet distributions, or produce SM fermions with a decay photon.

- The observation of neutrino oscillation [19, 20] (transition from one family to another) suggests neutrinos have non-zero masses, which is not predicted by the SM. The most widely adopted approach to explain small neutrino masses is the so-called seesaw mechanism [21], which can be embedded in the Left–Right Symmetric Model (LRSM) [22–24]. The LRSM contains SM singlet heavy neutrinos (left-handed neutrinos), and a right-handed gauge boson. The LRSM is thus augmented with violation of the global lepton number symmetry of the SM. Hence, the model can be tested by observing lepton number violating processes, such as the Keung–Senjanović process [25]. The neutrinos are classified as Majorana or Dirac depending on if the anti-particles are the same or different. Their final state signature is two or three

---

[2] Which is the minimum of the potential of the field being considered.

leptons and jets. If the heavy neutrino lifetime is longer, it can lead to a signature with a displaced vertex, as discussed later in this section.

Phenomenological model driven searches

- A vast array of models include dark matter particles [26] (like SUSY), or a family of dark particles, even with their own interactions, that comprise the dark sector. It can consist of particles that interact very rarely, termed weakly interacting massive particles [27] (WIMP). For a dark matter particle pair production, we require the presence of a jet from ISR to calculate momentum imbalance. If a dark matter particle is produced along with a high p_T jet or photon, then the striking final state signature will be only the high p_T jet or photon, leading to a large missing transverse momentum. These are referred to as mono-jet or mono-photon signatures. There are models predicting mono-*everything*, such as mono-W (leading to a mono-lepton), mono-Z, mono-top.

  From 2021, there is expected to be a dedicated detector [28], ForwArd Search ExpeRiment at the LHC (FASER), 480 m downstream from the interaction point within ATLAS to detect light and weakly interacting particles.

- The extended Higgs sector [29, 30] predicts singly or doubly charged Higgs boson-like particles, with decays to jets and isolated leptons or photons. Another class of models come under the umbrella of two Higgs Doublet models [31] (2HDM), which has two isospin doublets[3] ϕ_1 and ϕ_2, and five physical scalars. The latter consists of the two CP-even neutral scalar, one being the Higgs boson (h), the other being a heavier version (H), a CP-odd neutral (A) scalar, and two charged ($H^\pm$) scalars.

- Fourth (or higher) generation of quarks with spin-0 (scalar) or spin 1/2 or 3/2 (fermionic), with possible charges of 2/3, −1/3, 5/3 and −4/3 are postulated [32]. The current experimental and theoretical constraints rule out [33] the scalar variants, but fermionic variants remain a possibility. Since they are postulated to interact with rest of the fermions like third generation quarks (i.e. top and bottom quarks), they are often referred to as a top/bottom-partners. They can be singly (analogous to single top quark production) or pair-produced (with the spin 1/2 variant production mode being identical to SM $\bar{t}$), with decays to Wb, tZ, tH, Wt, Zb, hb, tg and $t\gamma$ final states (with the last two modes suppressed in certain models), constrained by (spin and charge) conservation principles. They are also referred to as *vector-like quarks* (VLQ).

- Leptoquarks [34] are composite objects consisting of a quark and a lepton, carrying non-integer charges (±5/3, ±4/3, ±2/3, ±1/3). Usually pair production is assumed. They decay into quarks and charged leptons or neutrinos, so the final state signature is jets along with leptons or missing energy.

---

[3] Isospin is a quantum number indicative of particles only differing by electric charge, such as neutron and proton form a doublet.

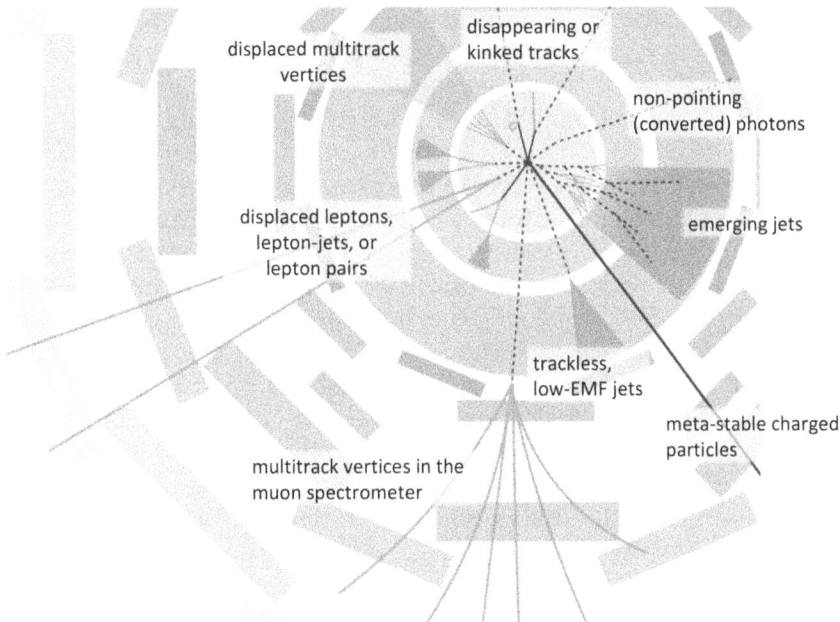

Figure 5.2. Schematic diagram showing different signatures for long lived particles, and how they manifest in different regions of the detector. Figure from [44] © Heather Russell.

- Heavy Z' and $W'^{\pm}$ bosons are postulated by a number of BSM models. They have decay modes like SM Z and $W^{\pm}$ boson or preferential decay modes to leptons or specific generation of quarks.
- Diboson resonance, with a new heavy particle producing WW, WZ or ZZ pairs, with hadronoic and leptonic decay modes are predicted by another class of models. Models predicting large extra dimensions and gravitons can lead to such signatures, among other scenarios.
- Lepton flavour violating (LFV) final states are predicted by many models as well [35]. The signature is events with exactly two different-flavour (charged) leptons.
- Long-lived particles (LLP) are predicted by many BSM scenarios, such as in different SUSY models, or UED models, for example. These are usually heavy particles which do not decay at the collision point where they are produced, but rather travel some distance inside the detector without interacting with detector material before decaying [36, 37]. The LHC experiments were not originally intended to detect these LLPs. Some of the common signatures are displaced vertices, truncated or kinked tracks, photons or leptons not pointing back to the primary vertex, jets appearing away from primary vertex (i.e. emerging jets), out-of-time energy deposits, as shown in figure 5.2. Since they travel with a speed slower than the SM particles, the SM background is almost absent in these searches. Depending on which sub-detector they decay in, the search strategies are different. Some charged long-lived particles can lose energy by ionisation in the detector

material, and energy loss per unit length, dE/dx is used to indirectly detect them. However, triggering is usually a challenge for such signatures. There are proposals for dedicated detectors far away from the collision points, such as the MAssive Timing Hodoscope for Ultra Stable neutraL pArticles (MATHUSLA) [38], or milliQan [39] to detect milli-charged particles.

- If new physics exists beyond the energy reach of the LHC, direct detection would not be possible. Moreover, their indirect effect may affect kinematic tails of distributions. However, to detect such minuscule effects, precision SM predictions are necessary, and usually global fits to all sensitive measurements are proposed. Different approaches have been adapted to extract new physics from precise SM measurements, and EFT [40, 41] (effective field theory) is one of them. The assumption in EFT is that the effect of new heavy particle fields in the energy scale we can probe can be encapsulated by modified couplings between SM particles, which are represented by higher-dimensional operators in field theory terms. This approach facilitates a model independent search using current experimental observations. In the Simplified Template Cross-sections [42] (STXS) approach, different kinematic regions are probed separately for different processes.

The searches for SUSY are almost always optimised for specific model being probed with specific model parameters (i.e. mass, coupling), while the other BSM searches (in ATLAS and CMS classified under exotics or exotica) are more driven by final state signatures. The other two LHC experiments have a somewhat different focus. The LHCb experiment aims to precisely measure bottom and charm hadron production, decays, and properties, as well as other particles with long lifetimes. Any discrepancies from SM can be indicative of new physics. The ALICE experiment primarily aims to establish quark–gluon plasma [43] (a new state of matter in which quarks and gluons are deconfined) in ultra-relativistic heavy-ion collisions.

Subtle difference from SM prediction can also be observed in signal strength μ, which is defined as the ratio of the number of events that were detected in all possible production and decay channels for a particular SM particle, over the expectations from the SM. So any deviation from $\mu = 1$ can potentially indicate the effect of new physics (later when discussing limit setting, μ will also be used to denote a related but different quantity, the signal strength modifier). Model independent searches are also performed, where events are subdivided based according to their final state objects into exclusive categories, and expected and observed number of events in each category is compared.

5.2 Observables/techniques

The objects reconstructed from measurements in the detector are typically charged particle tracks, jets (further classified based on whether originating from b-quarks or not), electrons, muons, and photons. However, in measurements and searches, the primary target is to select events corresponding to a particular process, extract the maximum amount of information of that process, and ignore the rest of the events.

Various observables are constructed using the previously mentioned objects which facilitate in globally characterising the events, or they are designed to be sensitive to specific physics aspects or processes. The simplest example is perhaps the previously mentioned MET, which is calculated from the imbalance of transverse momentum in the event, and represents the production of neutrinos in the SM. Some of the other commonly used observables and methods are described, but this list is far from being exhaustive.

5.2.1 Getting started

An intuitive way is to classify events depending on the multiplicity of reconstructed objects like jets or leptons, as particular processes result in specific final state configurations. For example, an event with top anti-top quark pair decaying semi-leptonically will have to have at least four jets (two of them b-tagged), one electron or muon, and a non-zero amount of MET. So to find such an event, these multiplicity requirements will have to be imposed (along with other requirements, to be discussed later).

Multiplicity is simply the number of that specific object in an event.

If a particle decays into two *daughter* particles, then the four-momentum of the decay particles can be added to obtain the four-momentum of the original particle, and from that the invariant mass or the individual components of the four-vector of the original particle can be determined. The most basic form of the four-momentum vector is (p_x, p_y, p_z, E), but it can be equivalently expressed in terms of (p_T, η, ϕ, E). A common example is Z boson decaying to two same-flavour, oppositely charged leptons. Adding the four-momenta of the leptons, the four-momentum of the Z boson can be obtained, and its mass can be seen to peak at the expected value.

A variable H_T is defined as the scalar sum of the transverse momenta of all jets, which is indicative of how energetic the event is in terms of jets. S_T is a more generalised version, which is defined as the scalar sum of the transverse momenta of all jets and leptons.

5.2.2 Cross-section

The most basic observable is cross-section, as it is directly comparable with theoretical predictions, both for measurements (to uncover shortcomings in the theory prediction) and searches (if new physics enhances the production rate). A large fraction of distributions are essentially cross-sections presented as a function of some observable. These can be considered counting experiments, where the number of events, N satisfying some requirements is measured, and it is related to the cross-section as:

$$N = L\sigma\varepsilon A + B$$

where L is the integrated luminosity (introduced in chapter 3), ε is the efficiency (introduced in chapter 4, it stands for product of individual efficiencies), A is the acceptance, and B is the number of background events. The acceptance refers to the fraction of particles measured in the detector considering the limited geometrical coverage of the detectors (often referred to as geometrical acceptance, when used in this context). When measuring energy or momentum, the acceptance also refers to the fraction of energy or momentum measured because of geometrical acceptance, as well as due to the measurement threshold. It is usually calculated using simulation. The background refers to the events which appear *signal-like* in terms of the final state configuration, but originated in a different process. A detailed discussion of backgrounds and how to estimate them is presented in section 5.3.

So the cross-section can be expressed as:

$$\sigma = (N - B)/L\varepsilon$$

in terms of the quantities that can be estimated from data and simulation.

To extract more information about the particles produced in a particular process, the concept of differential cross-section is used, which is essentially cross-section for process within a restricted range of some observable. Experimentally, it is impossible to measure the full production cross-sections, as we are limited by detector acceptance and efficiency. Usually the total cross-section value is obtained by extrapolating the measured value in a limited detector acceptance region and for a particular decay mode.

When using MC simulation, it needs to be *normalised* to the amount of data being used. The normalisation factor is:

$$N = Lumi_{data}/Lumi_{MC}$$
$$= Lumi_{data} \times \text{cross-section} \times \text{efficiency} \times k\text{-factor/sum-weights}$$

where the cross-section is given by the MC generator or external calculation for a sample, the filter-efficiency comes from the fact that often only a particular final state is chosen when running the generator. The k-factor is the correction for higher order. The sum-weights is the sum of all the weights of all the events in the sample, whether or not the events pass the analysis selection. The weights come from MC, trigger, reconstruction, identification efficiencies, calibrations, any other corrections, as discussed in sections 3.4 and 5.3.3.

5.2.3 Minimum-bias and underlying event

The simplest measurement in a collider is possibly the measurement of the number of tracks (corresponding to charged particles) and their transverse momentum. This is often referred to as minimum-bias (or minbias, MB) from the minimally biased trigger used to select the least biased, most inclusive possible event sample for these measurements. Of course any trigger introduces some bias, but selecting the events triggered by requiring one charged particle in the detector is usually most inclusive we can be, in principle they contain all types of interactions proportional to their natural production rate, and consequently it is characterised by having very few high p_T objects, since their branching fraction is much smaller. Usually the charged particle multiplicity, their dependence on transverse momentum and pseudorapidity, and the relationship between the mean transverse momentum and charged particle multiplicity are measured.

Often the term minimum-bias is used in conjunction with underlying event (UE). As explained in chapter 2, UE is defined as everything except the hard scatter, and measurements of UE usually involve measuring charged particle activities in the events with an identified high p_T object. However, sometimes, especially during the beginning of data taking a new center-of-mass energy, events selected with minimum-bias trigger are used to measure UE as well, with a requirement on the leading charged particle p_T.

In the traditional UE measurements, the leading high p_T object (track, jet, Z boson), is taken as the direction of dominant energy flow in the hard scattering, the event activity can be measured by (charged particle) tracks or by calorimeter energy clusters. As illustrated in figure 5.3, in each event, the azimuthal angular difference between tracks and the leading object $|\Delta\phi| = |\phi - \phi_{\text{leading}}|$, is used to assign each track to one of these regions (a concept first used in [45]):

- $|\Delta\phi| < 60°$, the *toward* region,
- $60° < |\Delta\phi| < 120°$, the *transverse* region, and
- $|\Delta\phi| > 120°$, the *away* region.

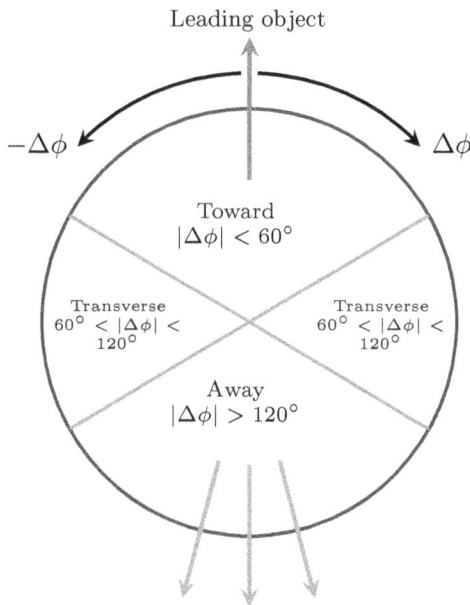

Figure 5.3. Definition of (azimuthal) angular regions to measure underlying event activity defined with respect to the chosen leading object.

While the underlying event populates the whole region, these azimuthal regions have different sensitivity to the UE, which is exploited to extract the UE activity. The transverse region is sensitive to the underlying event, since it is by construction perpendicular to the direction of the hard scatter and hence it is expected to have a lower level of activity from the hard scattering process compared to the away region, which in most cases would contain the objects which will balance the leading object.

In order to disentangle UE activity further, The max (min) transverse regions [46, 47] are defined on an event-by-event basis containing the larger (smaller) number of charged particles or to the region containing the larger (smaller) scalar p_T sum of charged particles. In the events with large initial or final-state radiation the trans-max region will usually contain the third jet in high-p_T jet production or the second jet in Z boson production while both the trans-max and trans-min regions receive contributions from the beam–beam remnants and soft MPI. Thus, the trans-min region is very sensitive to the beam–beam remnants. Also the event-by-event difference between trans-max and trans-min can be constructed, which is very sensitive to initial and final-state radiation. Certain results use the *trans-ave* or average transverse to denote the transverse region, and these two terms will be used interchangeably.

5.2.4 Double parton interaction

Another related class of measurement is the measurement of double parton interaction (DPI) cross-section, which is parametrised by a process independent effective cross-section σ_{eff} [48]. This quantity represents the probability of a second hard scatter from DPI, given that one from HS has occurred, and is given by:

$$\sigma_{eff} \equiv \frac{1}{1 + \delta_{ab}} \frac{\sigma_a \sigma_b}{\sigma_{ab}}$$

where it is assumed that the cross-section σ_{ab} for two processes a and b, where a is coming from the hard scatter and b is coming from DPI, both occurring simultaneously. The total cross-section can be factorised into cross-sections for both of these processes occurring separately. σ_{ab} can be interpreted as the difference between all events, denoted by σ_{ab}^{tot} and the events coming from where a is produced with b in the same hard scatter, σ_{ab}^{HS}: $\sigma_{ab} = \sigma_{ab}^{tot} - \sigma_{ab}^{HS}$. The quantity δ_{ab} is unity when $a = b$. Usually their measurements are done by estimating the fraction of DPI events by looking at angular observables sensitive to DPI topology, and using templates from MC models.

5.2.5 Asymmetry

Preferential particle production, in terms of detector region, lepton flavour or charge is usually expressed in terms of an asymmetry observable:

$$A \equiv \frac{N_1 - N_2}{N_1 + N_2}$$

where N_1 and N_2 are the numbers corresponding to the feature that is being probed. For example, they can be the number of particles produced with positive and negative η values (forward–backward asymmetry), or the number of positively and negatively charged particles (charge asymmetry), or the number of electrons and muons (lepton flavour asymmetry). Any deviation of A from zero indicates an asymmetry. It is normalised by the sum to enhance sensitivity to small values of the difference. Charge asymmetry in $t\bar{t}$ events is measured using number of events with $\Delta|y| > 0$ and $\Delta|y| < 0$, where $\Delta|y|$ is the difference between absolute value of top quark rapidity and the absolute value of the anti-top quark rapidity.

5.2.6 Gap fraction and azimuthal decorrelation

To probe properties of dijet events, a few observables are defined. The deviation of azimuthal angular difference between the jets $\Delta\phi$ from π indicates additional gluon or quark emission in between the jets, hence it is often referred to as azimuthal decorrelation. The same effect can also be probed by defining an observable called gap fraction:

$$f(Q_0) \equiv \sigma_{jj}(Q_0)/\sigma_{jj}$$

where σ_{jj} is the inclusive dijet cross-section, and $\sigma_{jj}(Q_0)$ is the cross-section for dijet production in the absence of jets with transverse momentum greater than Q_0 in the rapidity interval bounded by the dijet system. The variable Q_0 is referred to as the veto scale.

5.2.7 Event and jet shapes

Event shapes and jet shapes form a large class of observables which describe the geometric directions of energy flow in the whole event or inside the jets. Event shape observables [49] (common examples being thrust, sphericity) show the transition from isotropic events where the energy is distributed uniformly over the 4π solid angle to pencil-like or dijet events, shown in figure 5.4. In hadronic colliders, where the center-of-mass frame of the hard interaction may be boosted along the beam axis, event shape observables are defined in terms of the transverse momenta. The most intuitive version is to calculate the event shape from all particles in an event.

Pencil-like   Isotropic

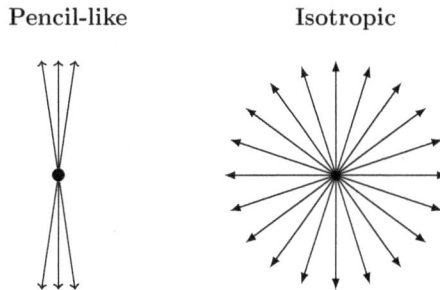

Figure 5.4. Schematic diagram illustrating the pencil-like and isotropic topology, which event shape observables try to identify.

These event shapes are called *directly global* event shapes. In an experiment, it is impossible to detect all particles in an event due to the detector coverage and the presence of beam-pipe. Event shapes which only include particles from a restricted phase space in η, e.g. all particles within $|\eta| < 2.5$, are called *central* event shapes. An example observable is (transverse) thrust which is defined as:

$$T_\perp \equiv \max_{\hat{n}} \frac{\sum_i |\vec{p}_{T,i} \cdot \hat{n}|}{\sum_i |\vec{p}_{\perp,i}|}$$

where the sum is performed over all the particles in an event and $\vec{p}_{\perp,i}$ are the transverse momenta of these particles. The unit vector $\hat{n}$ that maximizes the sum ratio is called the thrust axis, $\hat{n}_T$. Transverse thrust ranges from $T_\perp = 1$ for a back-to-back and to $T_\perp = 2/\pi(\langle |\cos\theta| \rangle)$ for a circularly symmetric distribution of particles in transverse plane, respectively. Because the majority of event shapes vanish in the case of a perfectly balanced dijet topology, it is convenient to define $\tau_\perp = 1 - T_\perp$, which shares this property. Several other event shape variables are also defined, such as transverse sphericity.

Jet shapes measure (transverse) energy flow as a function of the distance to the jet axis. The differential jet shape $\rho(r)$ is defined as the average fraction of the transverse momentum contained inside an annulus of inner and outer radius $r - \delta r/2$ and $r + \delta r/2$ around the jet axis. Alternatively, the integrated jet shape $\psi(r)$ is defined as the average fraction of the transverse momentum of particles inside a cone of radius r around the jet axis. They are schematically shown in figure 5.5.

The internal structure of the jet can also be probed with fragmentation function, defined in terms of the fractional transverse momenta carried by each charged particle associated with the jet:

$$z \equiv p_T^{\text{ch}}/p_T^{\text{jet}}$$

$$F(z, p_{\text{T}}) \equiv \frac{1}{N_{\text{jet}}} \frac{\mathrm{d}N_{\text{ch}}}{\mathrm{d}z}$$

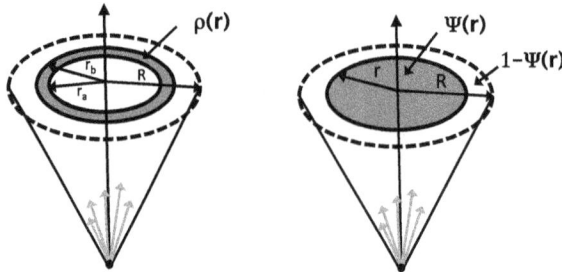

Figure 5.5. Schematic diagram illustrating the definition of differential (left) and integrated jet shape [50] (CMS Experiment © 2018 CERN).

where p_T^{jet} is the momentum of the jet, p_T^{ch} is the momentum of the charged particle, and N_{ch} is the number of charged particles in the jet, normalised to total number of jets. It gives distribution of momentum fraction carried by constituents.

5.2.8 Jet charge and jet pull

Apart from kinematic properties of jets, other observables can be constructed. Jet charge refers to the p_T-weighted sum of the electric charges of the particles in a jet, which can help to characterise the jet in terms of origin [51, 52]. It is defined in terms of all charged particles or tracks associated to the jet:

$$Q \equiv \frac{1}{(p_T^{\text{jet}})^\kappa} \sum_{i \in \text{jet}} q^i (p_T^i)^\kappa$$

where q^i is the charge (in units of the positron charge) of track i with transverse momentum p_T. The exponent κ controls the relative weight given to low and high momentum particles contributing to the jet charge.

Colour connection information between two jets can be probed by the jet-pull vector, which is a p_T-weighted radial moment of the jet [53]. For a given jet j with transverse momentum p_T, it is defined as:

$$\vec{P} \equiv \sum_{i \in \text{jet}} \frac{|\vec{\Delta r^i}| \cdot p_T^i}{p_T} \vec{\Delta r^i}$$

where the summation runs over the constituents of the jet that have transverse momentum p_T^i, and are located at the distance Δr^i from the jet axis. For two jets, the jet-pull angle is defined as the angle between the jet-pull vector of the first jet and the vector connecting the jets in the y–ϕ plane (often termed the jet connection vector), as shown in figure 5.6.

5.2.9 Angular observables

To gain understanding of Z boson production not just in terms of its p_T, an angular observable is often defined:

$$\phi^\star \equiv \tan(\pi - \Delta\phi)/2 \sin \theta_\eta^\star$$

where $\Delta\phi$ is azimuthal opening angle between the two decay leptons, and the variable $\theta_\eta^\star$ indicates the scattering angle of the dileptons with respect to the beam in the c.m. frame where the leptons are aligned. It is related to the pseudorapidities of the oppositely charged leptons by the relation $\cos \theta_\eta^\star = \tanh \Delta\eta_{ll}/2$, where $\Delta\eta_{ll}$ is the difference in pseudorapidity between the two leptons. Since $\phi^\star$ depends only on the angular variables, it is experimentally well measured compared to any quantities that rely on the momenta of the leptons.

For a two-body Drell–Yan like decay, the angular distribution of leptons follow $(1 + \cos^2 \theta)$, where θ is the angle of the lepton from the Z boson direction [55], which follows from helicity and parity conservation. That makes calculations in the

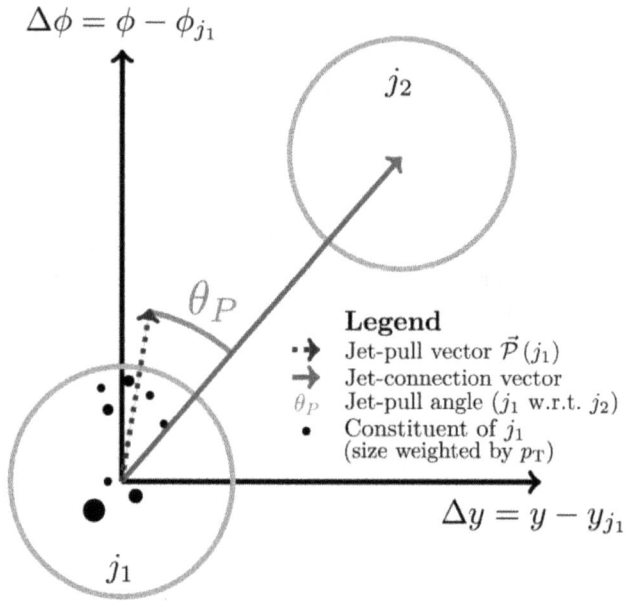

Figure 5.6. Schematic diagram illustrating the definition of jet pull observable [54] (ATLAS Experiment © 2018 CERN).

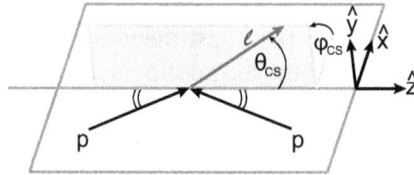

Figure 5.7. Schematic diagram illustrating the definition of Collin–Soper angle [57] (ATLAS Experiment © 2018 CERN).

dilepton rest frame easier, and Collin–Soper angles [56] are defined, which are sensitive to the spin and parity of the original particle. As shown in figure 5.7, the z-axis is chosen as the external bisector of the angle between the momenta of the two protons. Polar and azimuthal angles are calculated with respect to the negatively charged lepton and are labelled θ_{CS} and ϕ_{CS}, respectively.

For a similar two-body decay, of Higgs boson to diphoton, the production angle in Collin–Soper frame is defined as:

$$|\cos\theta^\star| \equiv \frac{|\sinh\Delta\eta^{\gamma\gamma}|}{\sqrt{1 + (p_{\mathrm{T}}^{\gamma\gamma}m_{\gamma\gamma})^2}} \frac{2p_{\mathrm{T}}^{\gamma_1}p_{\mathrm{T}}^{\gamma_2}}{(m_{\gamma\gamma})^2}$$

in terms of individual photon p_{T}, diphoton p_{T} and invariant mass. The $\Delta\eta^{\gamma\gamma}$ is the separation in pseudorapidity of the two photons. The concept can be generalised to cases like $H \rightarrow ZZ$ decay as well. For a spin-zero boson, it can be shown that the differential production cross-section will be independent of $\cos\theta^\star$.

5.2.10 *W* boson reconstruction

For a W boson decaying to a charged lepton and the corresponding (undetected) neutrino, the electron or muon momentum can be precisely measured, but not the missing energy from the neutrino. So the invariant mass of the W boson cannot be reconstructed as easily as the Z boson. It can be shown that if the W boson is produced at rest, the distribution of the p_T of the lepton peaks sharply at the m_W value. However at the LHC, W bosons are rarely produced at rest, rather they recoil against a quark or a gluon. That results in this peak being smeared, and measuring the m_W in this manner is not optimal. Instead, a slightly different approach is adopted.

The missing energy (or the momentum) carried by the neutrino can be expressed as: $\vec{p}_T^{\,\text{miss}} = -\vec{p}_T^{\,l} - \vec{u}_T$, where $\vec{u}_T$ is the vector sum of transverse energies of all detected objects except the charged lepton (l). This is also equivalent to the hadronic recoil against the W boson. Then the transverse mass is calculated by combining this information:

$$m_T \equiv \sqrt{2\left(p_T^l p_T^{\text{miss}} - \vec{p}_T^{\,l} \cdot \vec{p}_T^{\,\text{miss}}\right)} = \sqrt{2 p_T^l p_T^{\text{miss}}(1 - \cos\Delta\phi)}$$

where $\cos\Delta\phi$ is the azimuthal opening angle between the lepton and the p_T^{miss} vector. The momentum vectors of the lepton, $\vec{p}^{\,l}$ and of the neutrino, $\vec{p}^{\,miss}$ can be expressed in W boson rest frame as:

$$\vec{p}^{\,l} \equiv \frac{M_W}{2}(1, \sin\theta\cos\phi, \sin\theta\sin\phi, \cos\theta)$$

$$\vec{p}^{\,miss} \equiv \frac{M_W}{2}(1, -\sin\theta\cos\phi, -\sin\theta\sin\phi, -\cos\theta)$$

where θ and ϕ are the usual polar and azimuthal angles. Then transverse mass can be expressed by using the transverse components as [58] (as the angle between them in rest frame $\cos\Delta\phi = \pi$):

$$m_T^2 \equiv 2\frac{m_W^2}{4}(1 - \sin^2\theta\cos^2\phi - \sin^2\theta\sin^2\phi - \cos^2\theta)$$

$$= \frac{m_W^2}{2}(2\sin^2\theta) = m_W^2\sin^2\theta$$

So then using $\mu \equiv m_T/m_W$, we get $\mu^2 = \sin^2\theta$. Therefore, the differential cross-section in terms of m_T, or equivalently in terms of μ can be expressed as:

$$\frac{1}{\sigma}\frac{d\sigma}{d\mu} = \frac{\mu}{\sqrt{1 - \mu^2}}\frac{1}{\sigma}\frac{d\sigma}{d\cos\theta}$$

The factor of $1/\sqrt{1 - \mu^2}$ means a sharp peak at $m_T = m_W$, or in other words, $m_T^2 \leqslant m_W^2$. So the lower bound of the m_T distribution will approach the W boson mass, which is referred to as the Jacobian peak/edge [59]. The plots will be shown in section 7.2. For W bosons with non-zero p_T of the W boson, it can be shown that the smearing of this peak is governed by a factor of $p_T^W/\sqrt{s}$, so a fit of the m_T distribution gives an accurate value of m_W. Another related observable termed MET significance is often defined as: $S = p_T^{\text{miss}}/H_T$, which takes into account the resolution of objects in the MET calculation.

5.2.11 Stransverse mass

Stransverse mass [60], M_{T2} can be considered a more generalised version of M_T, where in the event, the two (pair-produced) particles decay to a final state containing an undetected particle X of mass m_X. For each particle, the visible system can be defined by $p_T^{vis,i}$, $E_T^{vis,i}$ and $m^{vis,i}$ obtained by summing the four-momenta of all detected particles from its decay (i can be 1 or 2). Each visible system is accompanied by an undetected particle with $p_T^{X,i}$. The (two) transverse masses can be defined, as before:

$$(m_T^i)^2 \equiv (m^{vis,i})^2 + (m_X)^2 + 2\left(E_T^{vis,i}E_T^{X,i} - p_T^{vis,i}p_T^{X,i}\right)$$

For the correct values of quantities on the right-hand side, the transverse masses cannot exceed the mass of the parent particles. However, experimentally only the total p_T^{miss} can be measured, not individual P_{TX}, i s. Therefore,

$$M_{T2}(m_X) \equiv \min_{p_T^{X,1}+p_T^{X,2}=p_T^{\text{miss}}}\left(\max\left(M_T^1, M_T^2\right)\right)$$

where the unknown mass m_X is a free parameter. The minimisation is performed over trial momenta of the undetected particles fulfilling the constraint on p_T^{miss}.

5.2.12 SUSY specific observables

There are a few methods dedicated to SUSY searches. To gain better sensitivity to pair production of heavy particles with subsequent direct or cascading decays to undetected particles along with jets, a couple of variables are defined:

$$M_R \equiv \sqrt{(|\vec{p}^{j_1}| + |\vec{p}^{j_2}|)^2 - \left(p_z^{j_1} + p_z^{j_2}\right)^2},$$

$$M_T^R \equiv \sqrt{\frac{E_T^{\text{miss}}\left(p_T^{j_1} + p_T^{j_2}\right) - \vec{p}_T^{\text{miss}} \cdot \left(\vec{p}_T^{j_1} + \vec{p}_T^{j_2}\right)}{2}}$$

where $\vec{p}^{j_i}$, $\vec{p}_T^{j_i}$, and $p_z^{j_i}$ are the momentum of the ith jet, its transverse component with respect to the beam axis, and its longitudinal component, respectively, with E_T^{miss} the magnitude of $\vec{p}_T^{\text{miss}}$. M_R describes the transverse momentum imbalance, and M_T^R quantifies the mass scale of new particle production in the event. The dimensionless ratio: $R \equiv M_R/M_T^R$ is then used in signal to background discrimination, and it is termed the R-frame razor [61]. For the signal process, the distribution of R peaks is near 0.5, since this is the ratio of two measurements of the same scale, while for multijet background, it is closer to zero. This is mostly used in searches involving squark and gluino pair production decaying to weakly interacting LSPs.

The concept of kinematic edge [62] is used to reduce the background contribution in final state signatures like $\chi_2 \to jj\chi_1$ or $\chi_2 \to \ell^+\ell^-\chi_1$, where χ_2 and χ_2 are the two lightest gluinos from a cascade decay. For the two opposite sign same-flavour lepton cases, if their mass difference, $\Delta m_\chi = \chi_2 - \chi_1$, is less than the Z boson mass (which usually is the dominant background), then the invariant mass distribution of the lepton pair gives a rising distribution with a kinematic endpoint (the so-called edge), the position of which is given by Δm_χ, below the Z boson mass peak. For the case with multiple jets, the right dijet combination should give an edge below the squark mass, if the gluinos came from the squark decay.

Another technique is known as the recursive jigsaw [63] (RJR), which exploits the expected long decay chain in SUSY models. Each decay step is translated to their rest frame, and information from all kinematic observables are combined and optimised to best reconstruct that step, which is termed a jigsaw or JR. Each event is then analysed through the recursive application of a series of JRs.

5.2.13 ISR tagging

ISR tagging [64] is a method to identify certain final states. Jets originating from initial state radiation (ISR) usually add extra jets to the event, leading to combinatorial problems in multijet final states. Such a jet can be tagged on an event-by-event basis exploiting the characteristics that it will not be p_T balanced with another jet, as in figure 5.8, and it will be separated in rapidity from other jets, mostly away from central region. A better idea of the event topology can then help in signal to background discrimination. Requiring a hard ISR object in the final state reduces the signal production rates, but allows it to trigger with high efficiency a third high-momentum object selecting dijet events with relatively low invariant mass.

In the mono-jet search mentioned earlier, if the high p_T jet comes from ISR (with a dark matter pair production), then the same technique can be applied. Similar methods exist for ISR (mono)photons as well.

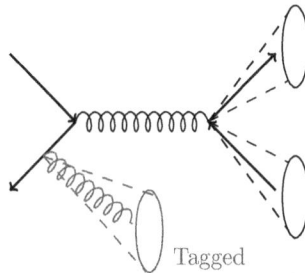

Figure 5.8. Feynman diagram depicting a process where a jet from initial state radiation can be tagged.

5.2.14 Trigger level analysis

High p_T threshold of jet triggers result in discarding of a large fraction of events. To gain more data, trigger level events containing limited objects are used for analysis that are not sensitive to other objects. This allows us to probe lower dijet resonance masses as well. Long-lived particle searches benefit from trigger level analysis as well, since standard triggers can throw away potentially interesting events. This is referred to as data scouting in CMS.

5.3 Analyses steps

In the previous section, we presented examples of observables we want to construct using the objects reconstructed in the detector. In most cases, our focus is on a particular physics process (signal) for which we want to look at such an observable (or a few of them), so we want to select events that originate in that process. However, it is usually difficult to unambiguously pick signal events only. Whenever we are looking for a specific final state, there can be other SM processes which produce the same final state, or similar enough final state which can be misidentified as a signal (background and fakes). So we employ object and event selections (colloquially referred to as *cuts*) to get a pure sample of signal events, and employ dedicated methods to remove contribution of background and fakes. Some of the background contributions can be estimated from simulation, but for the processes which are not known to be modelled well in simulation, data-driven methods are used. Common examples of the latter are multijet background and fakes. For searches (sometimes for measurements as well), we divide the phase space into control and signal (often also in validation) regions as well to optimise the cuts in an unbiased way. For measurements, we tend to correct for detector effects and present the results at particle level, using methods termed unfolding. In the following, these steps are described.

5.3.1 Object and event selection

The starting point for any analysis is event and object selection. Depending on what we are measuring, or what new physics signal we are looking for, we first determine the final state configuration. That not only includes the physics objects, but also takes into account the topology and kinematics. For example, if we are looking for events with Z boson decaying to a lepton pair without any extra jets, then not only would we want final states with two oppositely charged same-flavour leptons, but also the leptons would have to be balanced in azimuthal angle (often referred to as back-to-back) for momentum conservation. Additionally, invariant mass of the lepton pair has to be close to the Z boson mass.

The object selections are based on two considerations. First, the acceptance of the detector (geometric and threshold on energy or momentum measurement, as discussed in chapter 3) forces us to put requirements on p_T and η (or y, for massive objects) of the objects. Very low p_T leptons or photons may not make it to the calorimeter, and the energy deposits from them may be difficult to distinguish from noise or pile-up. In practice, events can be classified based on the trigger used, so even before the object level cuts, requiring an electron trigger will automatically result in choosing events with at least one electron with a certain p_T threshold based on the trigger setup. Secondly, these cuts are also chosen to ensure a high reconstruction or trigger efficiency (based on the trigger turn-on curve). Usually the object selection inherently include enforcing some isolation requirement on electron, muon or photons, along with requiring them to pass quality definitions (loose, medium or tight). For jets, quality requirements ensure that the jets are least affected by pile-up, or calorimeter noise.

The event selection ensures that only the events which contain the desired final state objects are kept, often requiring specific topology. That implies applying geometric cuts, such as requiring some minimum or maximum separation between two objects, demanding events to have some specific values of MET or H_T, or requiring the invariant mass of a pair of objects to be within a certain range. Any of the observables discussed in section 5.2 can in principle be used for this purpose, as well as the dedicated methods described there. More often than not, these are decided by comparing signal and background distributions obtained using simulated events.

Usually it is also required that the events to be analysed are recorded when the full detector is operational without any known problems. This is addressed by using a *good run list(s)* in the experiments.

Two other terms are often used to characterise collision events. An *inclusive* selection usually means a collection of processes are considered, which will include a particular process of interest. For example, an inclusive dijet selection will keep all events with at least two jets. On the other hand, an *exclusive* selection requires events only with the specific process of interest. So an exclusive dijet selection will keep events with exactly two jets.

5.3.2 Cut optimisation

The main aim of performing the selections is essentially to get as many signal events as possible, with the least contamination from background events. We must note that we cannot distinguish signal and background events on an event-by-event basis since all the selected events are equivalent. If they had some discernible difference, then we would have used that feature to discard more background events. So in the end, we will evaluate the probability that an event is signal-like, a topic that will be covered in section 7.1.3.

We need to define a measure to know when the selection is optimised. Let us denote by N the number of events we have at the end of our selection, that is the sum of S and B, the number of signal and background events respectively. Then we can write $S = N - B$. Assuming N is sufficiently large, which is usually the case, we can express the statistical uncertainty as $\sqrt{N}$ assuming a Poisson distribution (the different types of distributions encountered in particle physics will be discussed in section 5.4). We can further assume that the statistical uncertainty on the background estimation is minimal (but there will be systematic uncertainties, which will be discussed in chapter 6), then $S/\sqrt{N}$ gives the number of standard deviations away from 0 the signal is. It can be expressed as: $S/\sqrt{N} = S/\sqrt{S+B} \approx S/\sqrt{B}$, where the last expression is valid for small signals over large backgrounds (for $S \ll B$). This is often referred to as *significance*. A larger value of significance indicates that our cut choices are better. We note that the significance only increases as $\sqrt{N}$, so doubling the integrated luminosity will only result in $\sqrt{2}$ increase in significance. However, a better analysis methodology to increase signal selection efficiency will result in proportional improvement.

If the selection is too strict, that implies that the selection efficiency is low (i.e. many desired events will not be considered), while the purity (fraction of signal events in the chosen sample) will be high. Also it might lead to a small number of selected events, resulting in a large statistical uncertainty. On the other hand, a loose selection will lead to higher selection efficiency, but low purity, as many background events will be selected. A significant part of any analysis is to find the right balance, depending on that specific analysis. Cutflow tables, i.e., the number of events surviving each cut from signal and background is usually made in order to see the effect of changing the cuts. This can only be done with simulation. Multivariate techniques are used when this optimisation involves applying cuts on many variables simultaneously, a topic that will be discussed in chapter 9.

5.3.3 Reweighting

There can be different reasons why distributions from MC simulations may not describe the corresponding data distributions. It can be due to an inherent short-coming in the modelling of the process, or due to the mismodelling of some detector effect. The latter can be exacerbated by faults developed in detector or readout during data taking, which usually cannot be fixed instantaneously. It is not always possible to rerun the simulation with changed condition. So in order to make sure the MC describes the data as best as possible, the MC distributions are reweighted.

In each bin, the MC distribution is scaled upward or downward to match with the data, as shown in figure 5.9. Essentially this implies that each MC event contributes a non-unity entry in the histograms, as opposed to data, which always contribute unity. As MC distributions are used to derive other corrections to data, and often extrapolated other kinematic ranges (as we will see in the next section), this procedure ensures that we have a reliable correction or extrapolation when using simulation.

Reweighting the distribution of the number of vertices is necessary to address the inadequate pile-up modelling in MC. This is commonly referred to as pile-up reweighting, which is essentially an indirect way to correct the badly modelled visible cross-section in MC generators.

Reweighting reflects our incomplete knowledge about certain processes.

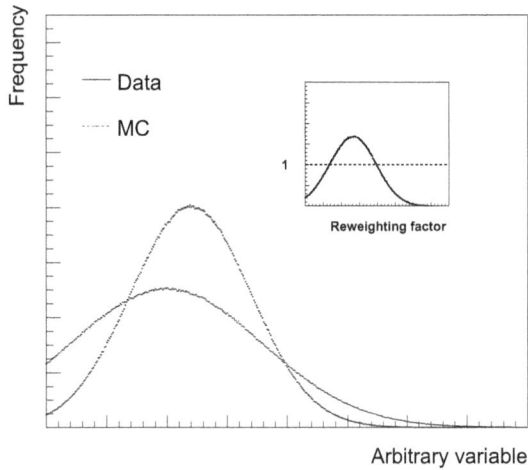

Figure 5.9. Schematic diagram showing the reweighting process to make simulation look like data.

Another related concept is smearing the MC distributions to reproduce the fluctuations in data better, which means multiplying the object four-momentum by a (typically small) random number. This is commonly used for distributions of p_T and angles of objects.

5.3.4 Signal, control and validation regions

In searches, the signal region (SR) is where we expect to detect the presence of new physics based on the model being probed. Even in some measurements, especially the ones for a rare process, a signal region is defined where the process of interest is expected to be seen. Usually the signal region is defined in terms of a particular observable which is sensitive to the signal being probed. For most searches, the observable of interest is usually the reconstructed mass of the new particle, and the signal region will be the mass range where the presence of the particle has not been ruled out (limits on new physics models will be discussed in chapter 7). For example, if we are searching for a new W' boson decaying to a top and bottom quark (where the top quark decays hadronically), and considering that previous searches have ruled out such a heavy boson with mass up to 3 TeV, then the SR can be defined by $M_{W'} > 3$ TeV, where $M_{W'}$ corresponds to the mass of the W' reconstructed using invariant mass of jets coming from top and bottom quarks. It is not always necessary that the signal region will lie at the edge of mass range. Before the discovery of the Higgs boson at the LHC [65], previous experimental results and theoretical calculations limited the mass range for existence between 114–150 GeV. So the searches considered this mass range (when reconstructing the Higgs boson mass in individual decay channels) as SR.

Obviously the main purpose of the object and event selection (i.e. cut optimisation) is to reduce the background as much as possible in the signal region (or define the signal region such that it is least polluted by background processes). In order to avoid bias, the cut optimisation is only performed using simulated samples. In other words, the data is *blinded* in the SR region during this step. Only after the selection is finalised, is the data in SR looked at (i.e. *unblinded*), and after this, no selection should be changed.

This approach ensures that the individual prejudices of the analysers do not affect the results. For example, if an analyser looks at data in the signal region, and sees no excess of events over background, the temptation may be to further tweak the selections. Conversely, an unexpected feature in data can be suppressed by an analyser (perhaps by overestimating uncertainties). Blinding avoids these potential pitfalls.

Cut optimisation without looking at the data in the signal region is a cornerstone of searches.

However, this implies that the estimation of background in the signal region needs to be as robust as possible. If the background is estimated from simulation, we have to ensure that the simulation is properly modelling the data, which is not possible in a blinded SR. Also, since the signal region is mostly chosen to reduce SM background contributions, the simulated samples tend to have fewer events there, making it statistically challenging to evaluate the background contribution there.

In order to address that, one or more control regions (CRs), orthogonal to signal region is defined (usually by inverting a signal selection, in order to find events which are topologically very similar to signal events but without any overlap). For example, in the previously mentioned W' search, the presence of a b-tagged jet is essential for the signal, so a CR can be defined by the exact same event selection, but requiring events with a non-b-tagged jet. Another example is loosening (or inverting) the isolation requirement on leptons, to get a fake enriched CR. The CR can also be defined just by inverting the mass threshold defining the signal region. Then a good description of data by the simulation gives us confidence in the simulation, and background estimation technique can be optimised there. The CR usually is expected to have good statistics for a thorough comparison of data and MC. Multiple CRs are defined in the case where there are multiple processes with significant contribution in SR, so a CR is defined for each of them.

Often additional validation regions (VR) are defined, which are also orthogonal to SR, where the background estimation or the application of normalisation on MC, obtained from CR is cross-checked against data. This assumes that there is no unexpected feature in the data in the CR/VR, so care must be taken to choose well understood regions as CR/VR. Usually the SR is identified first, and dominant backgrounds are ascertained. Then the CR and VR are determined to establish the modelling of these dominant background processes. In figure 5.10 (left), a schematic diagram of the different regions in a two-dimensional distribution of two example variables are depicted. Figure 5.10 (right) shows the actual data distribution

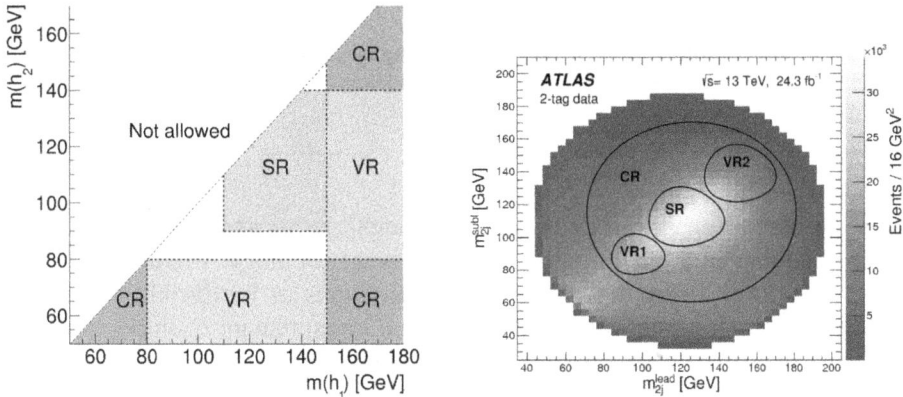

Figure 5.10. Examples of signal region (SR), control region (CR) and validation region (VR) definitions, schematically (left) and using real data (right) from ATLAS analyses [66] (ATLAS Experiment © 2018 CERN). The signal model here is assumed to predict two new particles, and observables corresponding to their masses are shown in the x and y axes in the two-dimensional distributions. The signal region is the area in the two-dimensional distribution which corresponds to masses of these two new particles which have not been ruled out by previous searches (or other theoretical constraints). Then the non-overlapping control and validation regions are defined. When the actual data was looked at, as expected most events were found to lie in the signal region.

(a detailed description of how to read these plots is in the next chapter), where as expected, most signal-like events were in the SR.

5.3.5 Background estimation

It is clear from the discussion so far that one of the most important aspects is the estimation of the background in any analysis. Commonly, background can be divided into three categories, based on their origin:

- Irreducible background: SM processes which produce the same final state. An example can be the SM production of two Z bosons in the search for Higgs boson decaying to two Z bosons.
- Reducible background: Similar enough final state which can be misidentified as signal, due to misidentified or misreconstructed objects. An example can be a dijet event appearing like a Z boson event decaying to an electron positron pair, where jets have been misidentified as electrons (and the invariant mass of the jets lie close to the Z boson mass).
- Combinatorial background: Incorrect combination of objects which happen to end up looking like the signal. For example, in all hadronic $t\bar{t}$ events, there are six jets, and a correct combination of jets is necessary to reconstruct the event. Since there is no obvious way of knowing the correct combination, there is a certain probability that a wrong combination will be chosen in some cases, which looks similar to signal.

Background estimation is usually performed using both simulation and data, the latter when the simulation does not model specific processes or objects well, such as multijet or fakes.

While the reducible backgrounds can in principle be rejected by enforcing strict selection requirements on the objects or the event topology, irreducible ones are more challenging. In order to calculate contributions of irreducible backgrounds with a high precision, the availability of precise measurement of these processes is necessary. So in that sense searches and measurements are complimentary. One of the ways to eliminate the combinatorial background, especially for jets, is to use large-radius jets, a technique that will be discussed in chapter 8.

Many different methods are employed to estimate the background, dependent on the specific analysis. The simplest way is obviously to use simulation to predict the background contribution in the SR. However, this approach has several short-comings. Firstly, we need to make sure we considered all possible SM processes that can contribute to the background, and assume that the simulation describes the data very well in a not so well-explored region of the phase space (as signal regions generally are). More importantly, the signal regions often lie at an extreme region of kinematic phase space, and generating enough simulated events to (statistically) cover that region often becomes very difficult. So for some processes, the background is estimated from simulation while for some other processes (like multijet, where it is hard to simulate a sufficient number of events to perform detailed MC studies), data-driven methods need to be used. Fakes are usually not modelled well in simulation, so usually data-driven methods are used for them as well. Often times, simulation predicts the shape of the background in CR well, but not the overall rate. Then a normalisation factor is derived in CR to address this mismatch. Then the derived normalisation factor is applied to simulation in SR while estimating the background contribution. Sometimes it is possible to determine the ratio of the signal and background normalisations as well.

Here, we give an overview of some commonly used techniques.

Inversion of cut(s): Some of the selection cuts are designed to remove specific backgrounds. So if those cuts are relaxed or inverted, then the selected events will be dominated by that background process, or in other words, a signal-free CR can be formed. For example, multijet or W+jets events are backgrounds for Z+jets (leptonic) signal, where one or both the leptons in the background are misidentified jets. Now the fake leptons from jets are expected to be positively and negatively charged at roughly the same rate, while the signal selection would have required two opposite sign leptons. So if we now select two same sign leptons, then that should give us an estimate of how many of these background events pass all the other selection requirements. This method is purely data-driven. Another method is to use different but well measured particles. For example, in SUSY searches with many jets and large missing energy, $Z(\rightarrow \nu\nu)$+jets is a major background. It is difficult to model

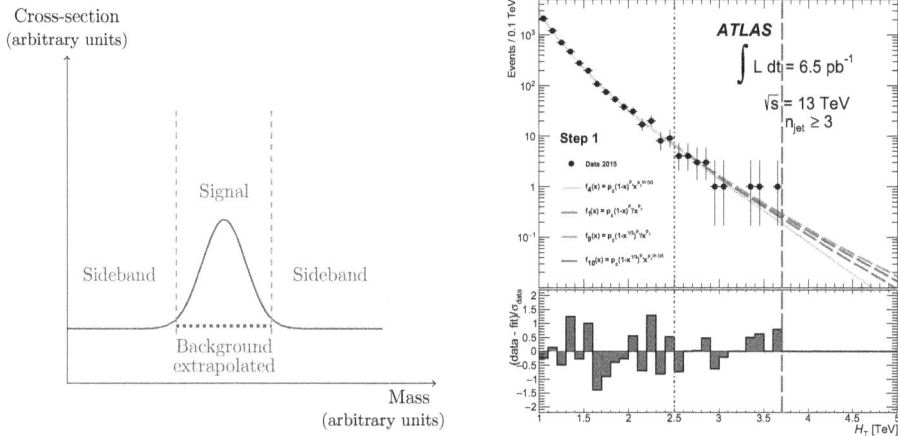

Figure 5.11. Schematic illustration of side-band (left) and shape extrapolation [67] (ATLAS Experiment © 2018 CERN) (right) methods of background estimation. For side-band method, once the signal region in the mass distribution is fixed, sidebands on either side are expected to contain only background contributions. So a smooth extrapolation under the signal region should indicate a background contribution. In some cases, where there are not enough events in the upper side-band, the shape from lower side-band can be approximated by a function and extrapolation of this function under the signal region gives the background estimate. Often multiple functions are tried.

MET well, but the contribution of $Z(\to \ell\ell)$+jets is easier to estimate. Then after correcting for acceptance (based on simulation), the yield from the latter can be scaled by branching ratio (six in this case) to give the background estimate for the former.

Side-band method: To estimate the background under a (prospective) resonance peak, side-band regions are defined away from the peak, where the background shape can be fitted by a function, and that fit can be extrapolated to signal region to estimate background there. An example is shown in figure 5.11 (left).

Shape extrapolation: If the distribution of interest can be approximated by a smooth function, then fitting the distribution with a suitable function in CR, and evaluating the background yield from this function in the SR is possible. If the fit can be based on data, that avoids dependence on the specific MC generator predilections. Usually different fit functions are tried, where quality of fit can be the criteria to choose the appropriate functional form. Different fit ranges in the CR are also tried, and sometimes these ranges can overlap (referred to as *sliding fit windows*). Many background distributions tend to be exponential (or damped exponential), and that fit can be extrapolated to the signal region to estimate the background there. An example is shown in figure 5.11 (right), where the fitted functions are extrapolated to the validation region and signal region. The control, validation and signal regions are delimited by the vertical lines.

Matrix element method: The matrix element method (MEM) [68–71] to discriminant signal events from background events is based on calculating the probability (technically the likelihood is calculated, which will be introduced in section 7.1.3) if the event is more signal-like or background-like. The probabilities are calculated from the matrix elements of the respective hard processes, accounting for detector effects. The shower deconstruction technique discussed in chapter 8 extends the method to include parton shower.

ABCD method: The ABCD method is another standard way of estimating background. The ABCD method relies on the fact that we have two independent distributions to distinguish between signal and background. Two uncorrelated variables (or variables with the least amount of correlation, the residual correlation is part of the systematic uncertainty then) need to be found. A two-dimensional scatter plot is made. Some requirement on the variable on x-axis and some other requirement on the variable on y-axis will correspond to the *cuts* on these two variables for signal optimisation. So then two lines are drawn through these cut values, and the plot is now divided into four regions, i.e. termed the ABCD regions, as in figure 5.12. The A region (top right) is the most signal-pure region, as it corresponds to the region chosen based on signal selection cuts. Now, since the variables are uncorrelated, the expectation is: $N_A \times N_D = N_B \times N_C$, where N_X is the number of events in region X, and we are multiplying diagonally opposite regions (and multiplication of diagonally opposite areas gives the same value). This is usually performed with simulation from all background processes without considering the signal. Then $N_A = (N_B \times N_C)/N_D$ gives the background contribution in the signal region.

Now for MC, this is performed with background MC samples, so the idea is to make sure non-signal regions are statistically viable. For data, of course, the additional complication is one has to blind the data first, then perform the background estimation. It can be the case, that the signal region would not be the top right corner. For example, if the y-axis values under the optimisation cut are used to select the signal, then right bottom will be the signal region. The diagonal

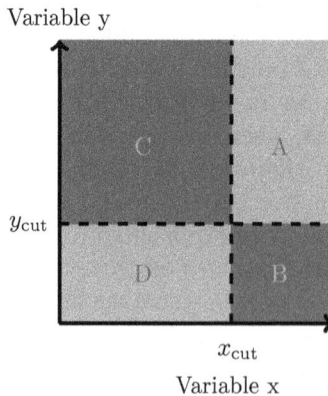

Variable y

y_{cut}

C   A

D   B

x_{cut}

Variable x

Figure 5.12. The ABCD method of background estimation. The x- and y-axis are two uncorrelated variables. The region A is dominated by the signal, while all other regions are dominated by backgrounds.

cross-multiplication still holds. Usually the last two variables in the selection stage are used. It must be noted that BCD regions are not really meant to be orthogonal CRs, rather regions with events that a looser signal selection would have accepted. It is not an abrupt reversal of a cut.

Matrix method for fake estimation: Misreconstructed (fake) leptons from a jet or a photon contaminates the selection of prompt isolated leptons from W and Z boson decays, therefore adding to the background for processes with real leptons. While the exact fraction of such fakes depends on the lepton identification criteria (tight, medium or loose), and also on its p_T, a rough estimate of the fraction of the fakes is 5%–10% for low p_T leptons (50 GeV), while a few percent beyond that.

One of the simpler methods to estimate lepton faking jet background is to select events with only a non-isolated electron, or if it is an opposite sign dileptonic event, with a same sign pair. The orthogonal samples are expected to contain the same level of background but no signal.

The so-called matrix method offers a more thorough approach. In addition to the nominal analysis selection (labeled tight sample), another sample with looser selection criteria (identification, or isolation) is chosen (let us call that the loose sample), where the tight sample is a subset of the loose one. Then we can express the number of loose and tight leptons in terms of real and fakes:

$$N^{\text{loose}} = N^{\text{loose}}_{\text{real}} + N^{\text{loose}}_{\text{fake}}$$
$$N^{\text{tight}} = N^{\text{tight}}_{\text{real}} + N^{\text{tight}}_{\text{fake}}$$
$$= \varepsilon_{\text{real}} N^{\text{loose}}_{\text{real}} + \varepsilon_{\text{fake}} N^{\text{loose}}_{\text{fake}}$$

where $\varepsilon_{\text{real}} = N^{\text{tight}}_{\text{real}} / N^{\text{loose}}_{\text{real}}$ and $\varepsilon_{\text{fake}} = N^{\text{tight}}_{\text{fake}} / N^{\text{loose}}_{\text{fake}}$, are respectively probabilities for a real or a fake lepton that satisfies the loose selection criteria, to also satisfy the tight one.

Then the quantity of interest can be expressed as:

$$N^{\text{tight}}_{\text{fake}} = \frac{\varepsilon_{\text{fake}}}{\varepsilon_{\text{real}} - \varepsilon_{\text{fake}}} (N^{\text{loose}} \varepsilon_{\text{real}} - N^{\text{tight}})$$

The efficiencies $\varepsilon_{\text{real}}$ and $\varepsilon_{\text{fake}}$ need to be determined. The $\varepsilon_{\text{real}}$ can be determined in a pure $Z \to \ell^+\ell^-$ sample by the tag and probe method, as discussed previously. For determination of $\varepsilon_{\text{fake}}$, a sample enriched in multijet events is used with non-isolated leptons, so that the contamination from real leptons from W and Z decays is negligible. Sometimes the $\varepsilon_{\text{fake}}$ is often termed as the *fake factor* (FF).

While this procedure in general works well for a single lepton selection (unless $\varepsilon_{\text{real}} \approx \varepsilon_{\text{fake}}$), for a dilepton selection, cross-terms corresponding to the case where one or both leptons are fake needs to be considered. Then the expression connecting them can be expressed in terms of a matrix (which gives the method its name, although even for the single lepton case, the formulae can be represented in a matrix):

$$
\begin{bmatrix} N_{\text{TT}} \\ N_{\text{TL}} \\ N_{\text{LT}} \\ N_{\text{LL}} \end{bmatrix} = \begin{bmatrix} \varepsilon_{\text{real}}\varepsilon_{\text{real}} & \varepsilon_{\text{real}}\varepsilon_{\text{fake}} & \varepsilon_{\text{fake}}\varepsilon_{\text{real}} & \varepsilon_{\text{fake}}\varepsilon_{\text{fake}} \\ \varepsilon_{\text{real}}(1 - \varepsilon_{\text{real}}) & \varepsilon_{\text{real}}(\varepsilon_{\text{fake}}) & \varepsilon_{\text{fake}}(1 - \varepsilon_{\text{real}}) & \varepsilon_{\text{fake}}(1 - \varepsilon_{\text{fake}}) \\ (1 - \varepsilon_{\text{real}})\varepsilon_{\text{real}} & (1 - \varepsilon_{\text{real}})\varepsilon_{\text{fake}} & (1 - \varepsilon_{\text{fake}})\varepsilon_{\text{real}} & (1 - \varepsilon_{\text{fake}})\varepsilon_{\text{fake}} \\ (1 - \varepsilon_{\text{real}})(1 - \varepsilon_{\text{real}}) & (1 - \varepsilon_{\text{real}})(1 - \varepsilon_{\text{fake}}) & (1 - \varepsilon_{\text{fake}})(1 - \varepsilon_{\text{real}}) & (1 - \varepsilon_{\text{fake}})(1 - \varepsilon_{\text{fake}}) \end{bmatrix}
$$
$$
\begin{bmatrix} N_{\text{RR}} \\ N_{\text{RF}} \\ N_{\text{FR}} \\ N_{\text{FF}} \end{bmatrix}
$$

where the double subscripts (L and T stands for loose and tight, as before) indicate the number of events satisfying the combined criteria. The composition of the signal sample can then be extracted by inverting the matrix. There are variations and extensions of this method, depending on how the different selections are chosen, and if there is overlap between them.

> The best background estimation method for a specific analysis is the one which performs the best in that analysis. That is to say, every analysis is unique, and the applicability and implementation of standard methods need to be checked carefully.

Non-collision background: Non-collision backgrounds refer to the events seen in the detector which have not been produced in collision at the center of the detector [72]. They mostly originate from beam-induced backgrounds (BIB), cosmic-ray showers, and collisions happening far from the interaction point. BIB at the LHC are due to proton losses upstream of the interaction point. These proton losses induce secondary cascades which can reach the ATLAS detector and become a source of background for physics analyses. The particles reaching underground LHC detectors are mostly muons from cosmic rays. Both BIB and cosmic-ray muons can cause a trigger by themselves, e.g. due to a large radiative energy loss in a calorimeter, or they can overlap with a collision event and together form a signature leading to a false trigger. To estimate and reject these non-collision backgrounds, we usually use timing from various detectors, as the particles from non-collision backgrounds will not be in-sync with the collision. For most of the high p_T analyses, these have negligible effects.

5.3.6 Unfolding

The measurements need to be corrected for detector effects, in order to make the results independent of the details of the particular experiment and comparable with theoretical predictions directly. *Unfolding* is the generic term used to describe this *detector-level* to *particle-level* correction for distributions. In general, this can account for limited detector acceptance, finite resolution, production of additional particles through secondary scatterings, and inefficiencies of the detector (if that is not explicitly corrected for during the analysis), as well as bin migrations. The concept of bin will be introduced formally in the next chapter, but for the current purpose it is sufficient to know that the histograms are divided into a finite number of bins, where each bin corresponds to a sub-range of the full range of the observable values. Bin migration occurs when in the same event, the value of the observable under consideration ends up in different bins in particle level and detector level distributions, presumably because of measurement inaccuracies.

Mathematically this is an ill-posed problem [73], as a unique answer cannot be assumed, and small changes in the measured distribution can often cause large changes in the corrected distribution. Simulated samples from MC event generators are used to perform the unfolding. Distributions obtained from the generated events correspond to particle-level (will be referred to as *gen*). Then the events are passed through a detector simulation programme, mimicking the behaviour of the intended detector as closely as possible, and the same distributions obtained using these events correspond to detector-level (will be referred to as *reco*). It must be noted that for unfolding, the particle level and detector level object selections must be independent of one another, to avoid any biases when correcting the data. Also the effect of multiple processes is difficult to account for, so the background contribution is usually subtracted before unfolding.

No unfolding method is ever perfect, they represent our best guess!

A few terms need to be defined, which are valid only for simulation. The term reco bin and gen bin will be used to denote a bin in detector level and particle level distribution respectively.

- Purity: fraction of events in a reco bin that were generated in the same gen bin, is denoted as:

$$P = \frac{N_{\text{reco}\in\text{samegen}}}{N_{\text{reco}}}$$

- Stability: fraction of the entries in a gen bin that are correctly reconstructed at the same reco bin, is denoted as:

$$S = \frac{N_{\text{reco}\in\text{samegen}}}{N_{\text{gen}}}$$

It follows,

$$S/P = N_{\text{reco}}/N_{\text{gen}}$$

- Fake/missed events: if a gen event does not have a corresponding reco event then it is called missed, while if a reco event does not have a corresponding gen event, it is called fake. These happen due to selection criteria applied individually at reco and gen level, and detector smearing moving events in and out of the selection range. To address these issues, over- and under-flow bins in the distributions are used in unfolding, although these extra bins are generally not published.

- Efficiency: fraction of events in a gen bin which have a corresponding reco event in any bin, denoted by $\varepsilon = \dfrac{N_{\text{genmatchedtoreco}}}{N_{\text{total gen}}}$

The binning of the distributions (separately for each observable) are often optimised to have high purity, and roughly the same number of events in each bin, which makes the result more robust and less dependent on bin migrations. Different methods have been used for unfolding, often after corrections are applied for detector acceptance and reconstruction efficiency.

Bin-by-bin unfolding: The simplest method is bin-by-bin unfolding, where the unfolding factor, u is the ratio of bin-content in gen and reco

$$u = \frac{\text{bin-value-particle-level}}{\text{bin-value-detector-level}}$$

is obtained for each bin in the distribution from MC, and then the data value in that particular bin is multiplied by the u for that bin to obtain the corrected data value in that bin. This simple method does not account for bin migrations, and for a steeply falling distribution, that is a significant shortcoming, but can give a reasonable approximation of the result for migration and statistics limited measurements. Figure 5.13 illustrates this process.

Bayesian iterative unfolding: This can be considered as an iterative method for calculating the maximum likelihood estimate of the matrix inversion [74]. Response matrices, $A_{ji} = P(R_j|G_i)$ for each distribution are derived from simulation, which states the probability for a given event at particle-level (G), contributing to bin i, to be reconstructed in a given detector-level (R) bin j. The inverse of the response

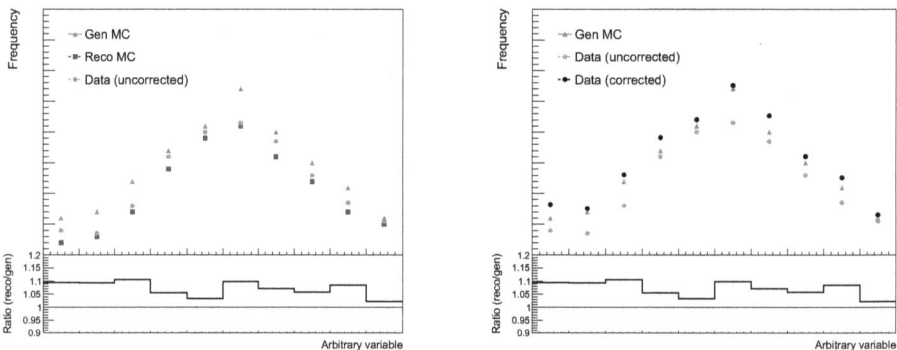

Figure 5.13. Illustration of the bin-by-bin unfolding procedure. The ratio (unfolding factor) in each bin between detector and particle level distributions is found (left), and the data distribution is corrected to particle level by multiplying each bin by the unfolding factor in the corresponding bin (right).

matrix is often referred to as the smearing matrix. Then using Bayes theorem (giving the method its name), an unfolding matrix U_{ij} is constructed:

$$U_{ij} = P(G_i|R_j) = \frac{P(R_j|G_i) \cdot P(G_i)}{\sum\limits_i P(R_j|G_i) \cdot P(G_i)} = \frac{A_{ji} \cdot P(G_i)}{\sum\limits_i A_{ji} \cdot P(G_i)}$$

where $P(G_i)$ can be considered the input prior. Then the number of data events in a particle level bin i, D_i' is given by

$$D_i' = \sum_j U_{ij} D_j$$

where D_j is the number of data events measured in bin j. The unfolding matrix can therefore be constructed using the response matrix obtained from simulation.

Using Bayes' theorem one can define the unfolding matrix as:

$$\theta_{ij} = P(T_i|D_j) = \frac{P(D_j|T_i) \cdot P(T_i)}{\sum\limits_i P(D_j|T_i) \cdot P(T_i)} = \frac{a_{ji} \cdot P(T_i)}{\sum\limits_i a_{ji} \cdot P(T_i)}. \tag{5.1}$$

where $P(T_i)$ is the input prior. An example of the response matrix is shown in figure 5.14. The more diagonal this matrix is, the less is the effect of bin migration.

To ensure that the final distributions are not biased by the shape predicted by simulation, the process is iterated, and each subsequent iteration uses the previous estimate for the corrected distribution (in other words, the reweighted gen level distribution) as $P(G_i)$. The difference between the results for different iterations indicates stability of the iterative procedure. The number of iterations is chosen such that this difference is minimised. Bayesian iterative unfolding is commonly used in LHC experiments.

This method can often result in large statistical fluctuations in the unfolded distribution. A standard way to address this is by so-called *regularisation*, which essentially means the resulting distribution is forced to be 'smooth', but this can add some bias to the result.

The singular value decomposition (SVD) [76] technique is also used, which is somewhat similar to the Bayesian iterative method, but focusses on suppressing statistical fluctuations (i.e. has an inbuilt regularisation). The software implementation (in ROOT) of Bayesian iterative unfolding and SVD used is called RooUnfold [77].

HBOM (Hit Backspace Once More) method: This method was developed during ATLAS Run 1 two-particle correlation analysis [78]. The effect of the detector on an observable is represented by an operator $\mathcal{A}$, thereby relating the corrected value, $V^{\text{corrected}}$ to the measured value, V^{measured} by: $V^{\text{corrected}} = \mathcal{A} V^{\text{measured}}$. Application of this operator an extra k times would result in an overcorrection:

$$V_k^{\text{measured}} = (\mathcal{A})^k V^{\text{measured}}$$
$$= \mathcal{A}^{k+1} Value^{\text{corrected}}$$

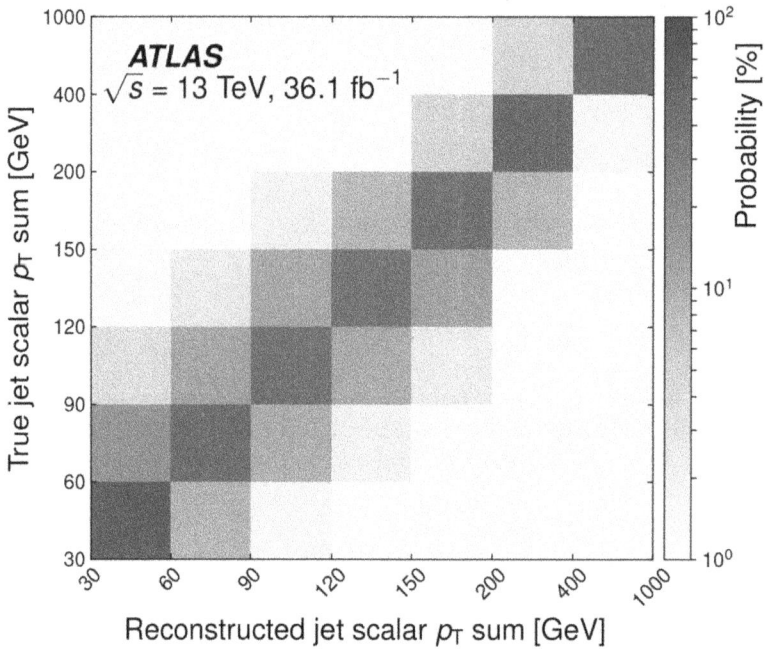

Figure 5.14. An example response matrix showing the scalar transverse momentum sum of all jets [75] (ATLAS Experiment © 2018 CERN). The x and y axes respectively represent the distributions corresponding to detector and particle levels, and each bin, (i, j) indicates the fraction of particle level events from bin i being observed in detector level bin j. The more diagonal the matrix is, the less the bin-to-bin migration is.

Assuming that the evolution of the value of the observable is a smoothly varying function under such iterations, an nth-order polynomial function $P_n(k)$ can be fitted to V_k^{measured} for $k > 0$. Then extrapolating this function back to $k = -1$ yields the $V^{\text{corrected}}$. This procedure is carried out independently for each bin of each distribution.

This method is applicable only in the cases where the detector effect can be parametrised in terms of its effect on the observables. For example, the effect of pile-up on an observable sensitive to pile-up can be simulated by adding more and more random tracks. The distribution without pile-up contamination can then be obtained from fitting to pile-up contaminated distributions with a different number of random tracks, and extrapolated to the distribution corresponding to zero pile-up tracks. This technique is completely data driven, unlike the other methods.

General considerations in unfolding: The performance of an unfolding procedure is evaluated by closure and cross-closure tests, where the unfolding is performed on the same or a different simulation sample, with which the corrections have been derived. Also, the performance of several methods is often evaluated. There is a school of thought that rather than performing unfolding, the theory predictions (or generator output) should be smeared based on the specific detector response. For measurements, this approach is not so prevalent, but for searches, experiments have started to provide the smearing numbers corresponding to some distributions.

The discussion so far has focussed on the case where a one-dimensional distribution needed to be unfolded. There may be cases where the distribution of interest is two-dimensional, or is a one-dimensional differential distribution (i.e. restricted to some range in another observable). In those cases, both the variables need to be unfolded simultaneously, which leads to a generalisation of the above methods.

Finally, rather than unfolding, *folding* the theory prediction, i.e. to smear them for detector effects to directly compare with data, avoids some of the complications and shortcomings associated with the unfolding process. However, specific experimental detector simulations are difficult to run outside of the experimental software framework, and are mostly not made publicly available. Approximate smearing functions for the detector response on a distribution-by-distribution basis, either provided by the experiments, or derived externally are often useful for a quick and easy comparison.

5.3.7 Summary of analysis steps

The above steps for measurement and search are summarised in figures 5.15 and 5.16, with the aid of two examples. These are, of course, overly simplified, with a lot of technical details omitted for ease of presentation, but helpful in understanding the complete process. For searches, optimising the selection and background estimation usually takes the most work, while for measurement, it is the unfolding and the estimation uncertainties which end up being the critical steps.

5.4 Detour: basic statistics

Statistics is the necessary tool to understand the precision and accuracy of the measurements, and to determine whether data is consistent with the current theory or any discrepancy observed is indicative of new physics.

5.4.1 Bayesian or frequentist?

Probability is a familiar concept, however often two different approaches are used. The classical or frequentist interpretation defines probability as the fraction of times a specific outcome will happen. The Bayesian interpretation introduces a subjective bias based on previous outcomes, essentially combining prior probability with the outcome. In other words, the frequentist approach starts with a hypothesis, and if the probability of actual outcome is very small, the hypothesis can be rejected. In Bayesian approach, the prior probability for a hypothesis will be updated based in the outcomes. The choice which approach to use is always based on the actual problem.

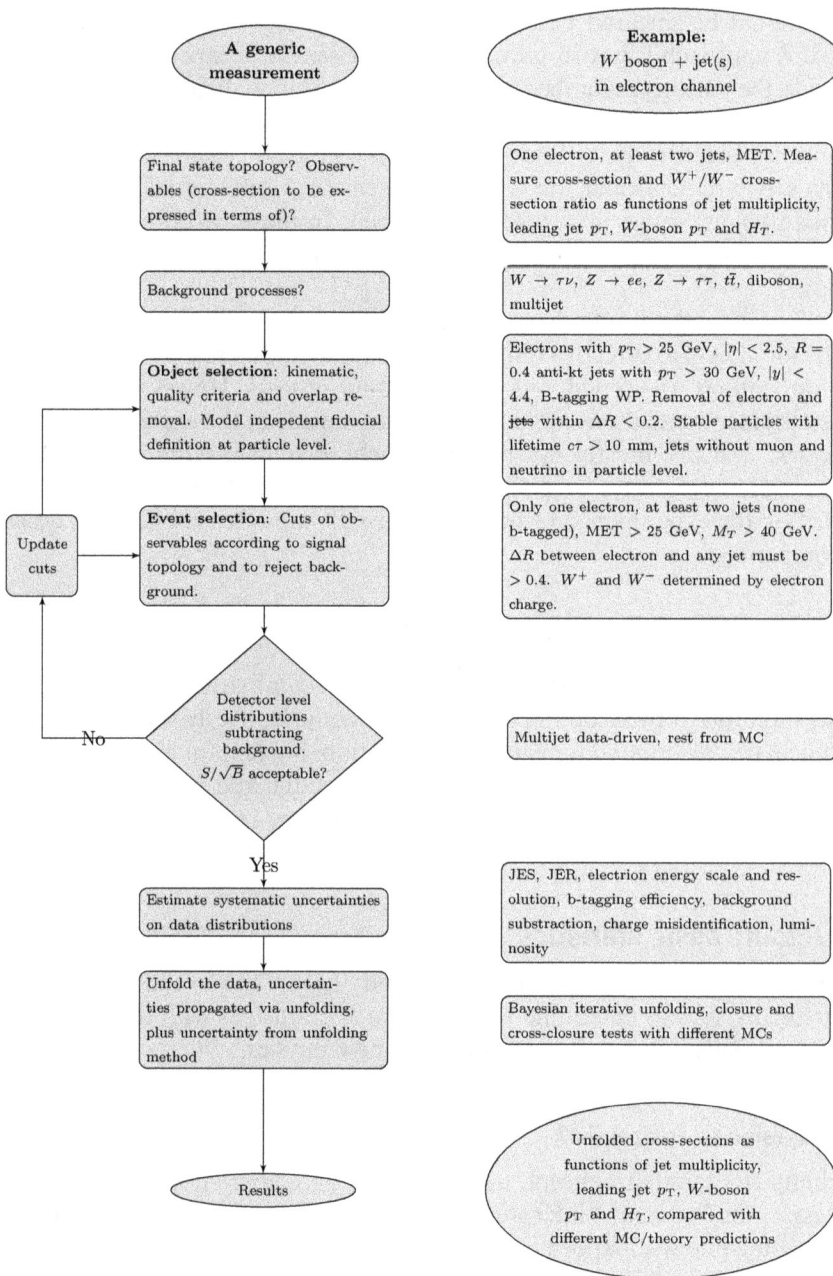

Figure 5.15. Simplified example of analysis steps in a measurement. The abbreviations are mentioned in the text.

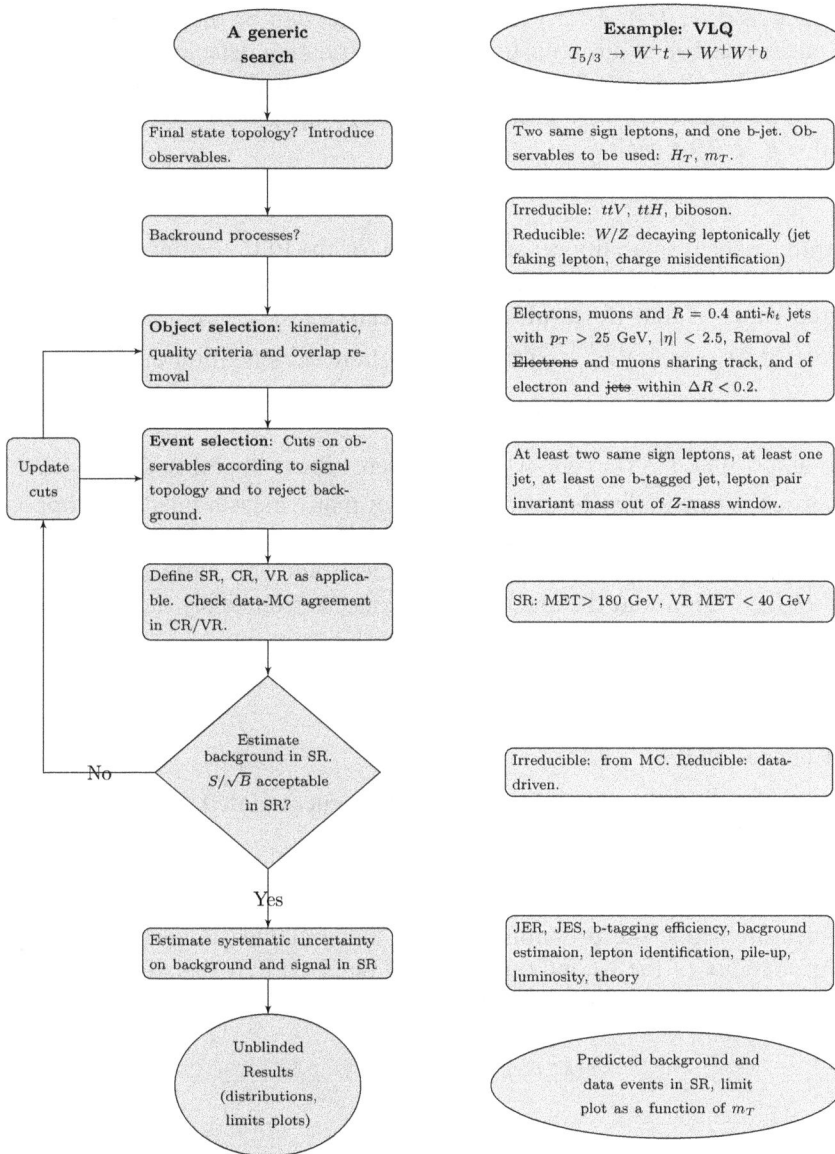

Figure 5.16. Simplified example of analysis steps in a search. The abbreviations are mentioned in the text.

5.4.2 Probability density function: PDFs again, but not partons

Almost all analyses involve measuring some observable. An observable in statistical terms is a random variable, usually denoted by X. For simplicity, the outcome of each measurement can be considered independent of the outcomes of other measurements. An event corresponds to a single measurement.

There are two types of random variables, discrete and continuous. A discrete random variable is one which may take on only a countable number of distinct values such as 0, 1, 2, 3, 4, ... Common examples can be the number of jets in an event, or the number of hits a particle makes in tracking detectors. Let us suppose a random variable X may take k different values, with the probability that $X = x_i$ defined to be $P(X = x_i) = p_i$. The probabilities p_i must satisfy the following:

- $0 < p_i < 1$ for each i
- $p_1 + p_2 + \cdots + p_k = 1$.

We introduce the probability density function (or the PDF, distinct from the parton density function, introduced earlier), $f(x)$, which denotes the probability that the measured value of that observable lies between x and $x + dx$. The mean value is then defined as: $\mu = \langle x \rangle = \int_a^b x f(x) \, dx$ and standard deviation as $\sigma = \langle x^2 \rangle - \mu^2 = \int_a^b x^2 f(x) \, dx - \mu^2$, within the range given by a and b. The cumulative PDF can be defined as: $F(b) = \int_{-\infty}^b f(x) \, dx$, which represents the probability that $x < b$.

The mean is essentially the average value of many measurements (as opposed to the median, which is the middle-most value of the measurements, and mode which is the value occurring the most number of times), and the variance represents the 'width' of the PDF about the mean. The variance is technically the second moment of the distribution, and except in rare cases, the higher moments (skewness, kurtosis) are not used.

5.4.3 Common PDFs

Some common discrete probability distributions, encountered in particle physics are mentioned below.

- Binomial distribution: a random process with only two possible outcomes of equal probability is represented as a binomial distribution. For n independent measurements, the probability of the first outcome is given by p, then the probability of the second outcome will be given by $1 - p$. The probability of getting the first outcome exactly k times is then given by:

$$f(k; n, p) = \binom{n}{k} p^k (1 - p)^{(n-k)}$$

where $k = 0, 1, \ldots, n$ and $\binom{n}{k} = \dfrac{n!}{k!(n - k)!}$.

It can be shown that for binomial distribution $\mu = np$ and $\sigma = \sqrt{np(1 - p)}$.

Binomial distributions are not very common in particle physics, because most distributions are continuous, so an abrupt true or false criterion is not often encountered. However, an example can be a hit (or no hit) in a particular layer of the detector, or number of events passing a certain cut.

The determination of efficiencies can be considered as an application of binomial sampling. Then the statistical uncertainty on the efficiency ε can be represented as: $\sqrt{\dfrac{\varepsilon(1 - \varepsilon)}{n}}$, where n is the total number of trials.

- Poisson distribution: when an observable can have discrete values over an interval but at a fixed rate, then it can be represented as a Poisson distribution. The probability of finding exactly k events between x and $x + dx$ is represented as:

$$f(k; \mu) = e^{-\mu}\frac{\mu^k}{k!}, \quad k = 0, 1, 2, \ldots$$

It can be shown that, μ is the mean, and $\sigma = \sqrt{\mu}$ is the standard deviation of the distribution. It is very commonly encountered when estimating number of events. In most cases the number of entries in a histogram bin follow a Poisson distribution. In the simplest case, then the variance can be taken as (symmetric) statistical uncertainty.

A continuous random variable is one which takes an infinite number of possible values. Continuous random variables are usually the result of measurements. Examples include measured mass, momentum, or energy. Generally two types of continuous probability distributions are encountered in particle physics:

- Gaussian distribution: this is the continuous version of the Poisson distribution. It is represented by:

$$f(x; \mu, \sigma) = \frac{1}{\sigma\sqrt{2\pi}}e^{-\frac{(x-\mu)^2}{2\sigma^2}}, \quad -\infty < x < \infty,$$ which is essentially a bell-shaped curve, with a peak at $x = \mu$ and the width described by σ. Increasing the mean shifts the density curve to the right, and increasing the standard deviation flattens the density curve. The probability of a particular measurement to be in 1, 2 and 3 σ can be calculated to be 68%, 95% and 99.7%. This is a good approximation for many typical measurements.

- Exponential distribution: this is encountered where the increase or decrease of a variable is proportional to the value of the variable, and the PDF is given by: $f(x; \lambda) = \lambda \exp^{-\lambda x}$, $x \geqslant 0$, where λ determines the steepness or the rate of the increase or decrease. In the cases where the time between the occurrence of an event follows exponential distribution with a rate λ, then the total number of events in a time period t follows the Poisson distribution with parameter λt. It can be shown that for exponential distributions, $\mu = \frac{1}{\lambda}$ and $\sigma = \frac{1}{\lambda^2}$. Many distributions, like transverse momentum spectrum of objects, exhibit this behavior.

In figure 5.17, the different types of distributions are shown. Often distributions arise which are convolutions of two different types of PDF. The convolution of multiple Poisson distributions are still Poisson, so separate distributions can be added and be treated as a single distribution. A Poisson distribution modified by a binomial results in a Poisson, and both behave as Gaussian for a large number of entries. Data yields are usually just one number, but they are often treated as the mean of a hypothetical Poisson distribution.

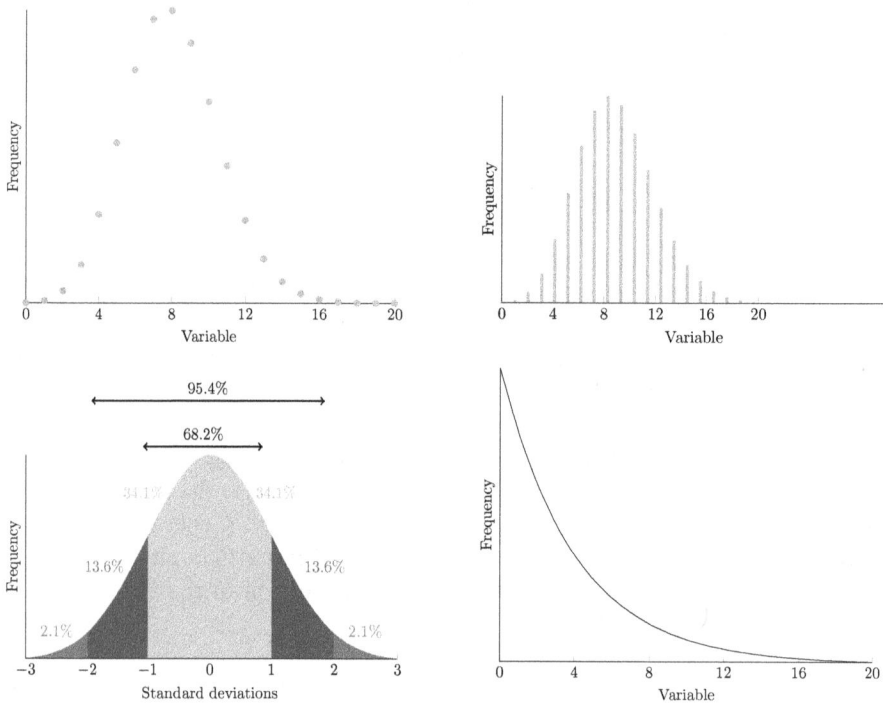

Figure 5.17. Examples of binomial (top left), Poisson (top right), Gaussian (bottom left) and exponential (bottom right) distributions.

5.4.4 Central limit theorem (CLT)

The central limit theorem states that the distribution of the sample mean $\bar{x}$, of N independent variables x_i with finite expectation μ and variance σ^2 in the limit $N \to \infty$, will approach a Gaussian distribution with mean μ and variance σ^2/N independent of the form of the distribution $f(x)$. As in most cases in particle physics we deal with large-sized samples, CLT is commonly used. This essentially implies that Poisson distribution of large n is very close to Gaussian.

5.4.5 Likelihood and parameter estimation

In many instances, experimental measurements need to be compared with some theoretical model predictions, but the model may depend on some free parameters that cannot be determined uniquely from theory. An example can be the mass of a particle. While it is known that repeated measurement of the mass will yield a Gaussian along the actual value, we will have to determine the mean and the standard deviation from the measurement.

A point estimate is a single number that is our best guess for the parameter. The probability density function based on the model can be expressed in the most general case as $P(x; \mu)$, where μ is (a set of) free parameter(s). In the above examples of PDFs, it can correspond with μ or λ. So PDFs essentially predict the expected distribution of the measured observable, assuming different values for the parameters μ.

Now to compare the observed data with the theory predictions, the concept of likelihood is introduced. If the continuous variable x follows a PDF $P(x; \mu)$, then the probability that a measurement of x will yield a specific value x_k is given by: $\mathcal{L}(\mu) = f(x; \mu)_{x=x_k}$. This is defined as likelihood. It depends on the value(s) of the parameter(s) μ, which essentially are the connection to the theory model being probed.

The likelihood that a sample which has N events to have measured a set of values $x_1, x_2, \ldots, x_n$ is the product of all likelihoods of the individual events:

$$\mathcal{L}(\mu) = P(x_1; \mu) \times P(x_1; \mu) \times \ldots P(x_N; \mu) = \prod_{i=1}^{N} P(x_i; \mu)$$

For computational purposes, the negative log-likelihood of a sample is preferred, which is defined as the negative of the sum of logarithms of all likelihoods:

$$-\ln \mathcal{L}(\mu) = -\sum_{i=1}^{N} \ln P(x_i; \mu)$$

Since the logarithm of a function is monotonically increasing function, the extremum points (maxima and minima) of the log-likelihood function will be located at the same extremum points of the likelihood function.

The aim is to estimate the true value(s) μ' of the parameter(s) μ (sometimes termed the point estimator) consistent with the measured value x_k. An estimator is unbiased if for a large number of measurements it gives the true value. Usually to obtain μ' the value of the parameter which maximises the likelihood is taken (MLE). The usual technique to find the maximum is to set the derivative $d\mathcal{L}(\mu)/d\mu$ equal to zero. In practice, $-d\ln\mathcal{L}(\mu)/d\mu$ (i.e. the negative log-likelihood as stated above) is set to zero. Assuming the CLT applies, the likelihood for a sample will have a Gaussian shape, which implies the negative log-likelihood $-\ln \mathcal{L}(\mu)$ curve has the shape of a concave upward parabola. The estimator μ' is the value of μ corresponding to the minimum of this curve.

For several estimators, the likelihood can be maximised with respect to each one individually. For example for Poisson distribution, the best estimator is the mean value, as expected. This can also be generalised for multivariate probability distributions, which is the case for most of the actual analyses. We simultaneously measure a set of k independent variables in an event, so the vector $\mathbf{x} = (x_1, x_2, \ldots, x_k)$ needs to be used in place of x.

5.4.6 Fit quality

Fitting the discrete data points with a continuous function is necessary in many instances. As discussed earlier, the fitted mean of the Gaussian distribution of measured mass values yield the best estimate of the mass. The background estimation relies on fitting the shape of the background with a function, as discussed in section 5.3. We need to judge the quality of these fits. Also, when a distribution from data needs to be compared with the same distribution from simulation, we need a measure of the compatibility.

The method of least squares is most commonly used, where the sum of the squares of the difference between the observed y-value, y_i and the fitting function evaluated at x_i is calculated: $\chi^2 = \sum_{i=1}^{N}(y_i - f(x_i))^2$. The fit with the least value of χ^2 will be preferred. In most cases, the data points have uncertainties associated with them, so a fit can be considered satisfactory if it passes through the points within the error bars, depicting the uncertainties. Since the uncertainties can have different values at different points, to take that into account, the χ^2 is redefined as: $\chi^2 = \sum_{i=1}^{N} \frac{(y_i - f(x_i))^2}{\sigma_i^2}$.
where σ_i is the uncertainty at point i.

If the fit function contains N_{fit} parameters, then choosing functions with increasing values of N_{fit} will in principle result in a better fit. If it is equal to the number of data points N, then clearly we can make the fit pass through every point, with $\chi^2 = 0$. So we define a parameter, degrees of freedom, $N_{\text{dof}} = N - N_{\text{fit}}$, and χ^2/N_{dof} becomes the criteria for judging the fit. In general, for an mth order polynomial, $N_{\text{fit}} = m + 1$, so $N_{\text{dof}} = N - m - 1$. For a good fit, we expect that the value of χ^2 should be about N_{dof}, resulting in $\chi^2 \approx 1$.

Another way to test if two distributions are similar is the Kolmogorov–Smirnov (KS) test, without relying on the shape of the distributions. It is based on creating so-called empirical distribution function (ECDF), which is a step function built using the ordered values. If $F(x) = \int_{-\infty}^{x} f(x') \, dx'$ is the cdf of x and $S(x) = \dfrac{\text{number of measurements with } x_i < x}{\text{total number of measurements}}$ is the ECDF of x. Then the test is based on the maximum distance between these two curves, given by $D = \sup|F(x) - S(x)|$, where sup is the supremum function[4].

5.5 ROOT terms

Particle physics experimental collaborations almost exclusively use ROOT software for data analysis and presentation of the results. The online user's guide [79] or many available tutorials can be good starting points for learning ROOT and like any other software, the best way to gain mastery is by using it.

The standard analysis flow in HEP consists of running an experiment-specific analysis framework on some data or MC samples, produced centrally inside the experiment. These sample formats and analysis frameworks are usually based on ROOT, but need experiment-specific software packages and databases. The output of these frameworks is often a *flat ntuple*, which analysers use as input to their *tree-analyser*, outputting the histograms they need to look at. Finally, the histograms are processed by a further (set of) code, which at the simplest case produces publication quality plots overlaying data and MC simulation results, or for searches produces

---

[4] Which essentially denotes the smallest upper bound of the function.

the limits plots described before. So what are these different objects in ROOT language?

- In order to save event-by-event information, data structure objects called *tuples* are used. An ntuple is an ordered list of *n* elements. Since in general many variables need to be saved for an event, ntuples are a convenient format to store event level information. They hold the copy of same structure for every entry/event.

- In ROOT, the implementation ntuples is done via *trees*. A tree can store different *branches*, which corresponds to a variable, and each branch can have *leaves*, for the cases that variable requires multiple entries for a single event. For example, the event-number or the missing energy can be saved in a branch, as they will have only one value for each event, whereas an event in principle can have multiple jets, so each jet p_T can be a branch, and individual p_T of each jet present in that event will be a leaf. The individual entries are double, float or integers. The words tree (technically TTree in ROOT syntax, as every object is TObject in ROOT) and ntuple are often used interchangeably.

- When a TTree contains container objects of non-ROOT classes, then the specific (experimental) framework used to create it is necessary to read it as well. It is then referred to as a non-flat ntuple, as opposed to a flat ntuple described above. Also, a ROOT file can contain multiple trees, usually for different analysis object selections, or to contain the systematic variations.

- The usual analysis strategy is to write a code to read the required branches of the ntuples, apply selection criteria as necessary, and fill histograms. ROOT provides a method called *Makeclass* as a starting point, which creates a header file containing information about all the branches and leaves in the TTree, and a C file with a skeleton example of how to loop over the events. While TBrowser or TTreeviewer can quickly display the contents of a branch, it just loops over all entries, which is normally not what an analysis would want.

- The output of the first step is usually another ROOT file, this time containing the histograms. Then depending on the aim of the analysis, the histograms can be further processed. At the least, a code for plotting is used, which makes sure all the necessary information to read the histogram is added to the plot. It should give the name of the experiment, the integrated luminosity corresponding to the data used, and the analysis channel clearly. Each axis should have clearly readable labels, with units. If two distributions are being compared, each should have different colours and markers, often an additional ratio panel under the main plot is useful. A useful mnemonics is to follow the *CULTS*, shorthand for (visible and distinct) Colours, Units, (axis) Labels, (informative) Text, and (readable) Size (of all the components of the plot).

Exercises

1. The LED models predict creation of mini black holes at the LHC. There were concerns that they can destroy the Universe, even a lawsuit was filed [80]. How can you judge if these concerns are valid?

2. Pile-up means extra tracks and energy deposits in the detector. In an analysis using tracks, obviously requiring one vertex will throw away most events, so rather than that, the usual criteria for tracks is that they must be associated with (or close to) the primary vertex. However, this is not guaranteed to fully get rid of tracks coming from pile-up interactions, can you think of why? If the analysis involves estimating the number of tracks and the sum of their transverse momentum, which among these will be more affected by the remaining tracks coming from pile-up?

3. In order to simulate the effect of pile-up, we add extra pp collisions to a simulated event with one collision. Since there is no way to know the exact number of extra collisions happening at one event, this pile-up *overlay* usually involves adding extra collisions from a Poisson distribution with the mean indicative of the expected average number of collisions in the data. However, pile-up interactions result in additional energy flow in the event, affecting calculation of MET, isolation variables, jet energy, and many other observables. So a mismodelled pile-up overlay in our simulation affects many facets of the analyses. How can we deal with this?

4. When MC generators give events with negative weights, how should you account for them in your analysis? What about large positive weights?

5. A student has prepared some comparison plots between data and simulation, say for Z boson p_T, in muon channel. A large discrepancy in the first few bins is observed. The kinematic cuts on muons are not different, and the simulation can be assumed to have the same muon identification and reconstruction efficiency as data. Can you think of a reason?

6. Assuming you are performing an analysis where you require four jets. You can either select all jets with the same p_T requirement (based on the considerations discussed so far), for example require all jets with $p_T > 50$ GeV. On the other hand, you can enforce a staggered p_T requirement, for example 50, 40, 30 and 25 GeV. Can you think of reasons why the first or the second approach will be preferred?

7. An observable A is constructed by $A = \frac{N^+ - N^-}{N^+ + N^-}$, where N^+ and N^- are the number of positively charged and negatively charged W bosons. It is found A is over unity as a function of W boson rapidity at the LHC. What does that tell you about the relative number of positively and negatively charged W bosons? If one is produced more, why? Would it be different at the Tevatron?

8. We have talked about cutflow tables in the text. An important step at the beginning of an analysis is to make sure the cutflow is identical for different analysers working on the same analysis. In order to understand possible sources of mistakes, let us consider a simple scenario, a final state of a Z

boson and two b-tagged jets, where the Z boson decays to two oppositely charged leptons. The cuts are shown:

Step	Selection
1	Electrons with $p_T > 25$ GeV and $|\eta| < 2.5$.
2	Muons with $p_T > 25$ GeV and $|\eta| < 2.5$.
3	Exactly two electrons or two muons with opposite charge (signal leptons), with 66 GeV $< M_{\text{lepton1, lepton2}} > 116$ GeV.
4	Jets with $p_T > 25$ GeV and $|\eta| < 2.5$.
5	Remove jets with $\Delta R(\text{jet, signallepton}) < 0.5$. This is a stricter criteria than standard OL, as b-jets can contain leptons.
6	Exactly two jets.
7	Two jets are b-tagged.

Now if we change the ordering of steps 1–4, will the final number of selected events change? What about steps 5 and 6?

9. Assume in an analysis, the background is estimated from an exponential functional fit to data from a steeply falling distribution. Is the validity of the approach dependent on the composition of the background? For example, the background may consist of one dominant process over the whole fit range, or may have two competing processes.

10. A recent trend is to turn the control region of a search into an unfolded cross-section measurement. This is particularly useful as the control region of searches is sometimes unique in terms of kinematics and configuration. What do you think are the most important criteria to determine if a certain control region can be turned into an unfolded cross-section measurement?

11. At a certain control region involving opposite sign dileptons, $t\bar{t}$ and Z+jets are the dominating background with much smaller multijet contribution. What will happen if the electron identification criteria is relaxed, i.e. loose electrons are chosen in place of tight?

12. For each of the following (BSM) signal processes given in the Feynman diagrams:
    (a) List the final state you expect.
    (b) State the most dominant standard model process which you think would be a background for that signal process.
    (c) What specific event selection requirement do you think will reject a large fraction of the background you stated above but will retain most of the signal?

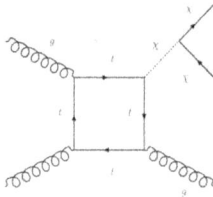

(a) Gluon gluon fusion producing a gluon and a new particle X, decaying to hypothetical χ particles, which do not interact with the detector

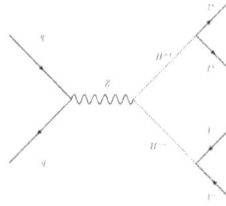

(b) The hypothetical charged Higgs bosons decaying to leptons

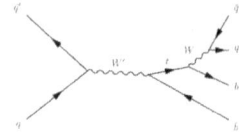

(c) The hypothetical W' decaying into a top quark and a bottom quark, and the top quark decays to another bottom quark and a quark anti-quark pair

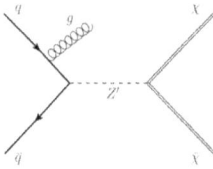

(d) The hypothetical virtual Z' decaying to hypothetical χ particles, which do not interact with the detector, along with QCD radiation

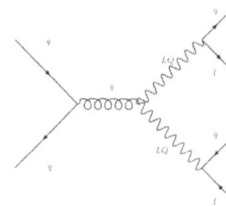

(e) Leptoquark pair production, decaying to quark and charged leptons.

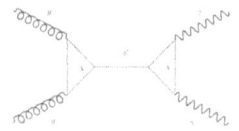

(f) production and decay of the SM-singlet scalar S at the LHC through coloured-scalar χ in the loop.

Question 12

13. If a signal process has a cross-section of 0.25 fb, and we have 100 fb^{-1} of data, what signal efficiency must the analysis have to be feasible?

14. What is the danger of tweaking cuts after the signal region has been unblinded in a search?

15. The selection of signal region is motivated by low background and high signal events. The control region is by definition orthogonal to signal region. For example, let us assume we are searching for dark matter particles being produced along with an SM Higgs boson, and we are focussing on the dominant $b\bar{b}$ decay mode of the Higgs boson. Then our signal region can be defined with at least two b-tagged jets, and the control region can be defined with zero or one b-tagged jets, which is orthogonal to signal region, as required. Do you foresee a problem with this? What can do you do to solve the problem?

References

[1] Brun R and Rademakers F 1997 ROOT: An object oriented data analysis framework *Nucl. Instrum. Meth.* A **389** 81–6

[2] Buckley A, Butterworth J, Lonnblad L, Grellscheid D, Hoeth H, Monk J, Schulz H and Siegert F 2013 Rivet user manual *Comput. Phys. Commun.* **184** 2803–19

[3] Conte E and Fuks B 2018 Confronting new physics theories to LHC data with MADANALYSIS 5 *Int. J. Mod. Phys.* A **33** 1830027

[4] Khachatryan V *et al* 2011 Charged particle multiplicities in *pp* interactions at $\sqrt{s} = 0.9$, 2.36, and 7 TeV *JHEP* **01** 079

[5] Rappoccio S 2019 The experimental status of direct searches for exotic physics beyond the standard model at the Large Hadron Collider *Rev. Phys.* **4** 100027

[6] Salam G P 2018 Theory vision talk at LHCP2018 *6th Large Hadron Collider Physics Conf. (LHCP 2018) (Bologna, Italy, June 4–9, 2018)*

[7] Lykken J D 1996 Introduction to supersymmetry *Fields, Strings and Duality. Proc. of the Summer School, Theoretical Advanced Study Institute in Elementary Particle Physics, TASI'96 (Boulder, USA, June 2–28, 1996)* pp 85–153

[8] Martin S P 1997 *A Supersymmetry Primer* pp 1–98
Martin S P 1998 *Adv. Ser. Direct. High Energy Phys.* **18** 1

[9] Aitchison I 2007 *Supersymmetry in Particle Physics: An Elementary Introduction* (Cambridge: Cambridge University Press)

[10] Kaluza T 1921 Zum Unitätsproblem der Physik *Sitzungsber. Preuss. Akad. Wiss. Berlin (Math. Phys.)* **1921** 966–72
Kaluza T 1921 *Int. J. Mod. Phys.* D **27** 1870001

[11] Klein O 1926 Quantum theory and five-dimensional theory of relativity (In German and English) *Z. Phys.* **37** 895–906
Klein O 1926 Quantum theory and five-dimensional theory of relativity *Z. Phys.* **76** (In German and English)

[12] Arkani-Hamed N, Dimopoulos S and Dvali G R 1998 The Hierarchy problem and new dimensions at a millimeter *Phys. Lett.* B **429** 263–72

[13] Arkani-Hamed N, Cohen A G and Georgi H 2001 Electroweak symmetry breaking from dimensional deconstruction *Phys. Lett.* B **513** 232–40

[14] Dimopoulos S and Landsberg G L 2001 Black holes at the LHC *Phys. Rev. Lett.* **87** 161602

[15] Lillie B, Randall L and Wang L-T 2007 The bulk RS KK-gluon at the LHC *JHEP* **09** 074

[16] Randall L and Sundrum R 1999 Large mass hierarchy from a small extra dimension *Phys. Rev. Lett.* **83** 3370–3

[17] Randall L and Sundrum R 1999 An alternative to compactification *Phys. Rev. Lett.* **83** 4690–3

[18] Eichten E J, Lane K D and Peskin M E 1983 New tests for quark and lepton substructure *Phys. Rev. Lett.* **50** 811–4

[19] Kajita T 2016 Nobel lecture: Discovery of atmospheric neutrino oscillations *Rev. Mod. Phys.* **88** 030501

[20] McDonald A B 2016 Nobel lecture: The sudbury neutrino observatory: Observation of flavor change for solar neutrinos *Rev. Mod. Phys.* **88** 030502

[21] Weinberg S 1979 Baryon and lepton nonconserving processes *Phys. Rev. Lett.* **43** 1566–70

[22] Pati J C and Salam A 1974 Lepton number as the fourth color *Phys. Rev.* D **10** 275–89
Pati J C and Salam A 1975 *Phys. Rev.* D **11** 703 Erratum

[23] Mohapatra R N and Pati J C 1975 A natural left-right symmetry *Phys. Rev.* D **11** 2558

[24] Senjanović G and Mohapatra R N 1975 Exact left-right symmetry and spontaneous violation of parity *Phys. Rev.* D **12** 1502

[25] Keung W-Y and Senjanović G 1983 Majorana neutrinos and the production of the right-handed charged gauge boson *Phys. Rev. Lett.* **50** 1427–30

[26] Bauer M and Plehn T 2017 *Yet Another Introduction to Dark Matter* (Berlin: Springer)

[27] Price P B, Snowden-Ifft D P and Freeman E S 1994 *Weakly Interacting Massive Particles as the Dark Matter of the Universe* (Boston, MA: Springer) pp 369–78

[28] Ariga A *et al* 2018 *Technical Proposal for FASER: ForwArd Search ExpeRiment at the LHC*

[29] Ivanov I P 2017 Building and testing models with extended Higgs sectors *Prog. Part. Nucl. Phys.* **95** 160–208

[30] Gunion J F 2002 *Extended Higgs sectors Proc. of 10th Int. Conf. of Supersymmetry and Unification of Fundamental Interactions SUSY'02 (Hamburg, Germany, June 17–23, 2002)* pp 80–103

[31] Branco G C, Ferreira P M, Lavoura L, Rebelo M N, Sher M and Silva J P 2012 Theory and phenomenology of two-Higgs-doublet models *Phys. Rep.* **516** 1–102

[32] Aguilar-Saavedra J A 2009 Identifying top partners at LHC *JHEP* **11** 030

[33] Lenz A 2013 Constraints on a fourth generation of fermions from Higgs Boson searches *Adv. High Energy Phys.* **2013** 910275

[34] Diaz B, Schmaltz M and Zhong Y-M 2017 The leptoquark Hunters guide: Pair production *JHEP* **10** 097

[35] Bernstein R H and Cooper P S 2013 Charged lepton flavor violation: An experimenter's guide *Phys. Rep.* **532** 27–64

[36] Lee L, Ohm C, Soffer A and Yu T-T 2018 Collider searches for long-lived particles beyond the Standard Model *Prog. Part. Nucl. Phys.* **106** 210–55

[37] Alimena J *et al* 2019 Searching for long-loved particles beyond the Standard Model and the Large Hadron Collider arXiv: 1903.04497

[38] Curtin D *et al* 2018 Long-lived particles at the energy frontier: The MATHUSLA physics case arXiv: 1806.07396

[39] Haas A, Hill C S, Izaguirre E and Yavin I 2015 Looking for milli-charged particles with a new experiment at the LHC *Phys. Lett.* B **746** 117–20

[40] Henning B, Lu X and Murayama H 2016 How to use the Standard Model effective field theory *JHEP* **01** 023

[41] Brivio I and Trott M 2017 The standard model as an effective field theory *Phys. Rep.* **793** 1–98

[42] Brooijmans G *et al* 2016 Les Houches 2015: Physics at TeV colliders - new physics working group report *9th Les Houches Workshop on Physics at TeV Colliders (PhysTeV 2015) (Les Houches, France, June 1–19, 2015)*

[43] Heinz U W and Jacob M 2000 Evidence for a new state of matter: an assessment of the results from the CERN Lead Beam Program arXiv:nucl-th/0002042

[44] *Graphics from Heather Russell* https://hrussell.web.cern.ch/hrussell/graphics.html (accessed: 2018-12-30).

[45] Field R D 2001 The underlying event in hard scattering processes *eConf* **C010630** P501

[46] Marchesini G and Webber B R 1988 Associated transverse energy in hadronic jet production *Phys. Rev.* D **38** 3419

[47] Pumplin J 1998 Hard underlying event correction to inclusive jet cross sections *Phys. Rev.* D **57** 5787–92

[48] Abe F *et al* 1993 Study of four-jet events and evidence for double parton interactions in $p\bar{p}$ collisions at $\sqrt{s} = 1.8$ tev *Phys. Rev.* D **47** 4857–71

[49] Banfi A, Salam G P and Zanderighi G 2010 Phenomenology of event shapes at hadron colliders *JHEP* **06** 038

[50] Chatrchyan S *et al* 2012 Shape, transverse size, and charged hadron multiplicity of jets in pp collisions at $\sqrt{s} = 7$ TeV *JHEP* **06** 160

[51] Field R D and Feynman R P 1978 A parametrization of the properties of quark jets *Nucl. Phys.* B **136** 1

Field R D and Feynman R P 1977 *Nucl. Phys.* B **763**

[52] Krohn D, Schwartz M D, Lin T and Waalewijn W J 2013 Jet charge at the LHC *Phys. Rev. Lett.* **110** 212001

[53] Gallicchio J and Schwartz M D 2010 Seeing in color: jet superstructure *Phys. Rev. Lett.* **105** 022001

[54] Aad G *et al* 2015 Measurement of colour flow with the jet pull angle in $t\bar{t}$ events using the ATLAS detector at $\sqrt{s} = 8$ TeV *Phys. Lett.* B **750** 475–93

[55] Drell S D and Yan T-M 1970 Massive lepton-pair production in hadron-hadron collisions at high energies *Phys. Rev. Lett.* **25** 316–20

[56] Collins J C and Soper D E 1977 Angular distribution of dileptons in high-energy hadron collisions *Phys. Rev.* D **16** 2219–25

[57] Aad G *et al* 2016 Measurement of the angular coefficients in Z-boson events using electron and muon pairs from data taken at $\sqrt{s} = 8$ TeV with the ATLAS detector *JHEP* **08** 159

[58] Smith J, van Neerven W L and Vermaseren J A M 1983 Transverse mass and width of the *w* boson *Phys. Rev. Lett.* **50** 1738–40

[59] Baur U 2003 Measuring the W boson mass at hadron colliders *Proc. of the Workshop of Electroweak Precision Data and the Higgs Mass (Zeuthen, Germany, February 28–March 1, 2003)* pp 47–59

[60] Lester C G and Summers D J 1999 Measuring masses of semiinvisibly decaying particles pair produced at hadron colliders *Phys. Lett.* B **463** 99–103

[61] Rogan C 2010 Kinematical variables towards new dynamics at the LHC arXiv:1006.2727

[62] Hinchliffe I, Paige F E, Shapiro M D, Soderqvist J and Yao W 1997 Precision SUSY measurements at CERN LHC *Phys. Rev.* D **55** 5520–40

[63] Jackson P and Rogan C 2017 Recursive jigsaw reconstruction: HEP event analysis in the presence of kinematic and combinatoric ambiguities *Phys. Rev.* D **96** 112007

[64] Krohn D, Randall L and Wang L-T 2011 On the feasibility and utility of ISR tagging arXiv:1101.0810

[65] Dissertori G 2015 The pre-LHC Higgs hunt *Philos. Trans. R. Soc. A: Math. Phys. Eng. Sci.* **373** 20140039

[66] Aaboud M *et al* 2018 Search for pair production of higgsinos in final states with at least three *b*-tagged jets in $\sqrt{s} = 13$ TeV *pp* collisions using the ATLAS detector *Phys. Rev.* D **98** 092002

[67] Aad G *et al* 2016 Search for strong gravity in multijet final states produced in pp collisions at $\sqrt{s} = 13$ TeV using the ATLAS detector at the LHC *JHEP* **03** 026

[68] Kondo K 1988 Dynamical likelihood method for reconstruction of events with missing momentum. I. method and toy models *J. Phys. Soc. Jpn.* **57** 4126–40

[69] Dalitz R H and Goldstein G R 1992 Decay and polarization properties of the top quark *Phys. Rev.* D **45** 1531–43

[70] Gainer J S, Lykken J, Matchev K T, Mrenna S and Park M 2013 The matrix element method: past, present, and future *Proc. of the 2013 Community Summer Study on the Future of U.S. Particle Physics: Snowmass on the Mississippi (CSS2013) (Minneapolis, MN, USA, July 29–August 6, 2013)*

[71] Martini T and Uwer P 2015 Extending the matrix element method beyond the Born approximation: Calculating event weights at next-to-leading order accuracy *JHEP* **09** 083

[72] Corti G and Talanov V 2006 Aspects of machine induced background in the LHC experiments *3rd LHC Project Workshop: 15th Chamonix Workshop Chamonix (Divonne-les-Bains, Switzerland, January 23–27, 2006)* pp 179–85

[73] Blobel V 2002 An unfolding method for high-energy physics experiments *Proc. of the Conf. of Advanced Statistical Techniques in Particle Physics (Durham, UK, March 18-22, 2002)* pp 258–67

[74] Agostinelli S *et al* 2003 GEANT4: A simulation toolkit *Nucl. Instrum. Meth.* A **506** 250–303

[75] Aaboud M *et al* 2018 $ZZ \to \ell^+\ell^-\ell'^+\ell'^-$ cross-section measurements and search for anomalous triple gauge couplings in 13 TeV *pp* collisions with the ATLAS detector *Phys. Rev.* D **97** 032005

[76] Hocker A and Kartvelishvili V 1996 SVD approach to data unfolding *Nucl. Instrum. Meth.* A **372** 469–81

[77] Adye T 2011 Unfolding algorithms and tests using RooUnfold *Proc. of the PHYSTAT 2011 Workshop (CERN, Geneva, Switzerland, January 2011, CERN-2011-006)* pp 313–8

[78] Monk J W and Oropeza-Barrera C 2013 The HBOM method for unfolding detector effects *Nucl. Instrum. Meth.* A **701** 17–24

[79] ROOT http://root.cern.ch/ (accessed: 2018-12-30)

[80] *LHC black hole lawsuit* https://phys.org/news/2010-09-lhc-lawsuit-case-dismissed-court.html (accessed: 2018-12-30)

IOP Publishing

Experimental Particle Physics
Understanding the measurements and searches at the Large Hadron Collider
Deepak Kar

Chapter 6

Uncertainties

Systematics … that's when you guys just vote, right?

—Lincoln Wolfenstein

The analysis techniques mentioned in chapter 5 are aimed at getting our results, be it a measurement or a search. The presentation of the results will be described in chapter 7. However, there is a missing ingredient in between, which is the estimation of uncertainty on the result. We start by discussing different sources of uncertainties in an analysis in section 6.1. Section 6.2 describes the different sources of systematic uncertainties. Then in section 6.3, some typical ways of estimating the uncertainties are covered. We end the chapter with some commonly used statistical techniques in section 6.4.

6.1 Types of uncertainties

Accuracy and precision are two important aspects of any experimental result. When the true value is known, accuracy is the degree to which a result agrees with the true value. Precision indicates the repeatability of the result. At the LHC, the latter is more important, because in most cases, the true value is not *a priori* known. The precision needs to be quantified by uncertainty, which is defined to be an interval around the measurement in which repeated measurements will fall. The term uncertainty is preferred over the often used term error, as it does not represent a mistake or flaw in the procedure. It is rather a limitation of the analysis design, or the incomplete knowledge of some aspects of the analysis. A new analysis following the same procedure (using the same amount data) may lead to a different measurement result, but usually the same uncertainty, and both the measured results should lie inside the uncertainty range. The result can in principle be different to true or accepted value of the observable, even in the cases where it is known, as inherent limitation of the procedure may not allow the measurement to get to it.

The uncertainties can be broadly divided into two categories. The statistical uncertainty arises because of inherently unpredictable fluctuations in measurement. Every measurement is limited by statistics, and using more data will give a more accurate result, but should not move the central value significantly. Although we have only one measurement, we can imagine that if we could have repeated the measurement N times with measured values x_i, then the standard deviation is: $\sigma = \sqrt{(x_i - \bar{x})/N}$, where $\bar{x}$ is the average. Then the σ is taken to be statistical uncertainty. This assumes that the distribution of measured values follows a Gaussian distribution. This also implies that the statistical uncertainty decreases by $1/\sqrt{N}$, as more data is collected.

The systematic uncertainties represent a bias on the measurement, so using more data does not improve the accuracy. Hence a systematic uncertainty represents a constant (not random) but unknown shift whose size is independent of N. In that sense, systematic uncertainties have origins in physics, not directly in the statistical limitations. Usually statistical and systematic uncertainties are assumed uncorrelated.

The errors are represented like:

$$m_{\mathrm{W}} = 80\,370 \pm 7(\mathrm{stat}) \pm 11(\mathrm{expt\ syst}) \pm 14(\mathrm{model\ syst})\ \mathrm{MeV}$$
$$= 80\,370 \pm 19\ \mathrm{MeV}$$

keeping statistical and systematic uncertainties separate (here, two different components of systematic uncertainty, experimental and model are shown separately as well, which will be discussed as we go along). This representation makes it easier to estimate how much the measurement could be improved with more data. The total uncertainty is the sum in quadrature (i.e. the square root of the sum of the squares) of the individual components. In the plots, to be discussed in section 7.1, usually error bars are used to denote statistical uncertainties, while error bands denote the total or the systematic uncertainties.

6.2 Sources of systematic uncertainties

It is necessary to know the sources of systematic uncertainties, in order to avoid them if possible, and estimate the ones that cannot be eliminated. Essentially there will be systematic uncertainties associated with every step of the analysis, as a result of measuring something, or estimating something that is not perfectly known because of certain limitations [1]. It is, therefore, important to keep in mind that systematic uncertainty does not arise because of (as an example) background estimation itself, but because of inherent limitation on background estimation technique employed.

An incomplete list (in no specific order) can be:

1. Luminosity: any cross-section measurement depends on the integrated luminosity, and for searches, if background contribution from SM processes is estimated from simulation, then that needs to be normalised to the amount of data taken. So uncertainty on the estimation of the luminosity plays an important role in measurements and in precise limit setting in searches.

2. Detector related: calibration of objects (i.e. energy scales), calculation of acceptance or efficiency (of identification or reconstruction), estimation of resolution (of energy or momentum), or simply an instrumental issue can lead to possible uncertainties. The idea is that any measurement in a detector comes with some imperfection, and when corrections are applied for these inadequacies an uncertainty on these corrections must be assessed. When no corrections can be applied, then the instrumental uncertainty must be considered.

3. Trigger related: almost all analyses use a trigger (or combination of them). It is therefore necessary to estimate the efficiency of the particular trigger with respect to the specific analysis selection. This estimation should have an associated uncertainty, depending on how it is performed.

4. Experimental condition: often experimental conditions change, which may not be known beforehand. The change of machine condition, or configuration of bunches can induce more pile-up, for example.

5. Theoretical inputs: like branching fractions or theoretical cross-sections of processes, come with their own uncertainties associated with their estimation.

6. Simulation: all Monte Carlo simulation programmes are based on phenomenological models to a certain extent. It is good practice to assess uncertainties on the aspects of the modelling to which the analysis is sensitive.

7. Background modelling: whether it is estimated from simulation or from data, it is always a guess in some sense (as we have no way of knowing the exact answer). The uncertainties encapsulate the inherent limitation of the background modelling.

8. Unfolding: similarly to background modelling, unfolding (for a measurement) is an ill-posed mathematical problem, as we have no way of knowing the idealised particle level result with absolute certainty.

9. Human bias: analysers can and do have an implicit bias. In many cases they have an *a priori* expectation of what the result will be. This can in some instances unconsciously shift the result toward the expectation.

10. We must be aware that there may be an analysis specific unknown effect, which has not been covered by the earlier points. This can be either because it was not thought of, or the effect was thought to be negligible during the analysis.

Once the list of possible sources of systematic uncertainties has been established, and procedures for estimating the effect of each on the distributions of interest has been decided (to be discussed in the next section), then they need to be applied to the analysis results. The systematic uncertainties can affect the shape of a distribution (such as those related to reconstruction efficiencies, energy scales and resolutions, background modelling) or the overall normalisation (such as luminosity, theoretical cross-sections) or both (such as parton shower). The shape variation can affect the event yield in a specific region of phase space, as shown in figure 6.1.

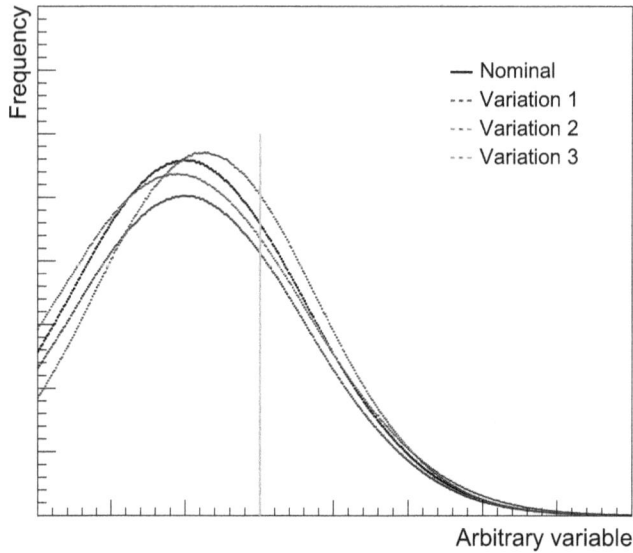

Figure 6.1. Schematic illustration of the effect of systematic uncertainties on a distribution. The black line represents the nominal shape, and the red, blue and green lines can be thought of as the same distribution obtained after applying systematic variations. The blue line represents a constant 10% downward shift, affecting the overall cross-section normalisation. The red and the green lines represent changes in the shape of the distribution, presumably because acceptance of jets or leptons was changed by application of the systematic effect. They affect the results differently. For example, if we are only interested in the yield to the right of the brown vertical line, even though the green line overall has a different shape as compared with the nominal, the yield difference will be negligible.

Measurements and searches differ in how the systematics are applied as shown in table 6.1. In measurements, the systematic uncertainties need to be applied to distributions obtained from data. In other words, the aim of a measurement is to present data distributions corrected for all detector effects, with the total systematic uncertainty shown at each data point. Usually this is obtained by constructing the data distributions individually for each systematic variation, and then unfolding each of the distributions. The nominal distribution is also unfolded using different MC samples to check the effect of varying the simulation model or some parameter in the model. Then the combined spread of these differently unfolded distributions from the nominal gives the resultant uncertainty. Then this data with the associated uncertainties are compared with MC distributions. Ideally, the MC distributions are expected to have orders of magnitude larger statistics than data, so no uncertainty on MC is assumed.

For the searches, the motivation is somewhat different. The aim is to find deviations between data and MC, so the systematic uncertainties on the MC distributions represent the possible spread of SM background or possible new physics signal shapes. The data distributions are shown only with statistical uncertainties.

6.3 Estimation of systematic uncertainties

The sources of systematic uncertainties were listed in the previous section. Now we discuss some standard methods to estimate their effects. It must be noted that each analysis is unique, and care must be taken to adapt the methods to specific analyses, or come up with new ideas in case the standard methods prove to be inadequate.

6.3.1 Luminosity

Dedicated working groups within the experiments provide the prescription to calculate the integrated luminosity from the good run list and trigger used, and associated uncertainty (usually a few percent up and down). These are based on nominal and alternative luminosity calculation methods, as described in section 2.2. Then the usual practice is to vary the luminosity within these given uncertainties and recalculate the cross-sections.

6.3.2 Experimental acceptance, efficiency, calibrations

Comparisons of data with simulation for as many variables as possible is a crucial step in every analysis, which is often performed in well-defined control regions. They can point to mismodelling in the MC generator, or can arise due to acceptance, efficiency or calibration difference between the data and simulation. If a disagreement is observed, the first approach is usually to try to understand the reason, and address it in the relevant corrections (i.e. by using scale factors) or by smearing the simulation. Any such correction or smearing introduces an uncertainty. In certain cases, a specific region of the detector may need to be excluded, or statistically limited tails of the distribution may be removed. These are usually checked with an independent data or simulation sample to avoid bias. Finally, an uncertainty is estimated based on the remaining difference between data and simulation. These are usually provided as forms of additional event weights.

Usually the effect of each source is estimated by varying the corresponding correction factor by one standard deviation (again, assuming a Gaussian distribution, so 1σ variation means that the values would lie within 68% of the mean) and observing the change in the distributions of interest. For example, we can repeat an

Table 6.1. Summary of how systematic uncertainties are applied for measurements and searches differently. The combination of different systematic uncertainties will be discussed at the end of the next section.

	Measurement	Search
Uncertainty on:	Data	MC simulation
How?	Unfold data distributions, corresponding to each variation	Recalculate final results, corresponding to each variation
Finally:	Differences from the nominal are the uncertainties. Combine them in quadrature or otherwise	

analysis by shifting the jet energy scale up and down by 1 σ from the central value, or by varying p_T of the leptons by momentum resolution scale factors. Then the shift in the distribution of interest is an estimate of the systematic uncertainty. Often this is done in individual bins. However, if the difference is asymmetric, or there are significant bin migrations resulting in correlations between the results in neighbouring bins, this approach is no longer robust.

An important sanity check for all experimental systematic uncertainties, is to verify the stability of the result with different data taking periods. Unless there are no known effects (like increase of pile-up), results for each sub-sample should be consistent.

6.3.3 Background estimation

Background estimation is one of the most crucial aspects of searches. Since the correct answer is not known, any background estimate needs to be presented with an associated uncertainty. The background estimation can be performed by different methods, yielding somewhat different results (if different methods give dramatically different results, then the methods should be revisited, rather than applying an uncertainty), and the differences from the nominal can be taken as the uncertainty. As every analysis is different, it is difficult to provide general guidelines, but some of the commonly employed checks are using different fit functions or fit windows, when the background is estimated by a fit. In ABCD method, the effect of correlation of the two variables considered can be used to calculate the uncertainty.

6.3.4 Unfolding

The uncertainty in unfolding is usually estimated by several checks:
- MC closure: a test where the detector-level distributions from a particular MC generator are unfolded using the smearing matrices (it is trivially true for bin-by-bin unfolding) derived using the same MC generator, and the results are compared to the original particle-level distributions.
- MC cross-closure: to test the modelling differences between two different MC generators. The distributions from an alternative generator are unfolded using smearing matrices (or bin-by-bin unfolding factors) derived from the nominal MC generator used in the analysis.
- Data-driven (cross)-closure: to test for the sensitivity of the unfolding method to the differences between the shape of the observable seen in data and in simulation. The detector-level and particle-level distributions from MC are reweighted in such a way so that the detector-level distributions match data. These reweighted distributions are unfolded using smearing matrices (or bin-by-bin unfolding factors) derived from the nominal MC generator used in the analysis.
- Iteration stability: in any iterative unfolding method, the results are expected to be stable with respect to the number of iterations after a small number of iterations are performed. Therefore, the dependence of the result on the number of iterations is tested.

For the MC closure and cross-closure tests, the difference between the original particle-level distribution and the unfolded distribution is indicative of the systematic uncertainty of the unfolding procedure. Similarly for data-driven (cross)-closure test, the unfolded result using reweighted MC is compared with the nominal result. Usually a specific number of iterations are used in iterative unfolding procedures, often based on the consideration of minimising the above mentioned closure and cross-closure uncertainties. The difference between the result using that nominal number of iterations and results using a few higher and lower number of iterations are considered as unfolding uncertainty as well.

Statistical uncertainty on unfolding is estimated by psuedo-experiments. The 68% IQR (inter-quartile range) of the variations generated by applying these toys to unfold the data are used as the band for the MC statistical uncertainty.

6.3.5 Theory/simulation

While in an ideal world, the simulation would have described the data perfectly, the phenomenological component of the generators, as well as the limited precision, means some uncertainties need to be considered.

For theoretical inputs, and for simulations in general, often more than one description or model is available, without a clear indication which is better. Then one description is used as default or nominal, and the discrepancy or the spread of discrepancies of the result obtained with other models is taken as systematic uncertainty. The model which describes the data best is usually taken as the nominal, but there are times when the decision is driven by practicality (which sample has the largest available statistics) or precedence.

The following are some of the common uncertainties assessed:

- Parton shower: Underlying event, hadronisation and fragmentation are modelled phenomenologically, and models are tuned to describe the data. Uncertainties are usually estimated by comparing different models (which is sometimes referred to as two-point systematics), and also by constructing suitable eigentunes covering the data-MC discrepancies. The construction of eigentunes is usually performed in the automated tuning tool, PROFESSOR.

- Parton distribution functions (PDF): There are several sources of uncertainty that affect the determination of P.DFs. Experimental uncertainties entering the datasets used in the PDF fits, and uncertainty on functional form used in the PDF fits are encapsulated in the PDF error eigensets. Usually the nominal value of the PDF, and these error eigenvalues are saved as weights in MC generator outputs (via the LHAPDF interface). Comparing different PDF sets (such as those provided by CTEQ, MSTW and NNPDF groups) usually accounts for the other effects, such as uncertainty of used α_S value, and general methodology used to obtain the PDF. The PDF4LHC group has a set of recommendations on PDF uncertainty estimation [2]. PDF uncertainties affect the total cross-section, unless there is a sharp rise at some x region, which can change the acceptance.

- Matching/merging: Uncertainties can be associated with choice matching/ merging scales values. For LO multi-leg samples, matching scale variation (like *xqut* in MADGRAPH) is an obvious way to estimate the effect. For NLO generators, varying the resummation scale, which defines an upper cutoff scale for the parton shower evolution, is a standard way. Certain generators contain specific parameters which can be varied. An example is the Powheg model resummation damping parameter, h_{damp}, which controls the matching of matrix elements to parton showers and regulates the high-p_T radiation.
- Normalisation: As defined in section 4.2.3, often k-factors are applied to simulated events to correct the cross-section to higher orders. If such k-factors are used, then an extra normalisation uncertainty comes from the determination of the k-factors.
- Scale variation: Missing higher orders in the perturbative expansion of the partonic cross-section is accounted for by varying the renormalisation and factorisation scales. Generally the scale variations will result in a variation of the cross-section and a variation of the differential shape of an observable. Commonly factors of 2 and 0.5 are used, but there is no guarantee that e.g. NNLO prediction will be contained within the NLO uncertainty band obtained by these variations.

6.3.6 Some remarks on uncertainties

Increasingly, systematic variations are being provided in the form of per-event (often referred as on-the-fly) weights in event generator programmes. That means using different weights each time will automatically result in producing the distributions corresponding to that variation.

Usually most systematic uncertainties cannot be avoided when doing a data analysis. However, some of them can still be suppressed. Every selection in principle introduces an associated efficiency loss, or results in performing the analysis in potentially more mismodelled phase-space, which can be avoided by keeping the selection simple. In general, if the analysis is dominated by one single source of systematic uncertainty, which is larger than the statistical uncertainty, then work is needed to understand if the origin of that can be understood better.

Each source of uncertainty is unique, so care needs to be taken to understand their origin, effects and correlation. A simple variation of cuts is neither a robust, nor a sensible way to estimate the uncertainties, even though it is used more often than not. A well motivated cut is employed to select more signal events than background events, or to perform the analysis in a well understood region of the detector defined in terms of kinematic cuts. Arbitrarily varying them changes the whole analysis, so that is not a source of systematic uncertainty, rather an expected difference. If the cut variation increases signal efficiency, then that probably indicates a problem with the chosen cut value, not an uncertainty. Only when a cut is designed to be robust, can this approach be used, with reasonable cut variations.

The other problem with this approach is that it can be difficult to isolate the effect solely from cut variation, as it can be correlated with a systematic uncertainty. For example, lowering the p_T cut on an object may result in selecting events with less than full trigger efficiency, which introduces its own systematic uncertainty from the estimation of this (in)efficiency.

6.4 Statistical methods used in uncertainty estimation

Here we review the statistical methods relevant for uncertainty estimation. The crucial aspect is again the CLT, as discussed in section 5.4, as systematic uncertainties can be propagated in a Gaussian manner via quadratic addition.

6.4.1 Nuisance parameters

Nuisance parameters (NP) are the parameter which are not being estimated, but introduced to take into account the uncertainties. The classic example of a nuisance parameter is the variance σ in a normal distribution, where the parameter of interest is the mean μ. The other examples are the parameters that take into account experimental uncertainties, such as luminosity, efficiency, resolution. These are accounted for by introducing NPs θ in the PDF, with the PDF now being be written as $f(x; \mu, \theta)$.

6.4.2 Profile likelihood

The nuisance parameters θ appear in the PDF along with μ, i.e. the likelihood function is a function of both μ and θ, written as $\mathcal{L}(\mu, \theta; x)$. The aim is to maximise the likelihood of finding μ, taking into account the NPs. For each μ, we have a different likelihood function $\mathcal{L}(\mu, \theta; x)$, and we can maximise the likelihood functions for each value of μ with respect to the nuisance parameters θ.

$$\hat{\theta}_\mu = \underset{\theta}{\mathrm{argmax}}\ \mathcal{L}(\mu, \theta; x).$$

Then we can choose the μ, which is the maximum over all likelihood functions. The new likelihood function is called the profile likelihood function, as the NPs have been *profiled out*, or in other words, the description of the systematic uncertainties has been incorporated into the likelihood function.

While this is a very powerful method, one needs to be careful when applying it. If the nuisance parameters have complicated but important features, then profiling artificially smooths them out. Another disadvantage of the profiling method is that it does not take proper account of the uncertainty on the nuisance parameter.

6.4.3 Bootstrapping

Bootstrapping is an alternate way to estimate uncertainties when the exact estimation is not possible. The term comes from the phrase 'pulling oneself up by one's bootstraps', which is a metaphor for accomplishing an impossible task without any outside help.

The assumption is the underlying distribution is not known, then *toy* or *pseudo experiments* are generated from the available sample. Let us consider a sample $\{x_1, \ldots, x_n\}$, then the following steps are performed for bootstrapping:

1. We take a bootstrap sample $\{x_1^*, \dots, x_n^*\}$—a random sample taken with replacement from the original sample $\{x_1, \dots, x_n\}$, of the same size, n as of the original sample
2. We then calculate the bootstrap statistic, say $a\{x_1^*, \dots, x_n^*\}$—a statistic such as mean, median, variance etc computed on the bootstrap samples
3. We repeat steps (1) and (2) say B (taking B to be large) times to create a bootstrap distribution—a distribution of bootstrap statistics, $a\{x_1^*, \dots, x_n^*\}$.

The envelope of the bootstrap distributions gives a robust estimate of uncertainty.

We can illustrate the method of bootstrap with the help of an example. Background estimation is often performed with fitting the data or simulation, and then the uncertainty corresponding to the fit needs to be estimated. The bootstrap distributions can be generated by varying each point randomly within the 1σ of the nominal, and fitting all the distributions. Then the spread of the fits can be considered the uncertainty on the fit.

6.4.4 Combining different uncertainties

If we have variables x and y, their means are given by $\langle x \rangle$ and $\langle y \rangle$, and the square of standard deviation, termed as variance, $V(x) = \langle x^2 \rangle - \langle x \rangle^2$ and $V(y) = \langle y^2 \rangle - \langle y \rangle^2$. However, if we want to construct a third variable from these two variables, then we need to define covariance: $Cov(x, y) = \langle xy \rangle - \langle x \rangle\langle y \rangle$, and correlation $\rho = Cov(x, y)/\sqrt{V(x)V(y)}$. The correlation $\rho = 0$ when the variables are not correlated, $\rho = 1$ for completely correlated variables, and $\rho = -1$ for completely anticorrelated variables. Correlation essentially means if both variables change similarly, then the uncertainty on the combined variable can be represented by:

$$\sigma_{xy} = (df(x)/dx)^2 V(x) + (df(y)/dy)^2 V(y) + 2(df/dx)(df/dy)Cov(x, y)$$

where $f(x)$ and $f(y)$ denote the PDFs corresponding to the variables. This means, for example, if the variables are completely correlated, the uncertainties will be added linearly. The most common case is when estimation of two different sources have the same statistical uncertainty, since they have been estimated from the same MC or data sample. Luminosity uncertainty when combining two different channels is another example.

Now let us consider two measurements, x_1 and x_2, with statistical uncertainties σ_1 and σ_2, and with a common systematic uncertainty of σ. The variances are given by: $V(x_1) = \sigma_1^2 + \sigma^2$ and $V(x_2) = \sigma_2^2 + \sigma^2$, with the covariance, $Cov(x_1, x_2) = \sigma^2$. This indicates that the statistical and systematic uncertainties can be added in quadrature. They can be represented in a covariance matrix:

$$\begin{bmatrix} \sigma_1^2 + \sigma^2 & \sigma^2 \\ \sigma^2 & \sigma_2^2 + \sigma^2 \end{bmatrix}$$

This can be generalised, if there are several sources of independent systematic uncertainties, they will be added in quadrature. In many instances, the uncertainties are expressed as a fraction of the nominal value, for example $\sigma = \varepsilon_1 x_1$, then the correlation matrix can be expressed in terms of those.

Often results from two different analysis channels, such as electron and muon channels need to be combined. While event yields or the distributions can be combined relatively easily, each systematic variation must be considered individually, taking into account their correlations. For example, while electron or muon efficiencies will be uncorrelated, the modelling uncertainty from MC generators will not be uncorrelated.

Exercises

1. The result of a certain measurement is 80 370 $\pm$ 19 MeV. Previously this was measured to be 80 385 $\pm$ 15 MeV. Are the two results compatible? Why or why not? Is the current measurement an improvement over the previous one? Why or why not? How would we know if more data will help in reducing the uncertainty?

2. The statistical uncertainty is estimated as $1/\sqrt{N}$ for N events. However, as we have discussed in chapter 4 many simulated programmes give weighted events. How should this calculation be modified in those cases?

3. One needs to be careful in assessing the impact of systematic uncertainties in an analysis. Let us assume in a search, the background estimation is completely data driven. Then how will the systematic uncertainties impact the result?

4. A general principle is, systematic uncertainty does not decrease with more data. Can you think of examples where this would not be true?

5. There are two ways of assessing systematic uncertainty on the reconstructed objects. In the top-down approach, we think of all possible sources affecting the reconstructed object, and in the bottom-up approach, we essentially consider uncertainties in each step while reconstructing the object. An example can be the uncertainty on jet observables. In the former approach (usually adopted), uncertainties on jet energy scale, jet energy resolution etc are estimated. In the bottom-up approach, the uncertainty is calculated for each input to the jet reconstruction. Can you think of advantages and disadvantages of the two approaches?

References

[1] Barlow R 2002 Systematic errors *Proc. of the Conf. on Advanced Statistical Techniques in Particle Physics (Durham, UK, March 18–22, 2002)* pp 134–44
[2] Butterworth J *et al* 2016 PDF4LHC recommendations for LHC Run II *J. Phys.* G **43** 023001

IOP Publishing

Experimental Particle Physics
Understanding the measurements and searches at the Large Hadron Collider
Deepak Kar

Chapter 7

Presenting and interpreting the results

*We have made the discovery of a new particle—a completely new particle—
which is most probably very different from all the other particles. It's nearly a
once in a lifetime experience, I would say.*

—Rolf-Dieter Heuer

The final step of an analysis is to present the results, and extract the maximum
amount of physics knowledge out of them. For measurements, it is usually the total
or differential cross-section as a function of the desired observables, or comparison
of shapes of distributions of some observables between data and simulation. The
final step in a search is either quantifying the deviation from SM, or setting an
exclusion limit on the new physics model being searched for. Almost always the
results are presented in histograms, colloquially referred to as *plots*. The previous
chapters have introduced different ingredients of analyses, and everything comes
together in this chapter. First we will describe the construction and different
elements of plots themselves in section 7.1, and then we will show how to interpret
the plots, using examples from published results in section 7.2.

7.1 Constructing the plots

7.1.1 Frequency distribution

The most common plot is the frequency distribution of an observable, where the
x-axis contains the value of the observable, and the y-axis shows in how many
instances that value was obtained. Since the number of events corresponds to the
cross-section (as discussed in chapter 5), these plots essentially present the cross-
section as a function of the specific observable. The plots are made in *bins*, as in the
x-axis range is divided into a number of sub-ranges. The number of sub-ranges (or in
other words, the size of a bin, bin-width) is chosen such that each bin has a
reasonable number of entries, where reasonable is very much dependent on the
specific analysis. We do not want to have large statistical uncertainty in a bin, but

doi:10.1088/2053-2563/ab1be6ch7

also prefer the bin size to be larger than the experimental resolution of the particular observable. The statistical uncertainty in a bin is given by $1/\sqrt{N}$, where N is the number of entries in that bin. This is stated by assuming Poissonian statistical distribution (for a random variable drawn from a Poisson distribution, it has a variance equal to the mean), which is usually the case for measurements with a large number of data points. The points are usually placed at the bin centre, i.e. at the average of low and high x-values of the bin. Sometimes a horizontal line is drawn to indicate the bin width. The vertical errors bars represent the (statistical and systematic) uncertainty for the bin values. Usually the uncertainties are symmetric (although certain systematic components can be asymmetric, for example uncertainty on trigger efficiency when the efficiency is close to unity, can have larger *minus* variations than *plus*, as the maximum cannot exceed unity).

If the data spans many orders of magnitude and there are important features to be identified at both large and small scales, then logarithmic axes are used. If the plot has logarithmic scale on an axis, the corresponding uncertainty bars will no longer appear symmetric. The cross-section plots are termed single or double differential according to how the measurement is restricted by other parameter ranges. Sometimes the y-axis contains the number of events per (for example) 5 GeV. This usually corresponds to the x-axis bin-width. More often that not, due to normalisation (by the number of events, or by the area under the distribution), these do not show the actual number of events, but the fraction proportional to the actual value.

Figure 7.1 (top left) shows the simplest example of an idealised one-dimensional plot. The y-axis represents the normalised frequency, while the x-axis shows the value of the observable. A real example is shown in figure 7.1 (top right), where the cross-section for $Z \rightarrow \ell^+\ell^-$+jets process (where $\ell^\pm$ stands for electrons or muons) is shown as a function of number of jets in the events. The data is compared to different MC model predictions of this signal process, and the ratios of MC/data at the bottom make the comparisons easier. Figure 7.1 (bottom left) shows an example where several different distributions are shown in the same plot, in this case dijet cross-section as a function of dijet invariant mass in different $y^\star$ ranges, where $y^\star$ is the average rapidity of the dijet system. This is an example of a double-differential cross-section plot, where the cross-section values in each bin are divided by the bin mass expanse, as a well as the rapidity range. We can also have single-differential (or commonly termed differential) cross-section plots, usually there each bin value is divided by the x-axis bin size. Sometimes, for ease of display, some of the lines are shifted, but that will be mentioned in the plot. Finally, in figure 7.1 (bottom right), we show an example of a stacked plot, where the leading jet p_T in events containing W boson and jets is plotted. Contributions of different standard model processes (from simulation), which survive the analysis selection are included. They are shown in different colours, one over another, after individually normalising each histogram. So the top-most line at any bin represents the total contribution of all processes considered in that bin.

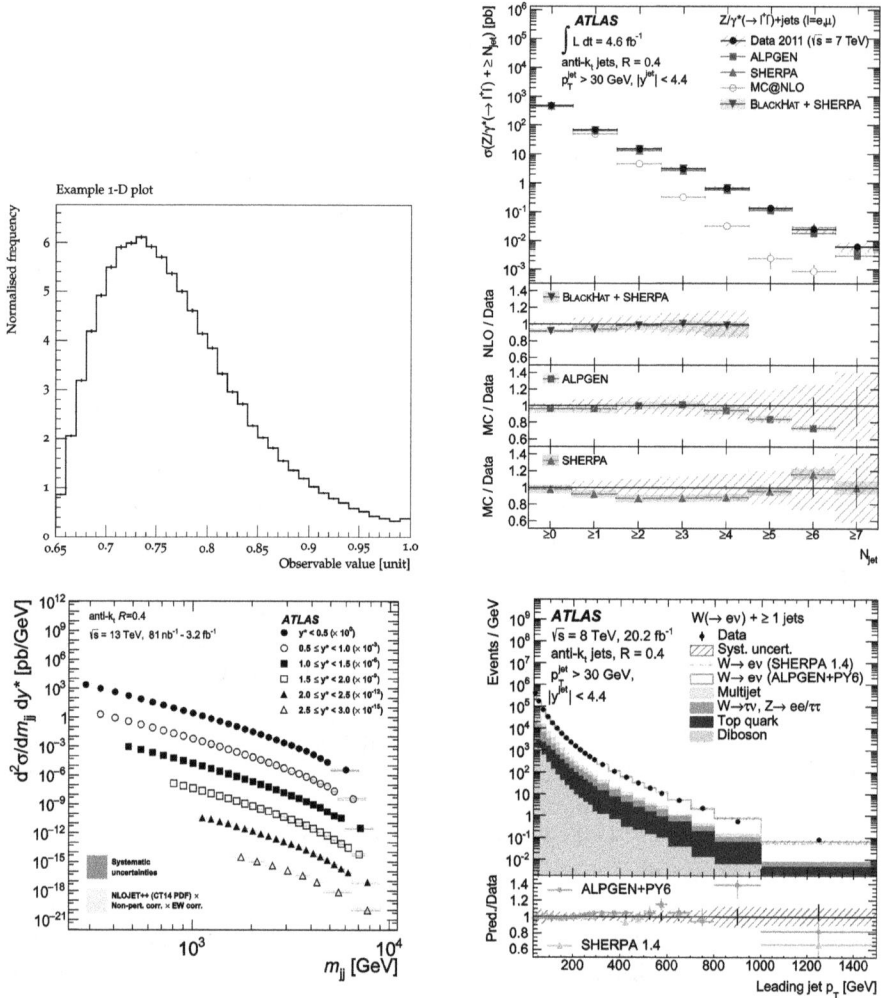

Figure 7.1. Examples of one-dimensional plots. A simple sketch (top left), distribution of number of jets with MC/data ratios [1] (top right), distribution of dijet invariant mass in different rapidity ranges [2] (bottom left), and distribution of leading jet transverse momentum, with different standard model process contributions stacked [3] (bottom right) (ATLAS Experiment © 2018 CERN). The y-axis in each case is a proxy for number of events.

7.1.2 Correlation between variables

The profile histograms present the dependence of the mean value of an observable as a binned function of another observable (plotted at the bin centre). While the event-by-event fluctuation of the dependent observable cannot be seen from profile plots, they are useful to assess the overall trend. An idealised example profile plot is in figure 7.2 (top left), where the mean value of observable2 is plotted as a function of observable1. The figure 7.2 (top right) shows an actual example, where observable2

Example profile plot

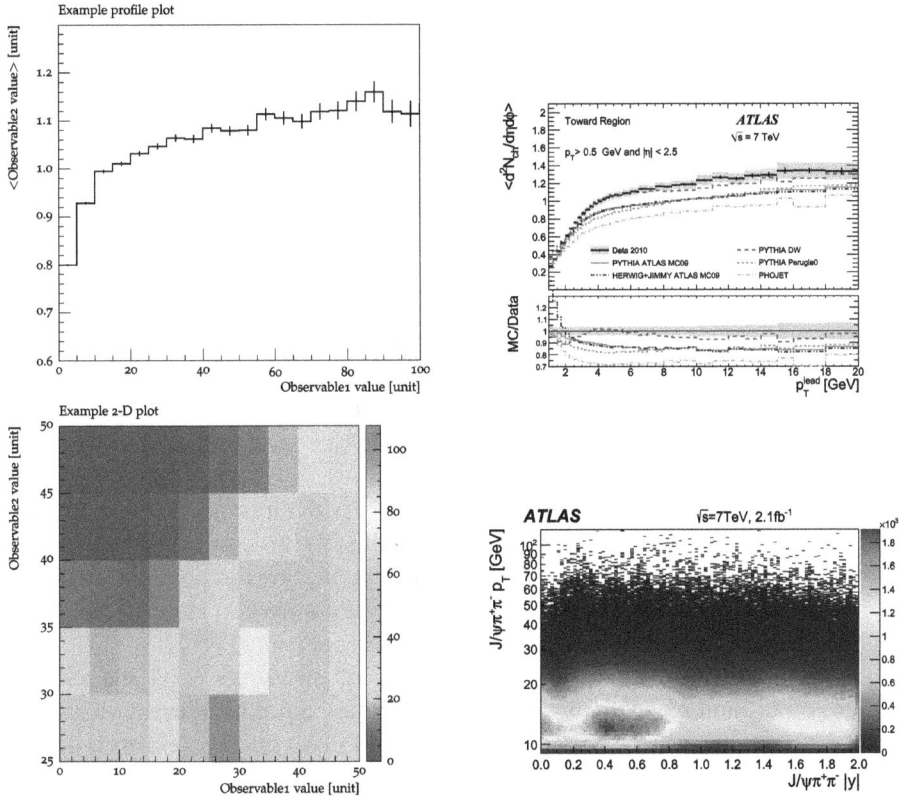

Figure 7.2. Examples of two-dimensional plots. A simple sketch of a profile (top left), a profile showing average charged particle multiplicity density against the p_T of the leading charged particle [4] (top right), a simple sketch of a heat map plot (bottom left), and a heat map plot showing correlation of absolute rapidity and p_T for a specific particle production [5] (ATLAS Experiment © 2018 CERN).

is the average number of charged particles, normalised by angular detector area where they are measured, and observable1 is the leading charged particle p_T.

When more fine-grained information between two observables is required, two-dimensional plots are employed, with values of the observables plotted along the x- and y-axes respectively, as shown in an idealised example in figure 7.2 (bottom left). These plots effectively have a hidden z-axis, to indicate the number of entries corresponding to a particular x and y value. In case an efficiency needs to be represented in the z-axis, the histogram is filled with the efficiencies as weights. In the simplest case (scatter plot), the density of points at a particular region of the plot indicates stronger or weaker correlation. A more sophisticated way (termed a heat map, as shown here) is to use different colours to indicate different range of frequencies. In the example two-dimensional heat-map plot in figure 7.2 (bottom right), the two observables are transverse momentum and rapidity of $\Psi(2S)$ particle reconstructed in $J/\Psi(\rightarrow \mu^+\mu^-)\pi^+\pi^-$ decay mode. Sometimes these plots are column-normalised, which allows us to focus on the statistically significant part of the distributions more clearly.

7.1.3 Limit plots

The observed number of events, N_{obs} in SR in data in a search in principle consists of signal and background events. We have discussed various methods of estimating the background, N_{bg} in SR. Our eventual aim is to check if the N_{obs} is consistent with N_{bg}, or if there is an excess of events, possibly indicating the presence of new physics. In the latter scenario, we need to quantify how confident we are about a discovery if such an excess is observed. In the former case, we construct the (exclusion) limit plots to indicate an upper bound on the signal production rate for the new physics model probed. In most cases, the limits are presented as a function of the hypothetical particle mass value, as the new physics model does not specify the new resonance mass *a priori*.

We can represent $N_{obs} = \mu N_{sig} + N_{bg}$, where N_{sig} is the number of expected signal events, and μ is referred to as signal strength modifier in this context. The presence of no new physics is represented by $\mu = 0$, which is the lowest possible value of μ. Higher values of μ would indicate evidence of new physics. Here μ can also be thought of as a parameter proportional to the cross-section.

Our aim is the estimate μ, given that we have measured N_{obs}, and estimated the values of N_{sig} and N_{bg} with associated uncertainties. Since most event yields follow Poisson distributions, we can write the probability of observing N_{obs} events is given by:

$$P(N_{obs}) = \frac{\lambda^{N_{obs}} e^{-\lambda}}{N_{obs}!} = \frac{\lambda^{\mu N_{sig} + N_{bg}} e^{-\lambda}}{(\mu N_{sig} + N_{bg})!}$$

where λ is a nuisance parameter (NP) signifying the width of the Poisson distribution. The consideration of uncertainties in the limit setting procedure is an important aspect. They are accounted for by NPs. The ranking or the pull plots are used to show the effect of different NPs on the determination of μ before (prefit) or after (postfit) profile likelihood fit. Then the likelihood function to estimate μ is given by:

$$\mathcal{L}(N_{obs}; \mu) = \frac{(\mu N_{sig} + N_{bg})}{N_{obs}!} e^{-(\mu N_{sig} + N_{bg})}$$

If the uncertainty in background estimation is quantified by the NP θ in the general case, then we can write:

$$\mathcal{L}(N_{obs}; \mu, \theta) = \frac{(\mu N_{sig} + \theta N_{bg})}{N_{obs}!} e^{-(\mu N_{sig} + \theta N_{bg})} \times \text{PDF}(\theta)$$

where $PDF(\theta)$ is the probability distribution of the NP θ. Additionally, often the background normalisation is performed by using a factor obtained in CR, so in the general case, the expression above will be multiplied by that factor as well.

We now need to compare two alternative hypotheses. The absence of new physics can be represented by null hypothesis $H_0 : \mu = 0$, which needs to be tested against an

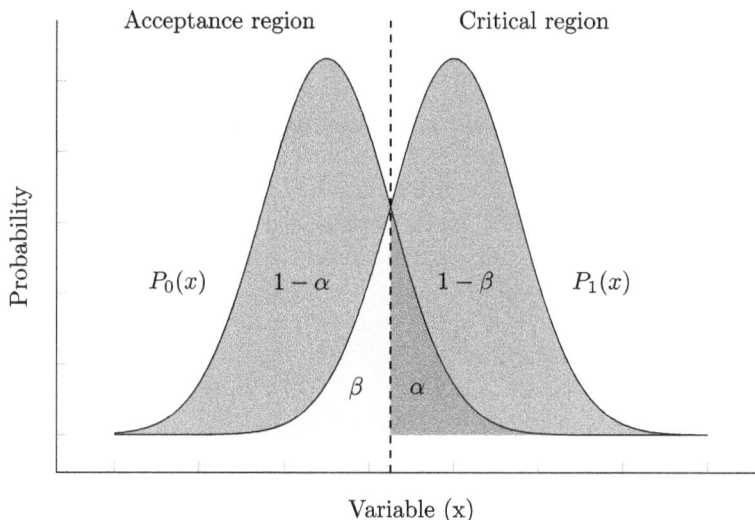

Figure 7.3. Illustration of hypothesis testing.

alternative hypothesis $H_a : \mu > 0$. We can alternatively term them as background only and signal + background hypotheses as well. These two hypotheses can be represented respectively by two PDFs $P_0(x)$ and $P_1(x)$. The PDFs $P_0(x)$ and $P_1(x)$ are usually of the same functional form, potentially with some overlap, but with different μ parameters. An example is shown in figure 7.3, where there is an overlap between the two PDFs.

The concept of confidence interval is introduced in order to perform the hypothesis testing. A confidence interval $[\mu_1, \mu_2]$ is an interval which contains the target value μ' of the parameter of interest μ obtained from the experimental measurement a certain fraction of times, denoted by $1 - \alpha$. The fraction $1 - \alpha$ is referred to as confidence level (CL). It can be represented by: $p(\mu' \in [\mu_1, \mu_2]) = 1 - \alpha$. Usually 95% (single sided) confidence level is used, which is the associated probability of that observation being correct 95% of the time. In other words, if the measurement is made repeatedly on independent datasets, the measured value will be obtained at least 95% of the time.

Then two figures of merit are defined. The probability of incorrectly rejecting H_0 in favour of H_1 when H_0 is true corresponds to α, which is termed as false positive or type I error. The probability of failing to reject H_0 in favour of H_1 when H_0 is false is denoted by β, which is termed as the false negative or type II error.

After obtaining N_{obs} from the analysis, we can have two distinct scenarios, discovery and exclusion. The N_{obs} can be termed as test statistic, which is essentially a single number from the measured data to be used for hypothesis testing. Usually we do not have just a single bin in the SR, so following the approach by Neyman and Pearson, the likelihood ratio (LR), which is defined as ratio of the likelihood under the null hypothesis and alternative hypothesis, is used to construct the test statistic, after profiling out the NPs. We can then construct PDFs for test static separately assuming H_0 and H_1 are respectively true.

Then the p-value is defined as the probability of observing a value of the test statistic that is $\geqslant N_{obs}$, assuming the null hypothesis is true. The p-value by itself does not translate into the probability of the null hypothesis being true, rather it is indicative of the probability of finding data at least as favourable to the alternative hypothesis if the null hypothesis is true. Therefore, the smaller the p-value is, the stronger the evidence against the null hypothesis. Since the test static considers the statistical uncertainty of the data, by construction p-values take into consideration the sample size. A large sample size increases the power of a hypothesis test without decreasing the confidence level $1 - \alpha$.

If the p-value is smaller than $\alpha = 0.05$ (corresponding to 95% CL), then the null hypothesis can be rejected. The usual threshold for discovery is 5σ, which corresponds to a p-value of 2.87×10^{-7}. If this condition is satisfied, one looks for an alternative hypothesis which can explain the data well. A 3σ deviation is often termed an observation, but not a discovery. On the other hand, the null hypothesis is verified if the opposite is true. Then the limit plot is constructed by evaluating for which values of test statistic this condition holds.

However, as the new physics is not known, H_1 cannot usually be a single hypothesis. In order to estimate the sensitivity of the analysis to different hypotheses, an MC sample is constructed, which gives median sensitivity over the whole range. This is termed the *Asimov data set*, which yields true parameter values when estimators for those parameters are evaluated. The name apparently was inspired by a science fiction short story by Isaac Asimov, where an ensemble of simulated experiments can be replaced by a single representative one. Often one wants to test H_0 without having a particular alternative hypothesis H_1, which is referred to as the model independent limit.

The above discussion assumes that we are searching for a signal with a specific mass, and the calculated p-value is referred to as the local p-value. However, in many scenarios, we *a priori* do not know where the signal can be, so we probe a mass range. When performing the search in a mass range, even if the null hypothesis is true, the chance of observing a spurious excess of events at some mass value increases because upward statistical fluctuation can occur anywhere. This is termed as the *look elsewhere effect* (LEE), as one needs to consider the whole mass range (i.e. look elsewhere!) to construct the p-value, termed the global p-value. In fact, when looking at a mass range ΔM for signal of width W, LEE results in up to $\Delta M / W$ factor of probability increase to observe an upward fluctuation.

In the example one-dimensional limit plot in figure 7.4 (top left), limits are set on the production of a hypothetical W'_R boson. The mass of it is reconstructed in hadronic decay mode and is plotted in the x-axis. The y-axis shows production cross-section of this process, multiplied by the branching fraction of the final state being probed. The production cross-section depends on the mass of the new particle probed. Every value is plotted in terms of 95% confidence level, as mentioned earlier. This plot contains a few different components.

- Expected line with 1σ and 2σ bands: this is obtained from MC simulation using SM processes only (i.e. corresponding to the background). The cross-section (times branching ratio) is estimated as a function of signal mass values

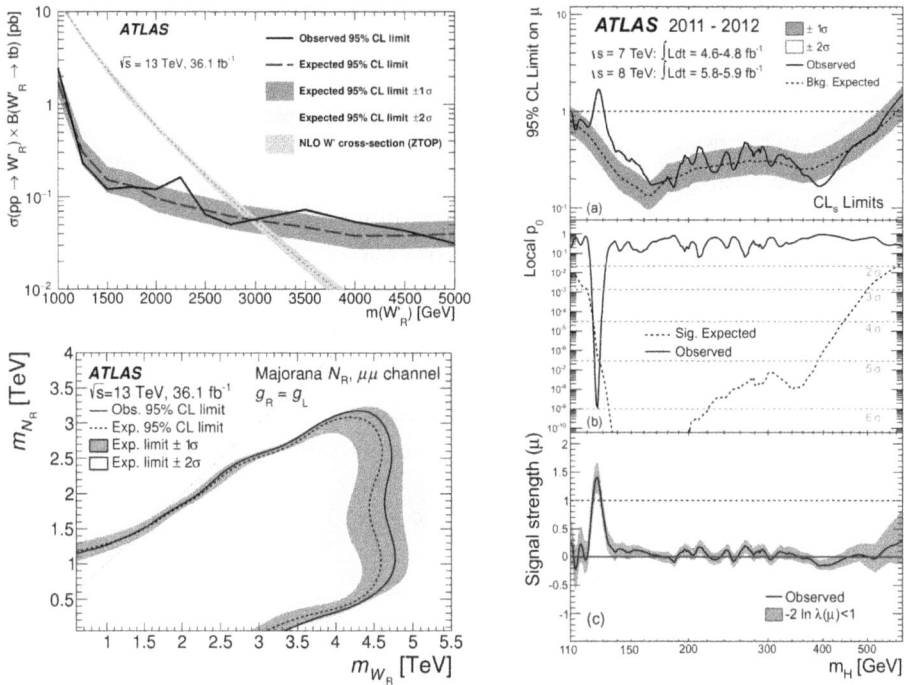

Figure 7.4. Examples of limit plot. An one-dimensional limit plot with the limit on cross-section on the y-axis [6] (top left), a one-dimensional limit plot with limit on the ratio of measured cross-section over the theory cross-section on the y-axis [7] (right), and a two dimensional limit plot [8] (bottom left) (ATLAS Experiment © 2018 CERN). The second one corresponds to the Higgs boson discovery, and it also contains the calculated p-value and the observed signal strength value as a function of the reconstructed Higgs boson mass.

(corresponding to the signal being searched for), using the analysis selection. So basically we are trying to find how events from SM processes will be BSM signal-like. Then a smooth line is drawn by extrapolating between the points.

The green and the yellow errors bands indicate $\pm 1\sigma$ and 2σ ranges, based on the uncertainties of the MC prediction, both from theoretical and experimental sources. Because of this choice of colours, the limit plots are often referred to as Brazil plots.

The bands are often computed using pseudo-experiments, considering statistical fluctuations as well as fluctuations of the systematic uncertainties. The median of the distribution of the limits from the pseudo experiments is taken as the expected limit, and 1σ and 2σ bands are defined as the ranges containing respectively 68% and 95% of the values obtained with the pseudo-experiments.

- Observed line: this is calculated from data, represents the number of events seen in the analysis, for each mass value. It is expected to stay within the expected bands if the simulation is accurate. The fluctuations represent the statistical uncertainty of the data.

- Theory prediction for the new model being searched for: the calculated values from the new model being probed are represented in the red line, with associated theoretical uncertainties. So this is essentially similar to the expected line, but using the BSM scenario rather than the SM used there. As long as the expected and observed lines are below the theory prediction, the conclusion is that no evidence of the new particle is seen. By this argument, the expected and observed limits are both 3 TeV, from where the theory prediction line intersects the expected and observed lines. If at any point, the observed line goes beyond the expected Brazil-bands, that may indicate that the data contains more events than SM predicts. The threshold for an observation is usually set at 3σ and for a discovery is at 5σ. Therefore, the *excess* at 2.25 TeV was not considered important, and that mass was still excluded. Only in the region where the observed line is higher than the theory line, and beyond the statistically allowed deviations from expected, can this particular new model be confirmed. With more data, the limits change, as the theory prediction will move up and the green and yellow bands are expected to shrink.

Often to simplify the plot, on the y-axis, the cross-section times branching ratio is divided by the theory prediction for the new model. In that case, the exciting parts are where the expected/observed lines drop below $y = 1$. That is the part where new physics can be observed if the observed line deviates upward from expected. If not, then it can be concluded that the new particle, as per this model, is not produced at the rate the model demands, so the mass range can be ruled out at 95% confidence level. An example is shown in figure 7.4 (right) for the Higgs boson discovery paper. The plot also shows local significance and signal strength.

Sometimes, limits on one parameter can be shown as a function of another model parameter, leading to a two-dimensional limit plot. The two-dimensional limit plots are essentially based on the same concept, but the exclusions are denoted by (inside of) the contours in the parameter space. They are drawn by testing if each point in parameter space is excluded or not (in practice drawn using interpolation as well from a discrete set of points). An example is shown in figure 7.4 (bottom left). Additional constraints can also exclude part of the parameter space as well.

7.1.4 ROC curves

Receiver operating characteristic (ROC) curves are used to show signal to background discrimination. These are usually constructed completely from simulation, where signal and background can be unambiguously identified.

An idealised example is shown in figure 7.5, detailing how they are constructed. We start with distributions of a certain variable (denoted by χ) for some arbitrary signal and background process in the top-left corner. higher values of χ contain most of the signal events, but background events also appear under the signal peak. If we expect the MC simulation to describe the data, then this plot for data should have

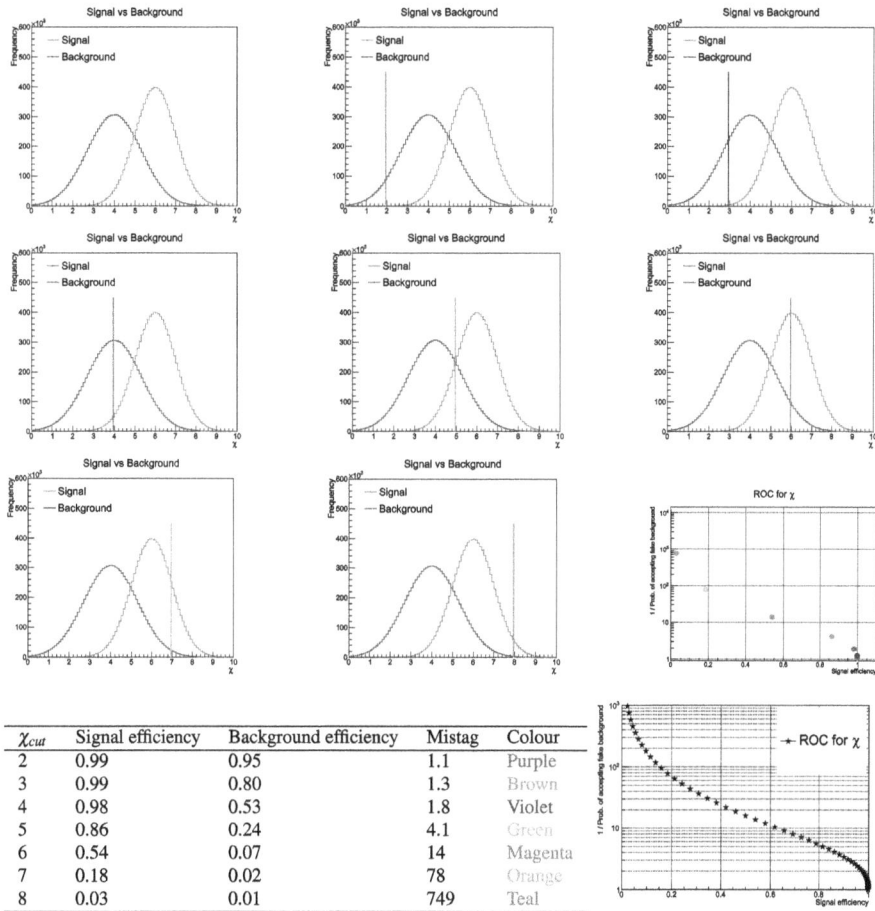

Figure 7.5. Steps for obtaining a ROC curve from a distribution of an arbitrary observable χ (top left) from signal and background events. The successive plots show how the χ_{cut} is placed at different values, and the table below shows the signal efficiency and background efficiency corresponding to that placement, with the points plotted in the last plot of the penultimate row. The final ROC is drawn by employing more fine-grained steps (bottom right plot).

χ_{cut}	Signal efficiency	Background efficiency	Mistag	Colour
2	0.99	0.95	1.1	Purple
3	0.99	0.80	1.3	Brown
4	0.98	0.53	1.8	Violet
5	0.86	0.24	4.1	Green
6	0.54	0.07	14	Magenta
7	0.18	0.02	78	Orange
8	0.03	0.01	749	Teal

the same characteristic, but we will just have one combined distribution, and would not know the fractional contribution of signal and background.

So if only the events over a certain value $\chi > \chi_{cut}$ are chosen, it will contain a certain fraction of signal (which we can term as signal efficiency), and a certain fraction of the background events (which we can term background efficiency). Since we actually want to reject the background, inverse of background efficiency (which is often termed the mis-tag rate or background rejection) is considered. Here the signal (or background) efficiency simply refers to integral of the signal (background) curve in the selected range divided by the total integral over the signal (background) curve.

By varying at χ_{cut} values from 2 to 8, as shown in the sequence of figures, the signal efficiency and background rejection can be tabulated for each cut, as shown in the table at the bottom. The χ_{cut} values in the table are represented in different colours. Now, each pair (signal efficiency, mistag rate) is plotted in the last figure of the penultimate row. This is essentially the ROC curve. It can be converted to the ROC curve shown at the bottom-right by not taking the discrete points as χ_{cut} values, but by a much more fine-grained scanning. The term AUC (area under the ROC curve) is often used to denote the two-dimensional area underneath the entire ROC curve.

While the ROC curve then contains no direct information about the individual χ_{cut} values, it gives us different working points (WPs), each corresponding to a certain signal efficiency and background rejection. Then, depending on what is more important in that particular analysis, a WP is chosen, and that corresponding χ_{cut} value is applied in data to get more signal-like events.

ROC curves are used extensively in areas where signal and background discrimination needs to be performed and jet substructure is one obvious example. The usage will be discussed in more detail in chapter 9.

7.2 Interpreting the plots

After obtaining the plots, the next step is to understand the features of these plots, and understand the physics message contained in them. We will use examples from published papers to illustrate this. We focus mostly on measurement results, as results of the searches are essentially represented by data and SM prediction in the signal region in terms of the sensitive variable (usually corresponding to the mass of the new particle being searched for), and exclusion limits on the BSM scenario probed are set. There are examples where distributions in well-understood control regions have been unfolded and presented as a measurement [9] to probe a new region of phase space.

This list is by no means exhaustive, nor it is intended to give a summary of the most interesting physics results obtained from the LHC so far. As our focus is on interpreting the different types of data distributions, and to keep the discussion general, we will not comment on the agreement or disagreement of data with different MC model predictions, which are of course a useful aspect of the measurements. Furthermore, distributions such as p_T or associated jet multiplicity of most objects show a very similar trend, we will not show them repeatedly. Many observables shown here have been introduced in chapter 5.

7.2.1 Total cross-section

Before measuring specific distributions, it is of interest to measure the total hadronic cross-section of the pp collisions, and see how it changes with centre-of-mass energy (the predictions from simulation were shown in chapter 2). It is easier to measure the total inelastic cross-section, as outgoing protons from elastic collisions are almost impossible to detect in ATLAS or CMS detectors, since they travel in a direction very close to the beam-pipe.

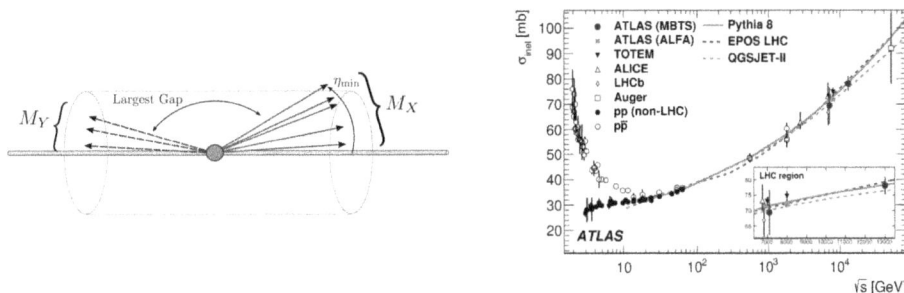

Figure 7.6. The definition of fiducial region for cross-section measurement [13] (left), and the measured inelastic cross-sections from different experiments as functions of centre-of-mass energy [14] (right) (ATLAS Experiment © 2018 CERN).

The total inelastic cross-section is usually measured by a counting experiment, accounting for non-collision background, trigger and event selection efficiencies. A fiducial region needs to be defined, which is done by finding the largest rapidity gap between any two final state hadrons, as in figure 7.6 (left). Then the invariant mass of the hadrons on either side is determined, and the larger of the two masses, denoted M_X, is used to define $\xi = M_X^2/s$, where s is the squared centre-of-mass energy. Then the measurement is restricted to over some certain values of ξ, and a correction is applied for the migration of events out of ξ acceptance to inside of it due to detector effects. The measured inelastic fiducial cross-section is then extrapolated by using MC model predictions to the total inelastic collision, and the results from different experiments and centre-of-mass energies are shown in figure 7.6 (right). A gradual increase in the cross-section can be seen. The total elastic cross-section is measured by forward detectors such as ALFA [10] (Absolute Luminosity For ATLAS, 240 m away from IP) and AFP [11] (ATLAS Forward Proton, 210 m from IP) associated with ATLAS, and TOTEM [12] (TOTal, Elastic and diffractive cross-section Measurement, 500 m from IP) associated with CMS.

7.2.2 Charged particle distributions

The most common examples of distributions with charged particles are so-called minimum-bias (MB) observables. In figure 7.7, several MB distributions constructed with charged particles are shown.

The p_T and multiplicity (which is a shorthand for number of charged particles) distributions, both normalised by number of events in figure 7.7 (top row) show exponential decrease, indicating most particles are produced with very low p_T, and most events have very few charged particles. To understand this, we can look at the nature of the collisions. A large fraction of interactions result in peripheral or semi-peripheral collisions (defined in chapter 2), where the small overlap between the protons results in a small amount of energy exchange in the collision. This leads to less sharing of the limited energy between a smaller number of particles, and those particles are produced with smaller p_T. This changes as collisions become more central.

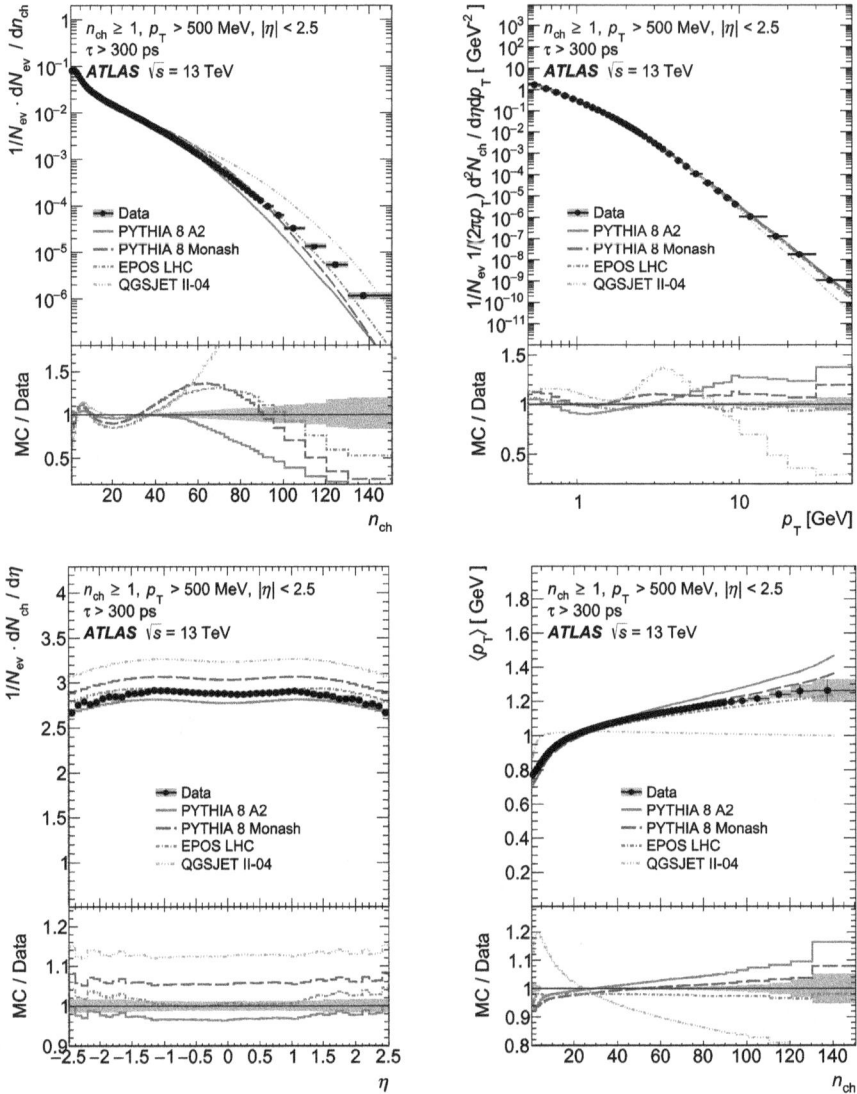

Figure 7.7. Distributions of charged particle multiplicity in each event n_{ch} (top left), their transverse momenta p_T (top right), their pseudorapidity η (bottom left), and their mean transverse momenta against multiplicity (bottom right) measured by ATLAS [15] (ATLAS Experiment © 2018 CERN). N_{ch} denotes the total number of charged particles in the full sample. In the first three cases the distributions are normalised by number of events, N_{ev}. In the second case, it is additionally normalised by the pseudorapidity and p_T range, as well as by a factor or $2\pi p_T$, the latter to make the cross-section Lorentz invariant. The ratio panels at the bottom are to look at data and MC model comparisons clearly.

The pseudorapidity distribution of charged particles (bottom left) is roughly uniform over the central part of the detector with a slight dip, but starts decreasing at the edge of the tracking detector, as more particles are produced in the central region (also the tracking efficiency decreases). The dip in the central region needs explanation. The charged particles in inclusive events are expected to be produced isotropically, and indeed if dN/dy distribution (i.e. against rapidity) was plotted, it will be flat. The actual plotted observable here is $dN/d\eta$, which can be related to dN/dy by following:

$$\frac{dN}{d\eta} = \frac{dN}{dy}\frac{dy}{d\eta}$$

Using equations obtained in chapter 3, section 3.3,

$$\frac{dy}{d\eta} = \frac{d}{d\eta}\frac{1}{2}\ln\frac{\sqrt{p_T^2\cosh^2\eta + m^2} + p_T\sinh\eta}{\sqrt{p_T^2\cosh^2\eta + m^2} - p_T\sinh\eta}$$

$$= \frac{\cosh\eta}{\sqrt{1 + (m/p_T)^2 + \sinh^2\eta}}$$

where m and p_T denote the mass and transverse momenta of the particles. Therefore, we have:

$$\frac{dN}{d\eta} = \frac{\cosh\eta}{\sqrt{1 + (m/p_T)^2 + \sinh^2\eta}}\frac{dy}{d\eta}$$

For $\eta \approx 0$, and for low p_T particles with non-zero mass (most charged particles are pions, and most of them have low p_T, as seen before), the first term gives a value smaller than unity, which corresponds to the dip.

Similar distributions are also obtained for different identified particle production (pion and K-mesons, lambda baryons, etc), by using vertex information and decay lengths.

Figure 7.7 (bottom right) shows mean p_T of charged particles per event plotted as a function of charged particle multiplicity. The gradual rise can be attributed to multiple parton interactions (MPI) and colour (re)connection effects. If charged particles were produced by single hard collision then higher multiplicity could have only come from very energetic events, as beam–beam remnants (BBR) are scale independent. This would have led to a sharper rise than seen. The colour reconnection leads to fewer additional charged particles produced with every additional MPI (same energy shared by fewer particles), therefore causing the rise.

There have been several efforts to describe the characteristics of inclusive particle production. Feynman postulated [16] that the total number of particles created in the collisions should rise logarithmically with $\sqrt{s}$. The width of the rapidity distribution was also expected to be proportional to $\log\sqrt{s}$ with the produced particles evenly distributed in rapidity. Then Feynman scaling implies that dN/dy distribution at mid-rapidity is independent of collision energy. The Koba–Nielsen–Olsen (KNO) [17] scaling suggested that the function $\Psi(z) = \langle N\rangle P(N)$ should be independent of collision energy, where $z = N/\langle N\rangle$ is the ratio of particle multiplicity in an event over average particle multiplicity, and $P(N)$ is the probability of N particles being produced in the event. However, experimental data supported neither hypothesis exactly [18, 19], and

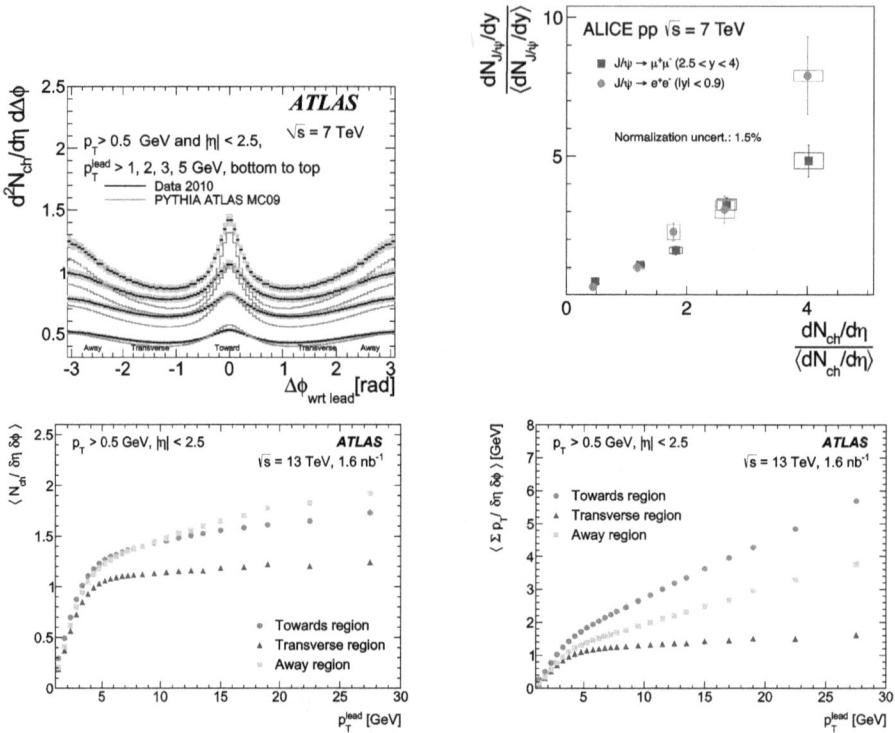

Figure 7.8. Angular distribution of charged particles with respect to the leading charged particle with different p_T thresholds [4] (top left), J/ψ multiplicity as a function of charged particle multiplicity [21] (top right). Average charged particle multiplicity (bottom left) and scalar sum of p_T (bottom right) as functions of leading charged particle p_T in three different angular regions [22] (ATLAS, ALICE Experiments © 2018 CERN). The charged particle multiplicity or scalar sum p_T distributions from ATLAS are normalised by the angular area (effectively converting them to densities), while for the ALICE plot, both quantities are *self-normalised*, as in divided by their average values.

theoretical shortcomings were found for high energy hadron colliders. It was seen that the multiplicity distribution more closely (but again not perfectly) follows a negative binomial distribution (NBD) [20], parametrised in terms of $\langle N \rangle$ and a parameter k determining the width. This can be phenomenologically explained by combining the probability of parton branchings from original hadrons, with the probability of them participating in collisions, before undergoing hadronisation.

Charged particle distributions in events with an identified hard scatter are sensitive to underlying event, as shown in figures 7.8 and 7.9. The angular distribution of charged particles shows the structure of the events in figure 7.8 (top left). The leading charged particle (with different p_T thresholds) is taken as reference direction (i.e. $\phi = 0$), and $\Delta\phi$ of the rest of the charged particles are plotted with respect to leading charged particle. So higher values at some $\Delta\phi$ ranges indicate higher activity. The plot effectively covers a full circle, as the left and right edges correspond to 2π. The leading p_T is usually aligned with the direction of the hardest energy flow, leading to an increased activity around $\phi = 0$. Conservation of momentum then requires the

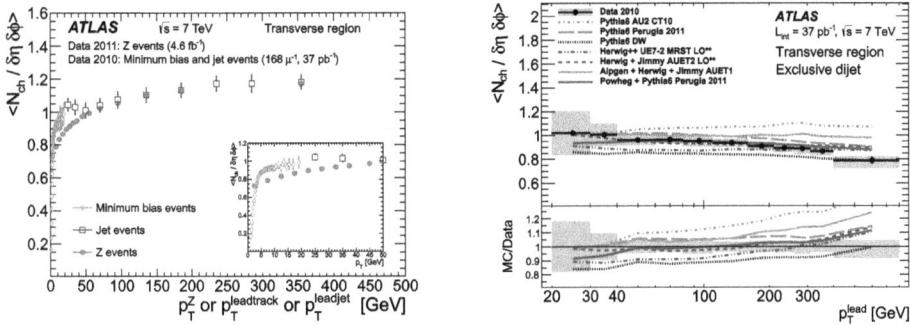

Figure 7.9. Charged particle multiplicity density as a function of the leading object p_T compared between inclusive events, in events with at least one jet, and in events with a Z boson [24] (left), and for exclusive dijet events [25] (right) (ATLAS Experiment © 2018 CERN).

hardest energy flow must be balanced by increased activity opposite to it, as seen by the higher activity at the edges.

The effect of MPI can also be seen in the plot of J/ψ yield as a function of charged particle multiplicity in figure 7.8 (top right). The number of J/ψ in each multiplicity interval (both normalised by η interval) is estimated from events with dileptonic invariant mass consistent with J/ψ mass (3–4 GeV) and associated charged particle multiplicity. However, rather than depicting the actual J/ψ yield or charged particle multiplicity, both values are normalised by corresponding average values for minimum bias events. The increase of multiplicity with yield indicates the effect of MPI, or associated hadronic activity with J/ψ production.

The mean charged particle multiplicity and mean charged particle (scalar) sum p_T (both normalised by angular area) as functions p_T of the leading (i.e. highest p_T) charged particle in the transverse region (defined in chapter 5) are the observables most sensitive to the UE, shown in figure 7.8 (bottom row). Some features of the distributions (measured in inclusive events, but requiring the leading charged particle to have $p_T > 1$ GeV) that are worth noting include:

- The activity (be it the multiplicity or p_T sum) initially shows a sharp rise, then saturates at a characteristic plateau height. This observation is consistent with the predictions by theoretical model [23], according to which MPI increases with increase of the event energy scale, as collisions become more central (i.e. proton and anti-proton overlap increases) from more peripheral at low p_T exchange. The plateau height is reached when MPI no longer increases, as the BBR level is roughly constant.

- While the transverse region exhibits the plateau-like behaviour, the toward and away region activities keep rising, consistent with the $\Delta\phi$ distribution earlier.

- One of the other distributions that was measured (but not shown here) was the standard deviation of charged-particle multiplicity and sum p_T profiles [4]. This showed conclusively, for the first time, that UE activity is dominated by event-by-event fluctuations, and a flat pedestal activity cannot be subtracted to remove the effect of UE.

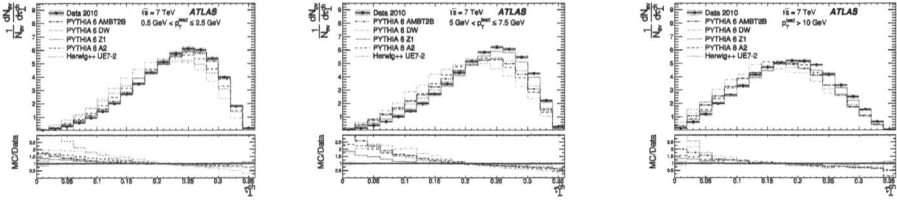

Figure 7.10. Distributions of transverse thrust constructed using charged particles with different leading charged particle p_T thresholds [26] (ATLAS Experiment © 2018 CERN).

The UE can be measured in many different processes using the same principle. In figure 7.9 (left), the charged particle multiplicity profile, comparing three different processes, the inclusive selection, Z boson production and high-p_T jet production (where the Z boson and the leading jet sets the leading direction and p_T scale). It is seen that the activities in Z boson and the leading jet events are similar, and activity continues smoothly from the leading charged particle to leading jet events. This leads us to believe that UE is somewhat universal. The advantage of measuring UE in Z boson events is that the toward region is also sensitive to UE by excluding the lepton pair. The right plot shows UE measurement as a function of leading jet p_T in an exclusive dijet topology. The advantage of this is to eliminate events with extra jets potentially contaminating the UE activity, resulting in much flatter distributions.

The event shape observables, constructed using charged particles gives a direct demonstration of the energy flow in the event, as shown in figure 7.10. The peak position moves to the left for transverse thrust distributions as the leading charged particle p_T is required to be higher, indicating more dijet-like events.

As an example of the final class of observables constructed with charged particles, we will discuss the so-called angular correlations. First, the correlation function for two charged particles, i and j is constructed in terms of their Dirac-delta function using azimuthal-angle $\Delta\phi$ and pseudorapidity $\Delta\eta$ differences, for a multiplicity interval N_{ch}:

$$R(\Delta\phi, \Delta\eta; N_{ch}) \approx \left\langle \sum_i \sum_{j \neq i} \delta(\eta_i - \eta_j - \Delta\eta)\delta(\phi_i - \phi_j - \Delta\phi) \right\rangle$$

where the normalisation depends on N_{ch}. The delta function ensures only the pairs yield a factor of unity. Then the correlation function is defined in terms of two such distributions:

$$C(\Delta\phi, \Delta\eta; N_{ch}) = R_1(\Delta\phi, \Delta\eta; N_{ch})/R_2(\Delta\phi, \Delta\eta; N_{ch})$$

where R_1 and R_2 usually denote the functions evaluated for charged particles pairs from the same events and charged particles pairs from random pairs of events. This normalisation helps to highlight the features of the distribution in the numerator.

One of the most striking results from the LHC was obtained by looking at this observable, as in figure 7.11 (top). For high multiplicity events, an enhancement at

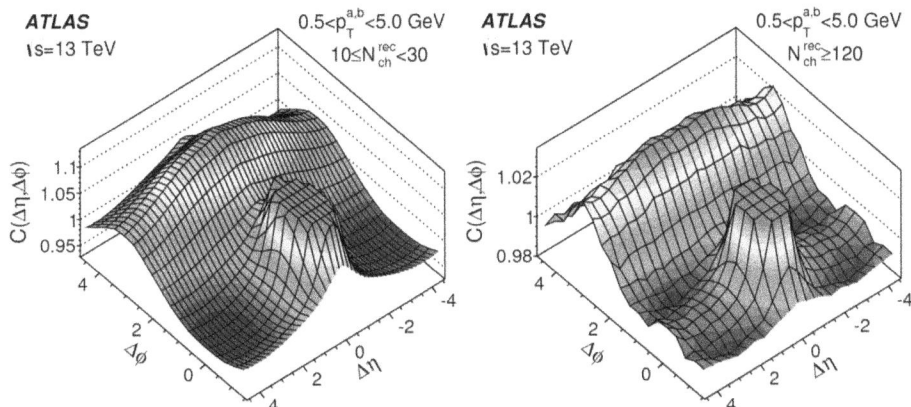

Figure 7.11. Distributions showing the absence (left) and emergence (right) of the ridge in two particle correlation distributions with increasing multiplicity [29] (ATLAS Experiment © 2018 CERN).

$\Delta\phi \approx 0$, extending over a wide range of $\Delta\eta$ was seen, which cannot be explained by the standard particle production mechanism so far (in heavy ion collisions, this is attributed to the formation of quark–gluon plasma [27], which is not expected in the proton–proton collisions considered here). This feature is referred to as a *ridge*. Of course this plot also contains the leading jet peak at $|\Delta\eta| < 2$, $|\Delta\phi| \approx 0$, and the away side balancing jet at $|\Delta\eta| < 2$, $|\Delta\phi| \approx \pi$. Another measurement of correlation is to probe the enhancement of the production identical boson pairs close by in phase space, which in the literature is known as Bose–Einstein correlations (BEC) [28]. The correlation function is expressed in terms of the relative momentum of the particles in the pair.

7.2.3 Jet distributions

Besides charged particle, jets are one of the most commonly encountered objects in colliders, and measurements of jets are essential to test predictions of perturbative QCD, and model the background for many of the searches. Figure 7.12 shows some jet cross-section measurements in terms of jet multiplicity, jet p_T and dijet invariant mass. The first plot shows how many jets are produced in events, and the second plot effectively shows how many jets within different p_T ranges were measured in seven ranges of (average) rapidity. The bottom plot shows the mass distribution for two leading jets in different ranges given by half of their rapidity difference. The cross-sections are multiplied by the factors indicated in the legend for ease of representation. They both show the same trend as seen for charged particles, most events contain few jets, and events with high jet multiplicity or jets with high p_T or mass are rarer. Each extra jet production can be considered suppressed by one order of α_S.

To probe the topology of the events with many jets, the gap fraction as a function of the mean p_T of the two (leading) jets, and azimuthal angle difference is measured, as shown in figure 7.13. For a purely $2 \rightarrow 2$ hard scattering, the jets should be back-to-back, while any additional quark or gluon emission alters the balance between the

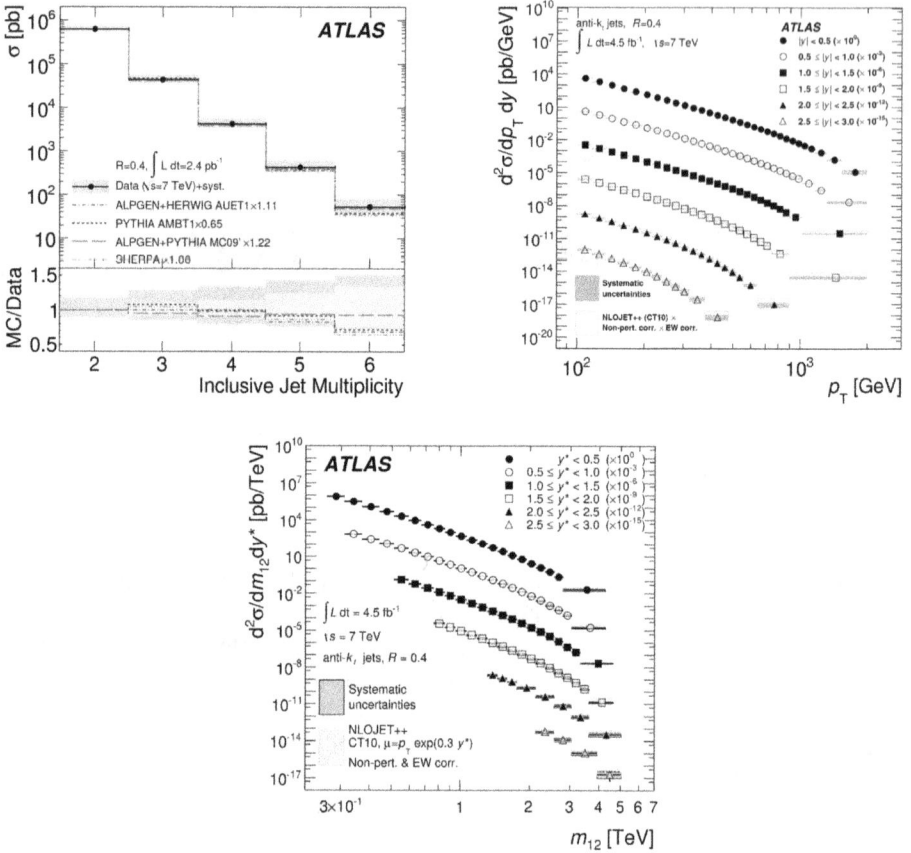

Figure 7.12. Jet cross-section as functions of inclusive jet multiplicity [30] (top left), inclusive jet p_T [31] (top right), and dijet invariant mass [32] (bottom) (ATLAS Experiment © 2018 CERN). The cross-sections are normalised by p_T or dijet invariant mass and rapidity range for the second and third plots.

partons and produces activity in the gap (thereby reducing the fraction of events with gaps), or resulting in an azimuthal decorrelation, as can be seen.

The internal structures of jets are probed with differential and integral jet shapes (more such observables will be discussed in chapter 8), as shown in figure 7.14. The dominant peak at small r for differential jet shape indicates that the majority of the jet momentum is concentrated close to the jet axis, which increases as the jet p_T increases. This indicates that the jets become narrower as p_T increases. This is also reflected in integrated jet shape, where the fraction of the jet transverse momentum outside a fixed radius $r = 0.3$, decreases as a function of p_T.

The strong coupling constant, α_S, (the only free parameter in QCD Lagrangian apart from the quark masses) can be measured from jet cross-section measurements. While a few different cross-section ratios (in order to cancel the dominant systematic uncertainties) have been used [35–37], perhaps the simplest is the ratio of inclusive 3-jet cross-section to the inclusive 2-jet cross-section, designated as R_{32}.

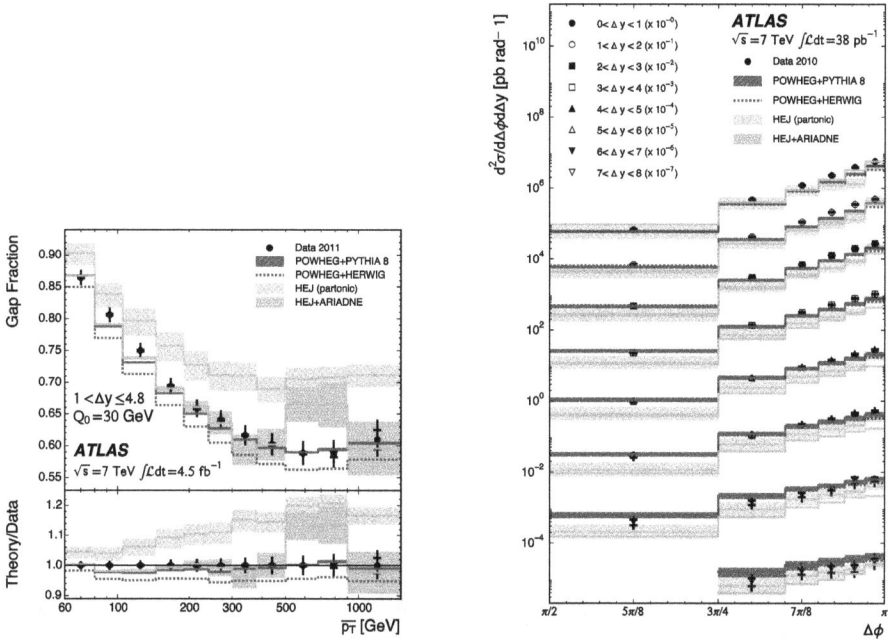

Figure 7.13. Gap fraction as a function of mean dijet p_T (left) and normalised cross-section as function of their azimuthal angle difference (right) for dijet events [33] (ATLAS Experiment © 2018 CERN).

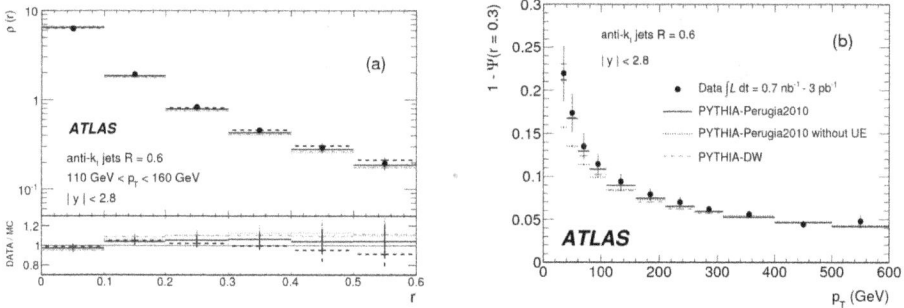

Figure 7.14. Differential jet shape as a function of distance from the jet axis (left) and (unity minus) integrated jet shape as a function of jet p_T (right) [34] (ATLAS Experiment © 2018 CERN).

Being able to distinguish light-quark jets from gluon jets on an event-by-event basis could significantly enhance the reach for many new physics searches at the LHC, as many certain signatures can have only one type of jet as signal. For example, VBF processes only have two forward quark-jets, and most SUSY models predict quark-jets as well. While at Tevatron the fraction of quark-jets was higher, the proton–proton collision at the LHC and the higher energy results in more gluon jets. The experimental discrimination (as well as the conceptual classification, as we will see in chapter 9) between quark and gluon jet is not trivial. On average, gluon jets have more constituents than quark-jets and the radiation pattern within gluon-

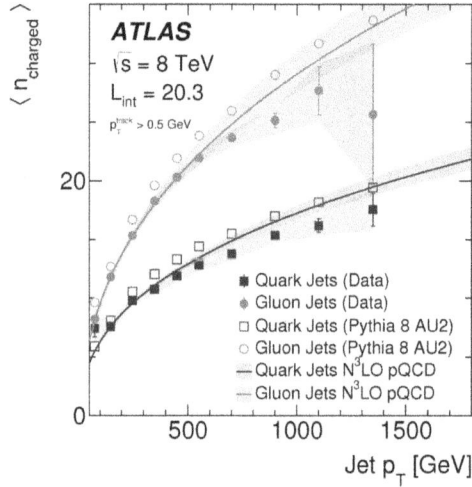

Figure 7.15. Average charged particle multiplicity inside quark and gluon initiated jets as a function of jet p_T [38] (ATLAS Experiment © 2018 CERN).

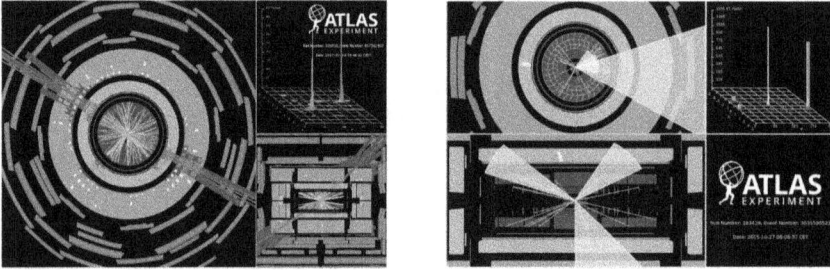

Figure 7.16. Event display of a well balanced djiet event (left) and of a photon produced with three jets (right) (ATLAS Experiment © 2018 CERN).

jets tends to be broader than quark-jets. Also, gluon-jets tend to be more forward in rapidity. Using these features, quark- and gluon-jets are categorised.

In figure 7.15, the average number of charged particles inside jets, as a function of jet p_T is shown separately for quark and gluon jets. Here the categorisation was performed by estimating the fraction of gluon jets in different rapidity ranges from simulation, and applying that information to data. While the number of charged particles increase with higher jet p_T, the increase is steeper for gluon-jets, consistent with expectations.

Finally, in figure 7.16, the event display for a balanced dijet event with invariant mass of 9.3 TeV (one of the highest observed so far), and a photon with $E_T = 1.3$ TeV produced along several jets is shown.

7.2.4 *W/Z* boson distributions

The vector boson productions result in characteristic features as well, and figure 7.17 show some selected results. They are probes of electroweak processes. Usually reconstruction of Z and W bosons in leptonic decay modes yield more precision than the dijet decay modes. The mass of the Z boson shows a peak at the PDG value (the background processes are heavily suppressed and not visible on the linear scale), but there is a large spread. This is a result of both theoretical and experimental reasons. If the Feynman diagram of the Z boson decay to leptons is calculated, the cross-section is seen to be proportional to $1/(s - m_z^2)^2 + (m_z\Gamma_z)^2)$, where $\sqrt{s}$ is the centre-of-mass energy is the collision frame, and Γ_z is the decay width. This means that the maximum cross-section occurs at $\sqrt{s} = m_z$, which causes the mass peak. However, the other terms cause the so-called Breit–Wigner lineshape or the spread of the peak. Experimentally, the momentum of the leptons can never be measured perfectly, so that also causes part of the smearing. This is characteristic of any resonant mass peak.

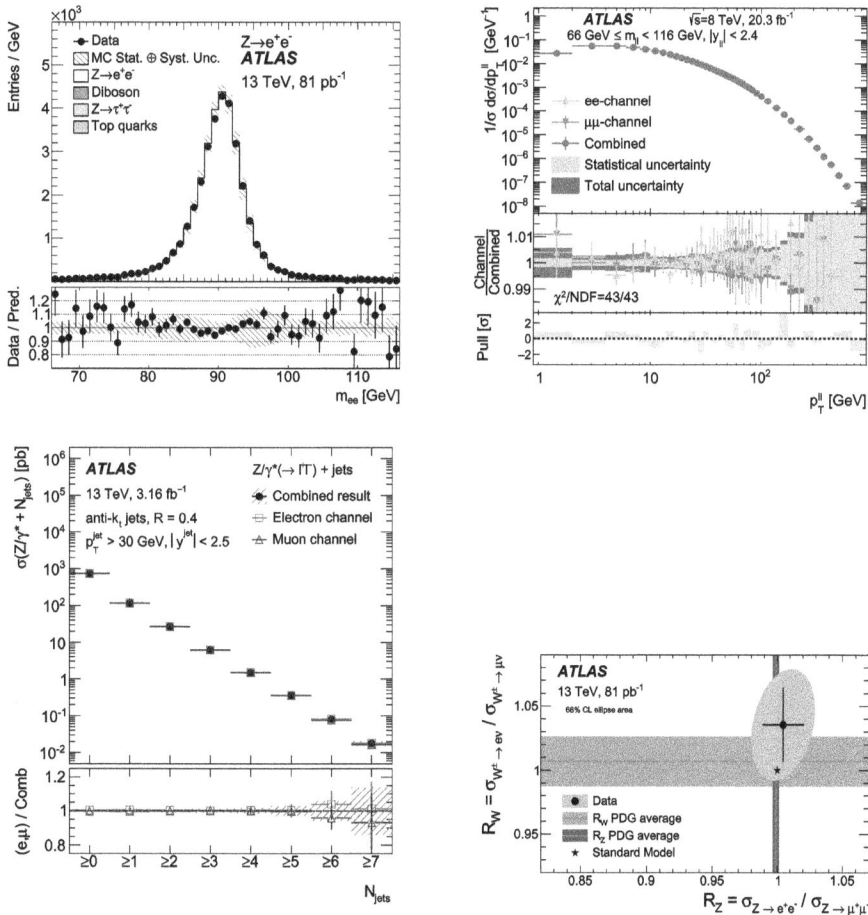

Figure 7.17. Invariant mass (top left), p_T (top right) of the dilepton system in Z boson events [39], jet multiplicity (bottom left) in such events [40], and a plot indicative of lepton universality [1] (bottom right) (ATLAS Experiment © 2018 CERN). The lepton universality is probed by constructing individual ratios of Z and W boson decay cross-sections to electron and muons.

The p_T distribution of Z bosons also has an interesting shape. It has the steeply falling shape as seen before for charged particles, but it also has a slight dip at very low p_T, leading to a peak at about 5 GeV. The initial dip is due to a selection bias, as electrons and muons can only be chosen with certain p_T threshold due to the detector resolution, resulting in some Z bosons with low p_T not being reconstructed. The exponential fall again results from the lack of available phase space to produce many energetic Z bosons. The number of jets produced in association with vector bosons, shown in figure 7.17 (bottom left) in an event is also a probe of QCD. It is shown for an event with a leptonically decaying Z boson. It shows a nice staircase-like structure, with each extra jet production suppressed by one order of α_S.

Theoretically, there should be no difference between the decay of the vectors bosons to electrons or muons, a concept known as lepton universality. The figure 7.17 (bottom left) confirms the lepton universality for both Z and W boson production. The ratio of cross-sections from electron and muon channels are consistent with unity. Another feature to note here is that when ratios are constructed, some experimental uncertainties cancel out.

The measurement of W boson mass is one of the most precise measurements in hadron colliders. As described in section 5.2, the distributions of p_T of the lepton, p_T^{miss} and m_T is measured, as shown in figure 7.18. The p_T of the lepton shows the expected Jacobian edge, while m_T peaks close to the actual m_W value. The last plot shows m_W values extracted from difference lepton and charge combinations, which are remarkably consistent.

Figure 7.19 (left) shows the distribution of the ΔR separation between the decay muon from a W boson event with the closest jet, for two different leading jet p_T intervals. The high ΔR peak corresponds to the topology of (balanced) back-to-back W boson and jet production. However, an enhancement is also seen for the region of small angular separation, which corresponds to nearly-collinear emission of a real W boson from a jet. As the leading-jet p_T increases, the fraction of events with the collinear W boson emission increases and the peak shifts to small values, while the fraction in back-to-back W + jets events decreases. This was interpreted as an increase in the collinear W boson emission probability as the jets become more energetic.

Diboson production along with two jets is interesting because it receives contributions from both electroweak and QCD processes, with the former dominated by VBS. The same sign production of $WWjj$ gets the largest contribution from VBS. This process is also sensitive to the existence of rarer quartic $WWWW$, $WW\gamma\gamma$, $WWZZ$, and $WWZ\gamma$ interaction vertices, apart from the more common WWZ and $WW\gamma$ gauge boson vertices. Possible physics beyond the SM can affect these vertices and introduce anomalous triple gauge couplings (aTGCs) [42] or anomalous quartic gauge couplings (aQGCs) [43], which are expected to be reflected in dimensionless anomalous coupling parameters α_4 and α_5, sensitive to this final state. In figure 7.19 (right), the invariant mass obtained from the four-momenta of two leptons and p_T^{miss} four vectors is shown, for events where the separation in rapidity between the two leading p_T jets is large, indicative of VBS process. The region >400 GeV is expected to have most sensitivity to the new physics parameters α_4 and α_5. While some excess

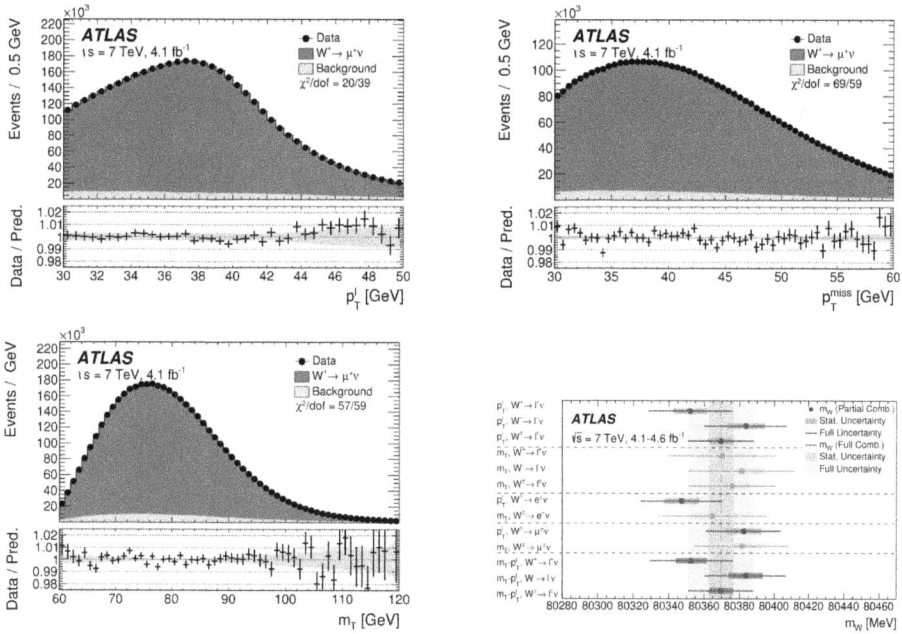

Figure 7.18. Distributions of lepton p_T (top left), missing transverse momentum (top right), transverse mass (bottom left), leading to estimation of W boson mass [41] (ATLAS Experiment © 2018 CERN).

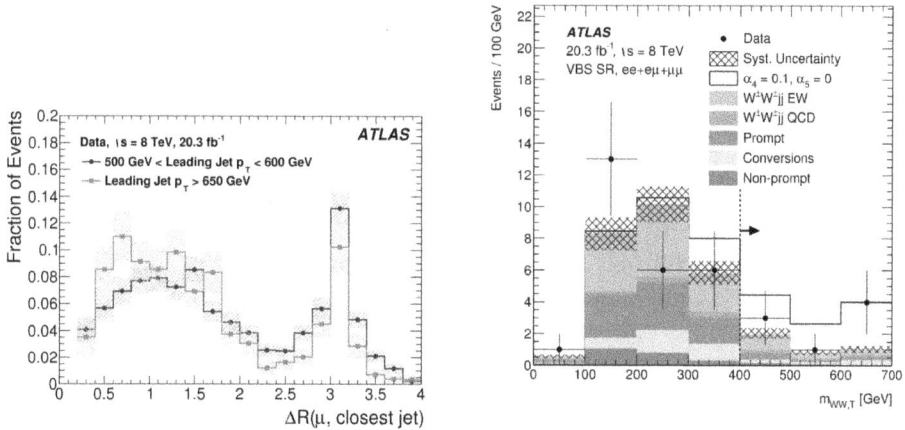

Figure 7.19. Distributions of angular separation between the muon with the closest jet in a W boson event [44] (left), and the invariant mass of two leptons and p_T^{miss} in events with two W bosons with two jets [45] (right), (ATLAS Experiment © 2018 CERN).

of events of the SM background is observed here, it is not enough to establish the presence of BSM effects.

Finally, in figure 7.20, the event display for a Z boson decaying to an electron–positron pair and a W boson decaying to a muon and neutrino is shown. In the first

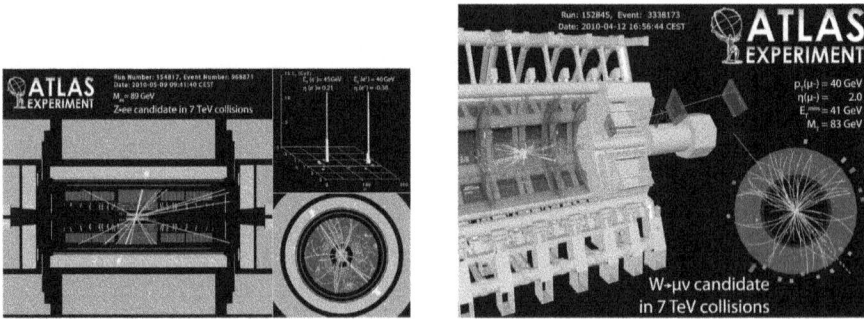

Figure 7.20. Event display of a Z boson to electron positron pair (left), and of a W boson decaying to a muon and neutrino (right) (ATLAS Experiment © 2018 CERN).

case, the invariant mass is 89 GeV and in the second case, the transverse mass is 83 GeV with missing energy of 41 GeV.

7.2.5 Top quark distributions

The top quark occupies a unique position, as the heaviest known quark, and it decays before hadronising unlike other quarks (with $\Gamma_{\text{top}} = 1.5$ GeV). As there are many new BSM scenarios predicting new heavy particles decaying to top quarks, the measurement of the top pair production cross-section, as well kinematics are an indirect probe of new physics. The decays of top quark also offer an indirect probe to BSM effects, as an enhancement in the branching ratio (BR) of relatively rare SM decay modes (i.e. $t \rightarrow Zq$, an example of FCNC process). The Wtb decay vertex coupling is sensitive to the V_{tb} element in CKM matrix. The usual cross-sections from $t\bar{t}$ events, as functions of top-quark p_T or jet multiplicity show the usual steeply falling nature.

Properties of top quarks themselves are interesting in their own right. Mass of the top quark, along with W and Higgs boson masses, constrain the other electroweak parameters in the SM [46, 47]. It also has implication of vacuum stability of SM [48, 49]. There is also an inherent ambiguity in the definition of top mass, as bare top quarks cannot be observed (unlike leptons, for example). There have been several approaches used to measure the top quark mass [50], most of these approaches are based on the position of the peak in the invariant-mass distribution of the decay products of the top-quark, a W boson (decaying hadronically) and a b-quark jet (the challenge is to pick the right combination of jets). In the literature this is known as the *pole mass* [51], as theoretically it corresponds to the mass value corresponding to the singularity in the top quark scattering amplitude integral (i.e. where the complex functions cease to be analytic). The pole mass is independent of energy scale at which it is evaluated, as well as the renormalisation scale chosen, as it corresponds to the intuitive physical mass of the particle. However, the theoretical calculation of quark masses diverge at λ_{QCD} (as quarks are confined), so the top quark mass needs to be calculated at some energy scale greater than (corresponding to a length scale smaller than) λ_{QCD}, defined in some specific renormalisation scheme. Commonly the so-called minimal subtracton scheme, or $\overline{MS}$ scheme [52, 53] is used, with the value of renormalisation scale $\mu_R = 2\,\text{GeV}$. For a different choice of renormalisation scale, the value will be different (at a renormalisation scale closer to λ_{QCD} the masses will of course be much bigger), hence this analytic top quark mass is referred to as *running mass*.

The mass definition used in MC models, which mimics the running mass evaluated with a certain scheme is measured experimentally by using simulated distributions of variables that are sensitive to m_{top} for different values of input m_{top}, termed templates. Then a fit to the distributions in data yields the value of m_{top} that best describes the data. These variables usually are the different kinematic variables in dileptonic decay mode, as the lepton measurements are more precise than jets. The identification of m_{top} with the pole mass suffers from the ambiguity described earlier, as well as from effects like non-perturbative connections and colour reconnection uncertainty [54]. The current theoretical calculations estimate a 1 GeV difference [55]. The decay width top quark is also measured using a template fit method from sensitive observables [56].

In figure 7.21, distributions sensitive to several properties of top quark are shown. Top quark pair production is parity invariant and hence the top quarks are not expected to be polarised, but their spins are expected to be correlated. The top left plot show the azimuthal angular difference between the charged leptons in dileptonic decay mode, which is sensitive to the spin of the top quark at production which is transferred to its decay products. In the absence of spin correlation, significantly more events are expected to have larger angular difference, where the actual distribution is seen to be flatter. This particular distribution actually shows that the slope of the data is higher than predictions based on SM, indicating stronger spin correlation than expected. Then the degree of correlation (or *helicity*) can be defined as the fractional difference between the number of events where the top and antitop quark spin orientations are aligned and those where the top quark spins have opposite alignment, which can be extracted again from template fits using $\Delta\phi$ distribution. The fraction of events with longitudinal, left-handed and right-handed polarisation (referred to as helicity fractions) of W bosons from top quark decay estimated from the distribution of $\cos\theta^\star$, which is the angle between the direction of the charged lepton and the reversed direction of the top quark, both in the rest frame of the W boson are shown in the top right plot with simulated events.

Top quark events also provide a useful testing ground for colour flow between different decay products. In a semileptonic $t\bar{t}$ event, four colour-charged final states can be identified: the two b-quarks produced directly by the decay of the top-quarks and the two quarks produced by the hadronically decaying W boson. As the W boson does not carry colour charge, its daughters must share a colour connection. The two b-quarks from the top-quark decays carry the colour charge of their respective top-quark parent, and are thus not expected to share a colour connection. This is probed by using the pull angle between the pairs of jets, as shown in the bottom row. A stronger effect is seen for the colour-connected case (W-decay products, left) compared to the non-colour-connected case (b-jets, right). A BSM

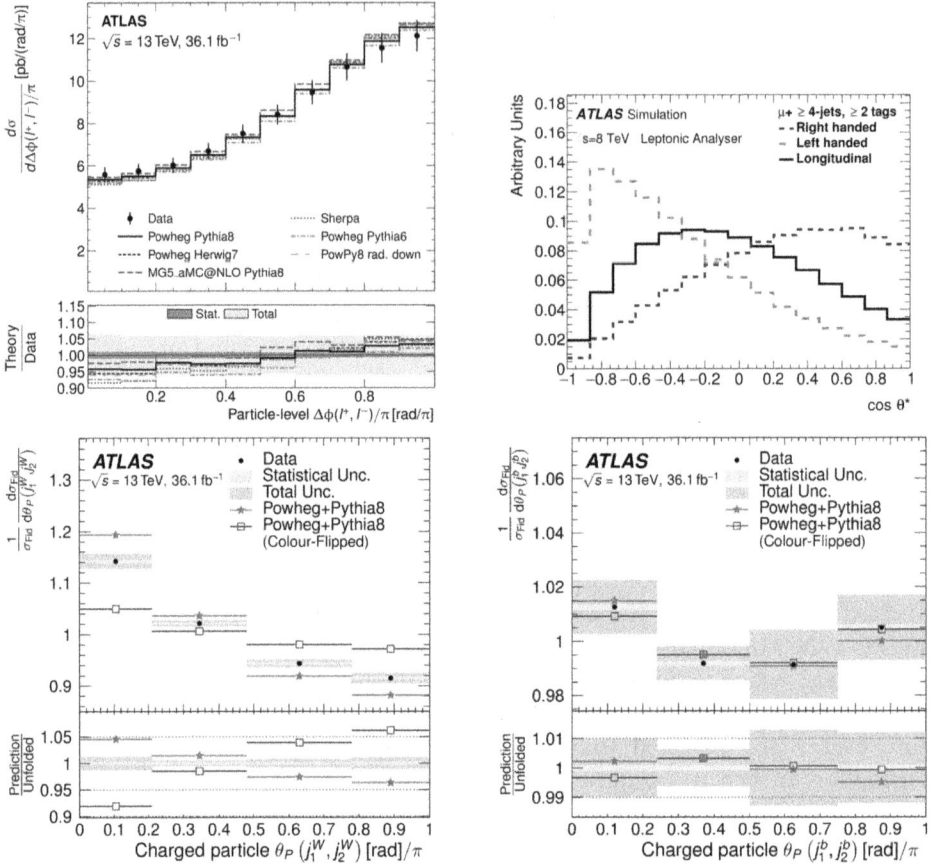

Figure 7.21. Distributions of the azimuthal angular difference between the leptons in dileptonic $t\bar{t}$ events [57] (top left), the angle between the direction of the charged lepton and the opposite direction of the top quark in the rest frame of the W boson [58] (right), and the pull angle between jets from W boson (bottom left) and between b jets (right) in semileptonic $t\bar{t}$ events [59] (ATLAS Experiment © 2018 CERN).

model with colour *flip* (where rather than SM colour-singlet W, a colour octet W was generated) was ruled out based on data.

Finally, in figure 7.22, several event displays are shown for top quark processes. The top row shows $t\bar{t}$ production dileptonic and semileptonic decay. In the first case, the event has electron, muon and two b-tag jets final state, and in the second case, a muon and four jets. In the bottom row, a rare event with four top quarks, with 13 jets (out of which four are b-tagged), one muon and missing energy of 288 GeV, and a t-channel single top quark production, decaying to muon and a b-tagged jet is shown.

7.2.6 Higgs boson distributions

The Higgs boson is the newest discovered (and so far the only one at the LHC) elementary particle and according to many, is the last missing piece in the SM. If and when a new particle is discovered, as the Higgs boson was discovered in 2012, the

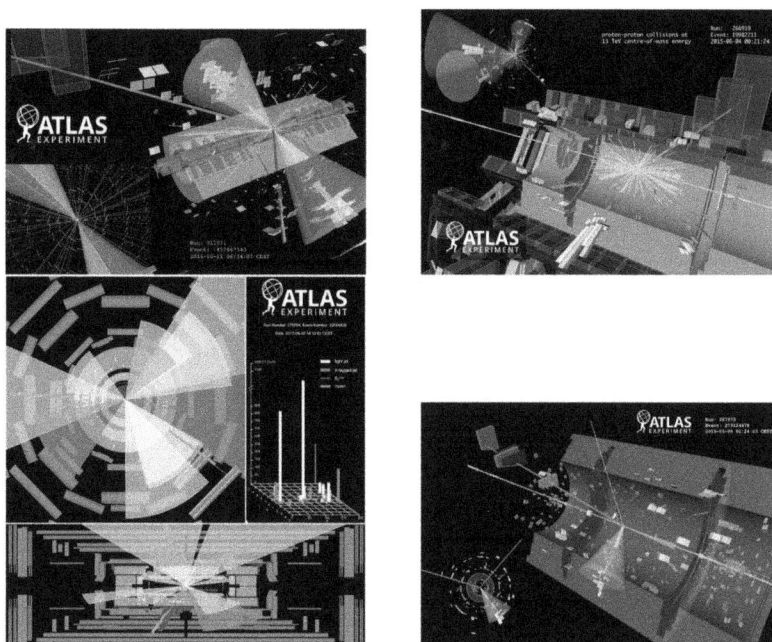

Figure 7.22. Event displays of a dileptonic $t\bar{t}$ event (top left), a semileptonic $t\bar{t}$ event (top right), an event with four top quarks (bottom left), and t-channel single top quark production (bottom right) (ATLAS Experiment © 2018 CERN).

measurements of properties of the new particle is critical to establish its identity. These are usually the measurement of mass, p_{T}, spin, parity, signal strength, coupling strength to other particles. Also compatibility of these measurements between different production and decay mode needs to be established. In figure 7.23, some of these results are shown.

The different decay modes of the Higgs boson was stated in chapter 4. Although not the most dominant, the four-lepton channel (i.e. from Higgs boson decaying to $ZZ^\star$) has the clearest and cleanest signature of all the possible Higgs boson decay modes (due to small background and good lepton reconstruction), therefore it is referred to as the *golden channel*. Many of the measurements are, therefore, performed in this channel.

The distribution of p_{T} and associated jet multiplicity (top right) for ZZ and $\gamma\gamma$ decay mode indicate that the measurement is still dominated by statistical uncertainties. The signal strength μ of the Higgs boson from four production and two decay modes is also shown. The μ here is defined as a ratio of cross-section of the Higgs bosons in a specified production and decay mode over the SM expectations in the same channel. So a value close to unity shows consistency with SM. The plot of the $\cos\theta^\star$, where $\theta^\star$ is the production angle of the leading Z boson in four lepton rest frame, shows behaviour consistent with spin-0 hypothesis.

The SM coupling of the Higgs boson to fermions is referred to as Yukawa coupling, with the coupling strength being proportional to the mass of the fermions.

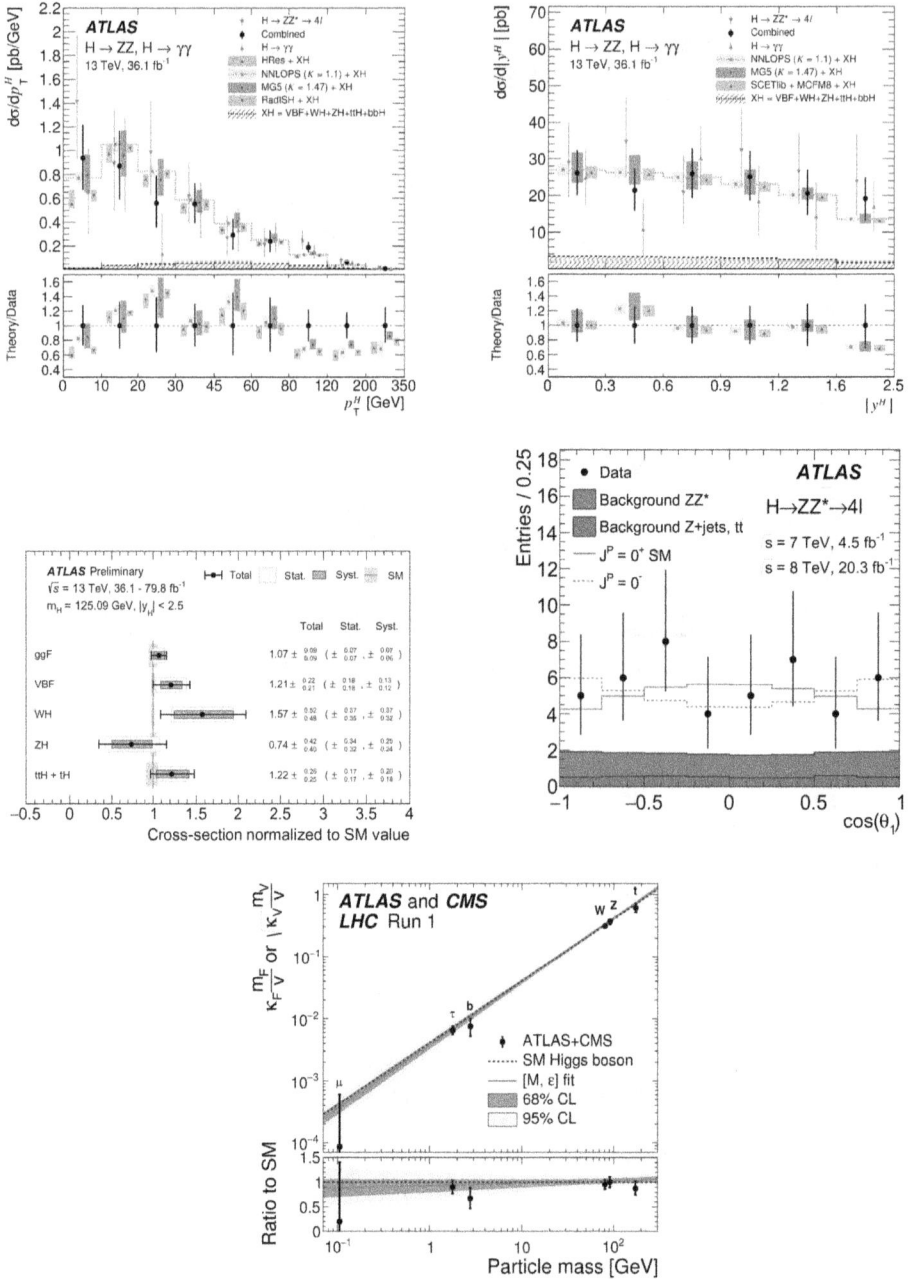

Figure 7.23. Distributions of Higgs boson p_T (top left), absolute value of rapidity (top right) in ZZ and diphoton decay modes [66], cross-sections in different production mode [67] (middle left), the production angle of the leading Z boson in four lepton rest frame in ZZ decay mode [68] (middle right), and the Higgs boson coupling to fermions and bosons as a function their masses [60] (bottom) (ATLAS and CMS Experiments © 2018 CERN).

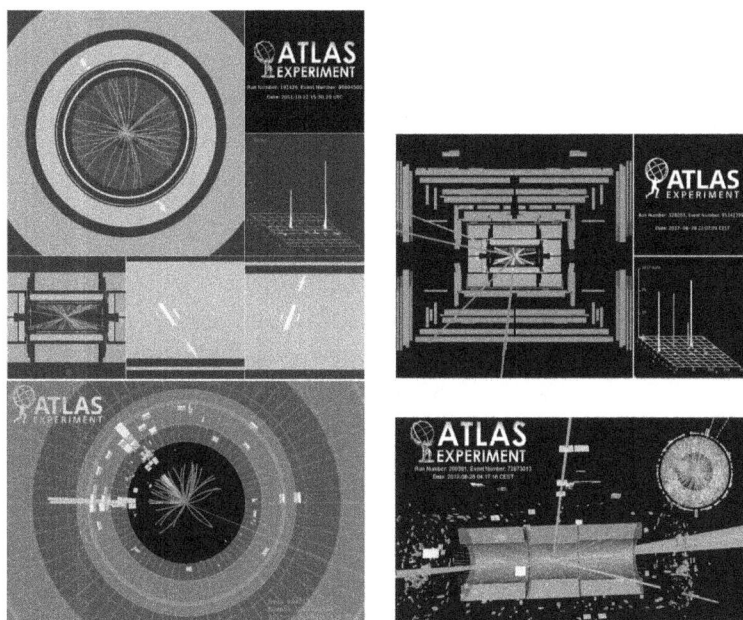

Figure 7.24. Event displays of a Higgs boson to diphoton event (top left), of a Higgs boson to four lepton event (top right), of a Higgs boson to two b jets event with a large missing energy indicative of associated production with a Z boson (bottom left) and of a Higgs boson to two τ leptons event (bottom right) (ATLAS Experiment © 2018 CERN).

The last plot shows parameters indicative of measured coupling as a function of mass of fermions and bosons [60], where the best fit is linear, as expected. A precise measurement of Yukawa coupling is not only important to confirm the SM but also to test new physics models which can affect this coupling. Fermionic decay modes of the Higgs boson provide direct measurement of the Yukawa coupling, observation of Higgs boson production in association with two top quarks [61], as well as the decay to tau lepton pair [62] provided direct evidence of this type of interaction. The most dominant decay mode, to a pair of bottom quarks, has also been observed experimentally [63, 64], at a rate consistent with the SM prediction. The SM also predicts Higgs self-coupling interactions [65], so analyses are being designed to look at the *hh* final state. The ggF production mode is sensitive to the trilinear Higgs coupling, while the VBF channel is sensitive to the coupling between the Higgs and the vector bosons. Much more data is needed to observe these processes.

Finally, in figure 7.24, several event displays are shown for Higgs boson processes. The first row shows resonant Higgs decays to two photons in the first case, and two electron and two muons in the second case, via $ZZ^\star$. The invariant masses of the reconstructed Higgs boson are 126.6 GeV and 124 GeV. The dielectron invariant mass is 80 GeV, while the dimuon invariant mass is 34 GeV, indicating the off-shell Z boson production in the latter case.

The left-bottom plot shows an event with Higgs boson to $b\bar{b}$ decay, represented by two b-tagged jets, and no leptons. The dijet invariant mass of 136.9 GeV. The missing energy is 294.4 GeV, indicating a Z boson decaying into a pair of neutrinos. The bottom-right plot shows a Higgs boson decaying to two leptonically decaying τ leptons, indicated by an electron and a muon. The calculated Higgs mass, using the missing energy of 43 GeV is 126 GeV. The event also has two forward jets, indicating a VBF production mode for the Higgs boson.

7.2.7 General search

The analyses discussed so far targeted a specific final state. General search is an approach which goes beyond that limitation, and compares a number of events in the data and in the SM background prediction for a large category of events. The events are categorised according to a multiplicity of different reconstructed physics objects, such as leptons, photons, jets, b-tagged jets, along with ranges of H_T or missing energy values. The categories tend to run into several hundreds, considering all different combinations. An example is shown in figure 7.25, where different event categories for events with two same-flavour leptons are shown. Each bin represents one category. This approach was initiated in CDF [69], where a model-independent approach (Vista) considered the overall features of the data, and was sensitive to large cross-section coming from new processes. Another quasi-model-independent approach (Sleuth) looked at high p_T tails of distributions.

7.2.8 Reinterpretation

LHC experiments have search programmes covering a wide array of new physics models, but it is impossible to cover all possible models, and the parameter spaces they span. Additionally, new models will continue to be proposed in the future, sometimes with similar but not identical final states as a current search. So a mechanism to *reinterpret* current results in terms of the models not probed originally allows us to make maximum use of our data. Reinterpretation, therefore, is an efficient way to re-use existing analyses that have good sensitivity for the new model

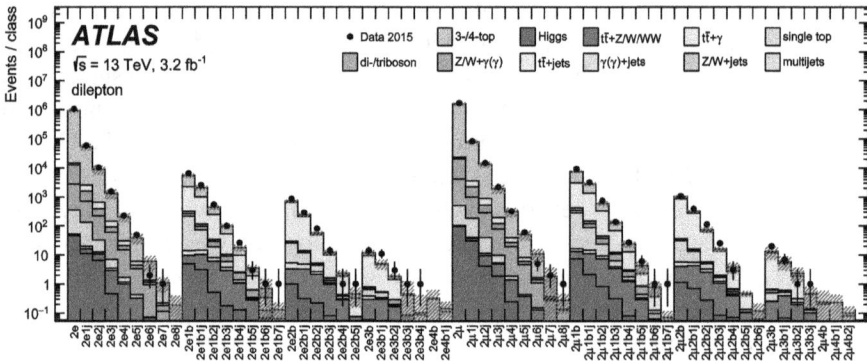

Figure 7.25. An example plot from a general search, where yields in different event categories between data and SM predictions are compared [70] (ATLAS Experiment © 2018 CERN). The x-axis here just indicates different (discrete) event categories.

without re-estimating background. If no analysis can exclude a model, that offers good motivation for a dedicated search.

The basic requirements are experimental acceptance and efficiencies, which can be used to convert particle level MC event counts (corresponding to the new signal) to what will be observed in the detector level analysis. The acceptance is essentially the analysis definition (simple cut based to complicated BDT), while efficiency is related to detector simulation. The selection needs to be conveyed clearly, with definition of the reconstructed objects, overlap removal used, sequence, and source code for calculation of complicated observables. Resolution and efficiency numbers for objects is necessary, and cutflow chart for some benchmark signal samples is useful for validation.

Most approaches build libraries of existing analyses, using this information, examples are SMODELS (Simplified Model Spectrum) [71], FASTLIM [72], CHECKMATE [73]. SMODELS and FASTLIM (the latter is mostly focused on SUSY) decomposes BSM signatures into simplified topologies in terms of mass spectrum and branching ratios. SMODELS also identifies the most important uncovered topologies in experimental searches. CHECKMATE takes simulated event files from a new model, and checks whether the model is excluded or allowed after performing a detector simulation based on DELPHES and testing various implemented analyses. Other programmes mostly use efficiency maps for objects.

There are other useful tools as well. RIVET [74] was originally developed as an analysis toolkit to compare particle level simulation output to published data, and it already includes hundreds of published unfolded analyses from LHC and previous experiments (with identical observable definition, event selection, binning of plots). However, now efficiencies and smearing for standard objects, based on experimental numbers are implemented, so a limited number of BSM analyses are included as well. MADANALYSIS [75] is a repository of published analyses as well. There is no specific format for the implementation of experimental analyses in these tools yet, but proposals have been made, classified as LHADA (Les Houches Analysis Description Accord for the LHC) [76].

CONTUR (Constraints on new theories using RIVET) [77] uses existing unfolded cross-section measurements to set limits on new models. Events generated from a new model are passed through relevant RIVET analyses, assuming SM background to be given by data. Finally, there are tools which rely on fitting a range of data, from LHC and beyond to detect signs of new physics, such as GAMBIT (Global and Modular Beyond-Standard Model Inference Tool) [78] or TOPFITTER [79], with the latter focused on top sector, using higher-dimensional effective operators.

Exercises

1. A plot showing a number of tracks originating from hadronic decay of a τ lepton candidate was seen to have peaks corresponding to one and three tracks. Explain why there are no peaks corresponding to even numbers of tracks?

2. Why does hadronic τ lepton decay result in a single jet?

3. In an analysis requiring two same flavour leptons, we usually restrict ourselves to *ee* or *μμ* final states, and ignore the *ττ* final state. Why is this a reasonable assumption?

4. Explain the difference in missing energy distributions between the two runs in the figure; the first one is for semi-leptonic $t\bar{t}$ decay, the second one is for hadronic $t\bar{t}$ decay.

5. The invariant mass of lepton pair is plotted for Drell–Yan events. If only the events with pair p_T with low values are selected, where will the mass peak be?

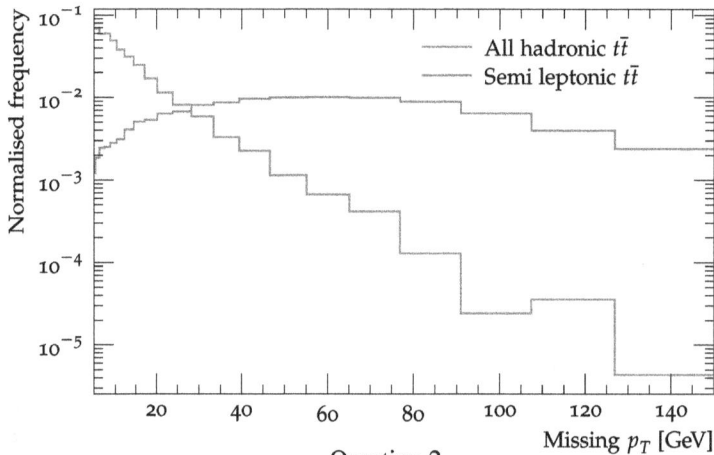

Question 2

6. Generally if the average number of charged particles in an inclusive event is compared to the average number of charged particles in transverse region, the latter is found to be twice or more. What does it tell you? Do you think the activity from the underlying event will be the same for different hard processes?

7. In an underlying event measurement with charged-particle jets, it was seen that the activity was initially higher (as a function of jet p_T) as the radius of the jet was increased. What could have caused the effect?

8. A one-dimensional histogram has 40 bins. You *rebin* the histogram, i.e. merge bin 1 and 2 into a single bin, and so on, resulting in a new histogram of 20 bins. How would the histogram change, will it move up or down or stay roughly similar?

9. When comparing shapes of distributions obtained with different numbers of events (i.e. do data from 2018 look the same as data from 2017? or let us assume we generated a few thousand simulation events, and trying to check how do they look compared to data), we need to *normalise* the histograms. What are some of the ways you can achieve that?

10. A critical consideration in any experimental analysis is the availability of sufficient MC events. It is usually said that MC should have at least a few times more statistics than in data. Why do you think more events than data is necessary?

11. Assuming an integrated luminosity of 150 fb^{-1} and average pileup of 50 at the LHC, in how many cases can one have two semileptonically decaying $t\bar{t}$ pairs from separate pp interactions?

12. If we expect 15 events from SM for a particular process, and we observe 20 events, what is the statistical significance of the observation in terms of σ? Assuming expected and observed event yields increase proportionally, how much luminosity do we need to reach 5σ, if the initial measurement was obtained with 10 fb^{-1}? Assume significance is given by $S/\sqrt{S+B}$, where S is the excess amount of observed events, and B the expected background events.

13. If in a search, we observe zero events in data, what is the minimum of number of signal events that must be expected to exclude that model?

14. In this question, we revisit several of the tantalising excesses which were published by collaborations, leading to a phenomenon termed as *ambulance chasing*, where hundreds of papers were put out on arXiv preprint server trying to explain the excess (the term presumably comes from lawyers trying to chase accident victims to sign up as clients).

    (a) CDF published [80] a bump at dijet mass spectrum at 140 GeV in 2010/2011 in events containing a leptonic W boson decay of about 4.5σ significance. However, no such excess was seen in $Z \rightarrow \nu\nu$ plus dijet events. Why was this a problem? Eventually [81] the excess was ascribed to mismodelling of multijet background in electron channel and not having dedicated jet energy calibration for quark and gluon jets. Why was the latter a problem?

    (b) ATLAS (and CMS, shortly thereafter) published [82, 83] bump at diphoton mass spectrum at 750 GeV, corresponding to 3.9σ significance in December 2015. Should we have considered the excesses seen by the collaborations statistically independent [84]? What role can human bias play, when one collaboration informally knows that the other collaboration is coming up with an excess, but still has not finished the analysis with due diligence?

    A new resonance compatible with the excess was calculated to have a cross-section of about 5 fb. However, later it was shown that a large fraction of events had zero jets. Why was that a problem? If the experiments saw the result remain even when photons were chosen to have *anti-isolation*, i.e. the isolation requirement was reversed, would that have been a cause for concern? Eventually [85, 86] the excess went away after adding more data in 2016, and thorough cross-check of photon isolation.

15. What results get cited in this field? The historically highest cited papers include papers describing the development of the Standard Model, and the Higgs boson discovery papers by ATLAS and CMS. The ATLAS and CMS detector papers are also very well cited, as most publications from the collaborations cite the respective detector papers. Can you guess some of the other most cited papers?

References

[1] Aaboud M *et al* 2017 Measurements of the production cross section of a Z boson in association with jets in pp collisions at $\sqrt{s} = 13$ TeV with the ATLAS detector *Eur. Phys. J. C* **77** 361

[2] Aaboud M *et al* 2018 Measurement of inclusive jet and dijet cross-sections in proton-proton collisions at $\sqrt{s} = 13$ TeV with the ATLAS detector *JHEP* **05** 195

[3] Aaboud M *et al* 2018 Measurement of differential cross sections and W^+/W^- cross-section ratios for W boson production in association with jets at $\sqrt{s} = 8$ TeV with the ATLAS detector *JHEP* **05** 077

[4] Aad G *et al* 2011 Measurement of underlying event characteristics using charged particles in pp collisions at $\sqrt{s} = 900$ GeV and 7 TeV with the ATLAS detector *Phys. Rev. D* **83** 112001

[5] Aad G *et al* 2014 Measurement of the production cross-section of $\psi(2S) \rightarrow J/\psi(\rightarrow\mu^+\mu^-)\pi^+\pi^-$ in pp collisions at $\sqrt{s} = 7$ TeV at ATLAS *JHEP* **09** 079

[6] Aaboud M *et al* 2018 Search for $W' \rightarrow tb$ decays in the hadronic final state using pp collisions at $\sqrt{s} = 13$ TeV with the ATLAS detector *Phys. Lett. B* **781** 327–48

[7] Aad G *et al* 2012 Observation of a new particle in the search for the Standard Model Higgs boson with the ATLAS detector at the LHC *Phys. Lett. B* **716** 1–29

[8] Aaboud M *et al* 2019 Search for heavy Majorana or Dirac neutrinos and right-handed W gauge bosons in final states with two charged leptons and two jets at $\sqrt{s} = 13$ TeV with the ATLAS detector *JHEP* **01** 016

[9] Aaboud M *et al* 2019 Searches for scalar leptoquarks and differential cross-section measurements in dilepton-dijet events in proton-proton collisions at a centre-of-mass energy of $\sqrt{s} = 13$ TeV with the ATLAS experiment *Report* CERN-EP-2018-262

[10] Abdel Khalek S *et al* 2016 The ALFA Roman pot detectors of ATLAS *JINST* **11** P11013

[11] Grinstein S 2016 The ATLAS forward proton detector (AFP) *Nucl. Part. Phys. Proc.* **273–275** 1180–4

[12] Latino G 2009 The TOTEM experiment at the LHC *Proc. of 44th Rencontres de Moriond on QCD and High Energy Interactions (La Thuile, Italy March 14–21, 2009)* pp 357–60

[13] Aad G *et al* 2014 Measurement of the total cross section from elastic scattering in pp collisions at $\sqrt{s} = 7$ TeV with the ATLAS detector *Nucl. Phys. B* **889** 486–548

[14] Aaboud M *et al* 2016 Measurement of the inelastic proton-proton cross section at $\sqrt{s} = 13$ TeV with the ATLAS detector at the LHC *Phys. Rev. Lett.* **117** 182002

[15] Aad G *et al* 2016 Charged-particle distributions in $\sqrt{s} = 13$ TeV pp interactions measured with the ATLAS detector at the LHC *Phys. Lett. B* **758** 67–88

[16] Feynman R P 1969 Very high-energy collisions of hadrons *Phys. Rev. Lett.* **23** 1415–7

[17] Koba Z, Nielsen H B and Olesen P 1972 Scaling of multiplicity distributions in high-energy hadron collisions *Nucl. Phys. B* **40** 317–34

[18] Alner G J *et al* 1984 Scaling violation favoring high multiplicity events at 540-GeV CMS energy *Phys. Lett.* B **138** 304–10

[19] Ansorge R E *et al* 1989 Charged particle multiplicity distributions at 200-GeV and 900-GeV center-of-mass energy *Z. Phys.* C **43** 357

[20] Blazek M 1986 Restoration of multiplicity scaling in high-energy proton proton and anti-proton-proton collisions *Z. Phys.* C **32** 309

[21] Abelev B *et al* 2012 J/ψ production as a function of charged particle multiplicity in *pp* collisions at $\sqrt{s} = 7$ TeV *Phys. Lett.* B **712** 165–75

[22] Aaboud M *et al* 2017 Measurement of charged-particle distributions sensitive to the underlying event in $\sqrt{s} = 13$ TeV proton-proton collisions with the ATLAS detector at the LHC *JHEP* **03** 157

[23] Sjostrand T and van Zijl M 1987 Multiple parton-parton interactions in an impact parameter picture *Phys. Lett.* B **188** 149

[24] Aad G *et al* 2014 Measurement of distributions sensitive to the underlying event in inclusive Z-boson production in *pp* collisions at $\sqrt{s} = 7$ TeV with the ATLAS detector *Eur. Phys. J.* C **74** 3195

[25] Aad G *et al* 2014 Measurement of the underlying event in jet events from 7 TeV proton-proton collisions with the ATLAS detector *Eur. Phys. J.* C **74** 2965

[26] Aad G *et al* 2013 Measurement of charged-particle event shape variables in $\sqrt{s} = 7$ TeV proton-proton interactions with the ATLAS detector *Phys. Rev.* D **88** 032004

[27] Shuryak E V 1978 Quark-gluon plasma and hadronic production of leptons, photons and psions *Phys. Rev.* B **78** 150–3

[28] Goldhaber G, Goldhaber S, Lee W and Pais A 1960 Influence of Bose–Einstein statistics on the antiproton-proton annihilation process *Phys. Rev.* **120** 300–12

[29] Aad G *et al* 2016 Observation of long-range elliptic azimuthal anisotropies in $\sqrt{s} = 13$ and 2.76 TeV *pp* collisions with the ATLAS detector *Phys. Rev. Lett.* **116** 172301

[30] Aad G *et al* 2011 Measurement of multi-jet cross sections in proton-proton collisions at a 7 TeV center-of-mass energy *Eur. Phys. J.* C **71** 1763

[31] Aad G *et al* 2015 Measurement of the inclusive jet cross-section in proton-proton collisions at $\sqrt{s} = 7$ TeV using 4.5 fb^{-1} of data with the ATLAS detector *JHEP* **02** 153
Aad G *et al* 2015 Measurement of the inclusive jet cross-section in proton-proton collisions at $\sqrt{s} = 7$ TeV using 4.5 fb^{-1} of data with the ATLAS detector *JHEP* **09** 141 Erratum

[32] Aad G *et al* 2014 Measurement of dijet cross sections in *pp* collisions at 7 TeV centre-of-mass energy using the ATLAS detector *JHEP* **05** 059

[33] Aad G *et al* 2014 Measurements of jet vetoes and azimuthal decorrelations in dijet events produced in *pp* collisions at $\sqrt{s} = 7$ TeV using the ATLAS detector *Eur. Phys. J.* C **74** 3117

[34] Aad G *et al* 2011 Study of jet shapes in inclusive jet production in *pp* collisions at $\sqrt{s} = 7$ TeV using the ATLAS detector *Phys. Rev.* D **83** 052003

[35] Warburton A 2015 Measurements of α_s in *pp* Collisions at the LHC *Proc. of 12th Conf. on the Intersections of Particle and Nuclear Physics (CIPANP 2015) (Vail, Colorado, USA, May 19–24, 2015)*

[36] d'Enterria D and Skands P Z 2015 *Proc. of High-Precision α_s Measurements from LHC to FCC-ee, (CERN, Geneva)*

[37] Britzger D, Rabbertz K, Savoiu D, Sieber G and Wobisch M 2019 Determination of the strong coupling constant using inclusive jet cross section data from multiple experiments *Eur. Phys. J.* C **79** 68

[38] Aad G *et al* 2016 Measurement of the charged-particle multiplicity inside jets from $\sqrt{s} = 8$ TeV *pp* collisions with the ATLAS detector *Eur. Phys. J.* C **76** 322

[39] Aad G *et al* 2016 Measurement of $W^{\pm}$ and Z-boson production cross sections in *pp* collisions at $\sqrt{s} = 13$ TeV with the ATLAS detector *Phys. Lett.* B **759** 601–21

[40] Aad G *et al* 2016 Measurement of the transverse momentum and ϕ_{η}^{*} distributions of Drell-Yan lepton pairs in proton-proton collisions at $\sqrt{s} = 8$ TeV with the ATLAS detector *Eur. Phys. J.* C **76** 291

[41] Aaboud M *et al* 2018 Measurement of the W-boson mass in pp collisions at $\sqrt{s} = 7$ TeV with the ATLAS detector *Eur. Phys. J.* C **78** 110
Aaboud M *et al* 2018 Measurement of the W-boson mass in pp collisions at $\sqrt{s} = 7$ TeV with the ATLAS detector *Eur. Phys. J.* C **78** 898 Erratum

[42] Falkowski A, Gonzalez-Alonso M, Greljo A, Marzocca D and Son M 2017 Anomalous triple gauge couplings in the effective field theory approach at the LHC *JHEP* **02** 115

[43] Eboli O J P, Gonzalez-Garcia M C, Lietti S M and Novaes S F 2001 Anomalous quartic gauge boson couplings at hadron colliders *Phys. Rev.* D **63** 075008

[44] Aaboud M *et al* 2017 Measurement of W boson angular distributions in events with high transverse momentum jets at $\sqrt{s} = 8$ TeV using the ATLAS detector *Phys. Lett.* B **765** 132–53

[45] Aaboud M *et al* 2017 Measurement of $W^{\pm}W^{\pm}$ vector-boson scattering and limits on anomalous quartic gauge couplings with the ATLAS detector *Phys. Rev.* D **96** 012007

[46] Osland P and Wu T T 1992 Parameters in the electroweak theory: 3. masses of the top quark and the Higgs boson *Z. Phys.* C **55** 593–604

[47] Kniehl B A 1996 Dependence of electroweak parameters on the definition of the top quark mass *Z. Phys.* C **72** 437–47

[48] Branchina V, Messina E and Platania A 2014 Top mass determination, Higgs inflation, and vacuum stability *JHEP* **09** 182

[49] Espinosa J R 2016 Implications of the top (and Higgs) mass for vacuum stability *PoS* **TOP2015** 043

[50] Hoang A H and Stewart I W 2008 Top mass measurements from jets and the tevatron top-quark mass *Nucl. Phys. Proc. Suppl.* **185** 220–6

[51] Smith M C and Willenbrock S S 1997 Top quark pole mass *Phys. Rev. Lett.* **79** 3825–8

[52] 't Hooft G 1973 Dimensional regularization and the renormalization group *Nucl. Phys.* B **61** 455–68

[53] Bardeen W A, Buras A J, Duke D W and Muta T 1978 Deep-inelastic scattering beyond the leading order in asymptotically free gauge theories *Phys. Rev.* D **18** 3998–4017

[54] Melnikov K and van Ritbergen T 2000 The Three loop relation between the MS-bar and the pole quark masses *Phys. Lett.* B **482** 99–108

[55] Hoang A H 2014 *The top mass: interpretation and theoretical uncertainties Proc. of 7th Int. Workshop on Top Quark Physics (TOP2014) (Cannes, France, September 28–October 3, 2014)*

[56] Aaboud M *et al* 2018 Direct top-quark decay width measurement in the $t\bar{t}$ lepton+jets channel at $\sqrt{s} = 8$ TeV with the ATLAS experiment *Eur. Phys. J.* C **78** 129

[57] ATLAS Collaboration 2019 Measurements of top-quark pair spin correlations in the $e\mu$ channel at $\sqrt{s} = 13$ TeV using pp collisions in the ATLAS detector *Technical Report* ATLAS-CONF-2018-027, CERN, Geneva arXiv:1903.07570. (see also https://cds.cern.ch/record/2682109)

[58] Aaboud M *et al* 2017 Measurement of the W boson polarisation in $t\bar{t}$ events from pp collisions at $\sqrt{s} = 8$ TeV in the lepton + jets channel with ATLAS *Eur. Phys. J.* C **77** 264

[59] Aaboud M *et al* 2018 Measurement of colour flow using jet-pull observables in $t\bar{t}$ events with the ATLAS experiment at $\sqrt{s} = 13$ TeV *Eur. Phys. J.* C **78** 847

[60] Aad G *et al* 2016 Measurements of the Higgs boson production and decay rates and constraints on its couplings from a combined ATLAS and CMS analysis of the LHC pp collision data at $\sqrt{s} = 7$ and 8 TeV *JHEP* **08** 045

[61] Aaboud M *et al* 2018 Observation of $H \rightarrow b\bar{b}$ decays and VH production with the ATLAS detector *Phys. Lett.* B **786** 59–86

[62] Sirunyan A M *et al* 2018 Observation of the Higgs boson decay to a pair of τ leptons with the CMS detector *Phys. Lett.* B **779** 283–316

[63] Aaboud M *et al* 2018 Observation of Higgs boson production in association with a top quark pair at the LHC with the ATLAS detector *Phys. Lett.* B **784** 173–91

[64] Sirunyan A M *et al* 2018 Observation of $t\bar{t}H$ production *Phys. Rev. Lett.* **120** 231801

[65] Dolan M J, Englert C and Spannowsky M 2012 Higgs self-coupling measurements at the LHC *JHEP* **10** 112

[66] Aaboud M *et al* 2018 Combined measurement of differential and total cross sections in the $H \rightarrow \gamma\gamma$ and the $H \rightarrow ZZ^{\star} \rightarrow 4\ell$ decay channels at $\sqrt{s} = 13$ TeV with the ATLAS detector *Phys. Lett.* B **786** 114–33

[67] ATLAS Collaboration 2017 Combined measurements of Higgs boson production and decay in the $H \rightarrow ZZ^{\star} \rightarrow 4\ell$ and $H \rightarrow \gamma\gamma$ channels using $\sqrt{s} = 13$ TeV pp collision data collected with the ATLAS experiment *Technical Report* ATLAS-CONF-2017-047, CERN, Geneva http://cdsweb.cern.ch/record/2273854

[68] ATLAS Collaboration 2018 Combined measurements of Higgs boson production and decay using up to 80 fb^{-1} of proton-proton collision data at $\sqrt{s} = 13$ TeV collected with the ATLAS experiment *Technical Report* ATLAS-CONF-2018-031, CERN, Geneva https://cds.cern.ch/record/2629412

[69] Aaltonen T *et al* 2009 Global search for new physics with 2.0 fb^{-1} at CDF *Phys. Rev.* D **79** 011101

[70] Aaboud M *et al* 2019 A strategy for a general search for new phenomena using data-derived signal regions and its application within the ATLAS experiment *Eur. Phys. J.* C **79** 120

[71] Kraml S, Kulkarni S, Laa U, Lessa A, Magerl W, Proschofsky-Spindler D and Waltenberger W 2014 SModelS: a tool for interpreting simplified-model results from the LHC and its application to supersymmetry *Eur. Phys. J.* C **74** 2868

[72] Papucci M, Sakurai K, Weiler A and Zeune L 2014 Fastlim: a fast LHC limit calculator *Eur. Phys. J.* C **74** 3163

[73] Drees M, Dreiner H, Schmeier D, Tattersall J and Kim J S 2015 CheckMATE: confronting your favourite new physics model with LHC data *Comput. Phys. Commun.* **187** 227–65

[74] Buckley A, Butterworth J, Lonnblad L, Grellscheid D, Hoeth H, Monk J, Schulz H and Siegert F 2013 Rivet user manual *Comput. Phys. Commun.* **184** 2803–19

[75] Conte E and Fuks B 2018 Confronting new physics theories to LHC data with MADANALYSIS 5 *Int. J. Mod. Phys.* A **33** 1830027

[76] Brooijmans G *et al* 2016 Les Houches 2015: Physics at TeV colliders - new physics working group report *9th Les Houches Workshop on Physics at TeV Colliders (PhysTeV 2015) (Les Houches, France, June 1–19, 2015)*

[77] Butterworth J M, Grellscheid D, Krämer M, Sarrazin B and Yallup D 2017 Constraining new physics with collider measurements of Standard Model signatures *JHEP* **03** 078

[78] Athron P *et al* 2017 GAMBIT: The global and modular beyond-the-standard-model inference tool *Eur. Phys. J. C* **77** 784
Athron P *et al* 2018 GAMBIT: The global and modular beyond-the-standard-model inference tool *Eur. Phys. J. C* **78** 98 Addendum

[79] Buckley A, Englert C, Ferrando J, Miller D J, Moore L, Russell M and White C D 2016 Constraining top quark effective theory in the LHC Run II era *JHEP* **04** 015

[80] Aaltonen T *et al* 2001 Invariant mass distribution of jet pairs produced in association with a W boson in $p\bar{p}$ collisions at $\sqrt{s} = 1.96$ TeV *Phys. Rev. Lett.* **106** 171801

[81] Aaltonen T *et al* 2014 Invariant-mass distribution of jet pairs produced in association with a W boson in $p\bar{p}$ collisions at $\sqrt{s} = 1.96$ TeV using the full CDF Run II data set *Phys. Rev. D* **89** 092001

[82] Aaboud M *et al* 2016 Search for resonances in diphoton events at $\sqrt{s} = 13$ TeV with the ATLAS detector *JHEP* **09** 001

[83] Khachatryan V *et al* 2016 Search for resonant production of high-mass photon pairs in proton-proton collisions at $\sqrt{s} = 8$ and 13 TeV *Phys. Rev. Lett.* **117** 051802

[84] Fowlie A 2017 Bayes factor of the ATLAS diphoton excess: Using Bayes factors to understand anomalies at the LHC *Eur. Phys. J. Plus* **132** 46

[85] Aaboud M *et al* 2017 Search for new phenomena in high-mass diphoton final states using 37 fb^{-1} of proton-proton collisions collected at $\sqrt{s} = 13$ TeV with the ATLAS detector *Phys. Lett. B* **775** 105–25

[86] Sirunyan A M *et al* 2018 Search for physics beyond the standard model in high-mass diphoton events from proton-proton collisions at $\sqrt{s} = 13$ TeV *Phys. Rev. D* **98** 092001

IOP Publishing

Experimental Particle Physics
Understanding the measurements and searches at the Large Hadron Collider
Deepak Kar

Chapter 8

Advanced topic: jet substructure

More is less. Just because you know the QCD Lagrangian doesnt mean you know all of its physics.

—Andrew Larkoski

Jets have been already introduced in chapter 3, and measurements and searches using jets have been discussed extensively in chapters 5 and 7. Jets are defined in terms of the reconstruction algorithm and the radius parameter used to construct them. Let us now look at two searches with many jets in the final state.

The first one involves a hypothetical W' boson decaying into a top quark and a bottom quark, and the top quark decays to another bottom quark and a quark–anti quark pair (of course the top quark can decay leptonically, and that is usually dealt with in a separate analysis). The process is shown in figure 8.1 (right). In this case, there are four jets (of which two can be b-tagged) in the final state, resulting in a large multijet background. The traditional analysis technique involves reconstructing four jets, and combining three of the four jets to form the hadronically decaying top quark candidate. However that leads to the possibility of picking the wrong combination of three jets, either because of selecting the wrong b-tagged jet, or picking a jet which does not reconstruct the top quark. The latter can happen when a jet not coming from the decay of a top quark (from underlying event or pile-up) or a jet coming from extra radiation (which still may be from top quark decay products, but in that case not all decay products will be considered) is selected.

Another example is the associated production of a Higgs boson with a top–anti-top quark pair (tthbb), shown in figure 8.1 (right). The largest branching fraction of a Higgs boson is to a bottom–anti-bottom quark pair. The top quarks can decay leptonically or hadronically, leading to a dileptonic (10.5%), semileptonic (43.8%) or hadronic (45.7%) $t\bar{t}$ final state. The largest cross-section is for a final state with eight jets, four of which can be b-tagged. This leads to a combinatorial nightmare to

doi:10.1088/2053-2563/ab1be6ch8

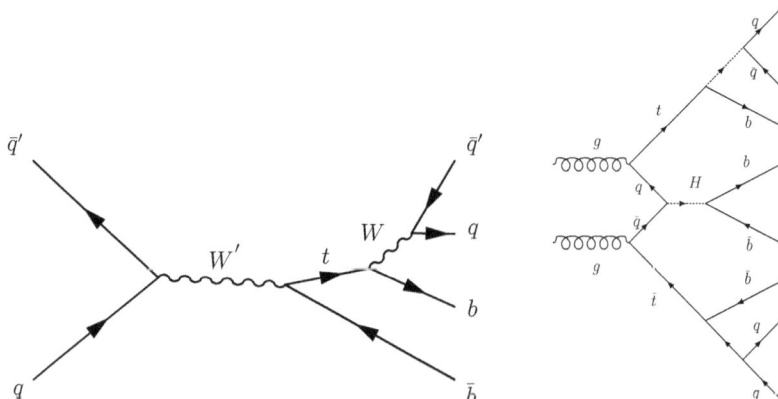

Figure 8.1. Feynman diagrams depicting a hypothetical W' boson decaying to a top and a bottom quark with the top quark decaying hadronically (left), and a Higgs boson produced in association with a top quark pair with the Higgs boson decaying to a bottom quark pair, and both the top quarks decaying hadronically (right).

reconstruct the top quark and Higgs boson candidates. Even for the semileptonic channel, there are six jets, of which four can be b-tagged.

In order to reconstruct these types of final states efficiently, a different approach is adopted. Rather than using individual jets, if a larger-radius jet is constructed which contains all the decay products of the heavy hadronically decaying object, then the combinatorial problem can be avoided, and the large-radius jet kinematics will mimic that of the heavy particle. Then the challenge becomes identifying which large-radius jet actually originated from an interesting heavy particle, and which originated from light quark and gluon splittings. The use of jet substructure involves looking at the internal (sub)structure of large-radius jets. We will first motivate the use of these large-radius jets and their substructure with the first such analysis in section 8.1. An obvious pitfall of constructing a jet sweeping a large area is that they also include contributions from pile-up and underlying events, section 8.2 describes how that is addressed. Then, section 8.3 discusses different observables and techniques to identify the large-radius jets based on their origin. In section 8.4, we discuss some experimental results. Finally, we will conclude the chapter by mentioning some theoretical and experimental aspects related to jet substructure in section 8.5. For completeness, we also provide a list of more technical references [1–4].

It must be noted that the energy deposits from high p_T electron or even photons can form large-radius jets as well, often referred to as leptonic (large-radius) jets. The former arises in the case of a leptonic top quark decay, where only the electron, or the electron and the bottom quark decay products can be inside the large-radius jet. While leptonic jets give rise to an interesting set of experimental signatures, we will mostly restrict our discussion to hadronic jets in this chapter.

8.1 Large-radius jets

For a massive particle with four-momentum $(E, \vec{p}\,)$, and mass m, decaying into two daughter particles A and B with four-momentum $(E_A, \vec{p}_A)$, $(E_B, \vec{p}_B)$, we can write:

$$m^2 = (E_A + E_B)^2 - (\vec{p}_A + \vec{p}_B)^2$$
$$= (E_A)^2 + (E_B)^2 + 2E_A E_B - |\vec{p}_A|^2$$
$$- |\vec{p}_B|^2 - 2\vec{p}_A \cdot \vec{p}_B \cos\theta$$

where θ is the angular separation between the vectors. Now assuming that the masses of the decay products are small compared to their energy (which is a realistic assumption in the case of heavy hadronically decaying objects), we can write, $E_A \approx |p_A|$ and $E_B \approx |p_B|$. Furthermore, only the transverse component of momentum is relevant for colliders, so the magnitude of three-momenta can be replaced by the magnitude of p_T. Finally, we can assume a symmetric decay (which is again a reasonable approximation for these cases), with both the decay particles having the same transverse momentum p_T. With these simplifications, we get:

$$m^2 = \frac{p_T^2}{2}(1 - \cos\theta) = \frac{p_T^2}{2}\frac{\theta^2}{2}$$

by making one more simplifying assumption, that the angles between the two decay products are small, the motivation for which will become clear as we go along. So finally we have:

$$m^2 \approx \frac{p_T^2 \theta^2}{4}$$

The angular separation between the two decay particles can be equivalently represented in terms of ΔR, the distance measure in collider coordinates. That leads us to:

$$\Delta R = \frac{2m}{p_T}$$

which means that the light particles from a symmetric two-body decay of a massive particle will be within a distance of ΔR of each other, the value of which depends only on the mass and the p_T of the massive particle. For example, for a top quark with p_T of 350 GeV (the mass of a top quark is 175 GeV), the decay products, W boson and the bottom quark will be roughly separated by $\Delta R \approx 1$. The corresponding W boson decay products will also be within $\Delta R < 1$ for a W boson p_T of about 150 GeV (the mass of a W boson is 80 GeV). This can be seen in figure 8.2, where the angular separation between the decay products is shown as a function of the original heavy particle p_T in a simulated sample. It can be seen when the original particle has more p_T than the threshold indicated above, their decay products can be contained in a single jet with $\Delta R = 1$, as seen in figure 8.3. This is what we meant by the term large-radius jets before. The *standard* jets we encountered before usually have $R = 0.4$–0.6, which we will refer to as small-radius jets in this chapter to avoid confusion. The jets with even smaller radius, inside a large-radius jet will be referred to as *subjets or prongs*.

A high p_T particle corresponds to a particle produced with a large Lorentz boost (i.e. values of higher β) in the transverse direction, as opposed to being produced

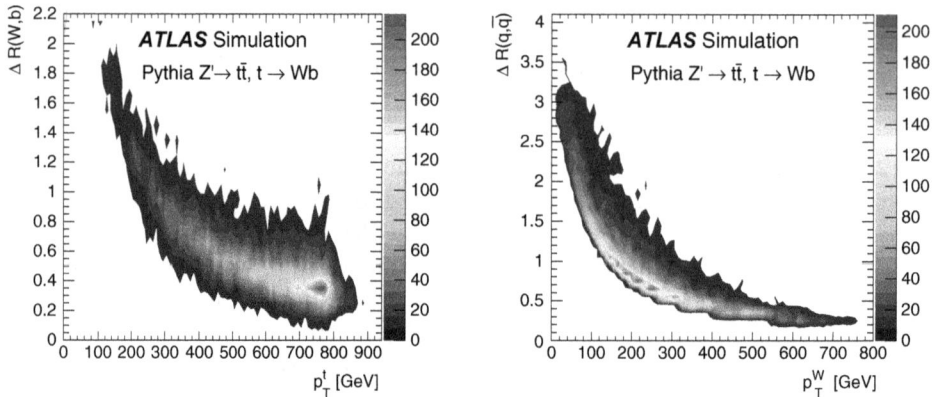

Figure 8.2. The angular separation between the W boson and b quark in top quark decay (left) and between the quarks from W boson decay (right), as functions of the p_T of the decaying particle from simulation [5] (ATLAS Experiment © 2018 CERN).

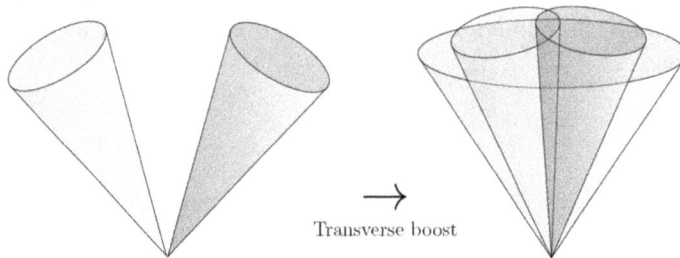

Figure 8.3. A schematic diagram showing two jets from a hadronically decaying particle overlap when it has sufficient transverse boost, and a larger-radius jet can contain both of them.

almost at rest in the centre-of-mass frame. With the increase of the LHC centre-of-mass energy, more particles are being produced with appreciable *boost*. In fact many analyses using the large-radius jets are therefore referred to as boosted analysis, and the standard analyses using small-radius jets are termed resolved[1].

This discussion is relevant for hadronically decaying bosons (Z, W, Higgs), top quarks, as well as for any new heavy particle decaying to partons, which subsequently hadronise, so their experimental signature can look quite similar to standard jets. It is worth noting that the substructure techniques are only applicable for particles with a high p_T (and p_T of particles almost always fall exponentially), so by design they are not sensitive to full process cross-section. That is why usually a (complementary) resolved analysis is also performed to cover the full p_T range, identifying jets with smaller radii. We should also note that boosted objects are usually more central in the detector, leading to a better acceptance than in the resolved case.

---

[1] The biggest meeting in this area is also named Boost!

Large-radius jets can originate[2] not only from heavy resonances as described above, but also from light quarks (usually means other quarks apart from the top quark), gluons and even charged leptons (if they deposit energy in the calorimeters, which electrons do more than muons). Therefore, the challenge becomes discriminating large-radius jets coming from signal (heavy particles of interest) to large-radius jets coming from background (mostly light quark, gluon or lepton jets).

There are different approaches for achieving this. Possibly the first to look at the boosted objects was Mike Seymour (1991) [6], where he stated: 'The W-finder used in this study utilises this cut by running a jet-finder twice, with a cone-size of $\Delta R = 0.75$ and $\Delta R = 0.4$, and then demands a big jet containing two small jets $\cdots$'. The first look at what we can term a substructure observable was perhaps by Butterworth *et al* (2002), [7], where they stated: 'The scale of the splitting is indeed high in the signal and softer in the W+ jets background $\cdots$'. The development of IRC-safe jet algorithms (and their implementation in computationally feasible fastjet programme) during those years led to the much acclaimed BDRS paper (2008, where the acronym stood for the authors, Butterworth, Davison, Rubin, Salam) [8], which in some sense kick-started the field. We will describe their method in some detail in order to point out the salient features of a substructure based analysis, which we will encounter repeatedly.

This was before the LHC started, so the Higgs boson was not discovered at that time. They looked at the final state where the Higgs boson was produced in association with a W or a Z boson, which helps in distinguishing the events due to the presence of one or two leptons. The (most dominant) decay mode of the Higgs boson, to a pair of bottom–anti-bottom quarks was considered. The presence of lepton(s) made this analysis somewhat viable, as otherwise the Higgs decay to two b-tagged jets gets buried under a large multijet background. In the boosted regime, the W/Z boson is produced almost back-to-back with the large-radius jet containing Higgs boson decay products. The leading CA with $R = 1.2$ and $p_T > 200$ GeV was used to reconstruct the Higgs boson.

The first step was to remove the constituents coming from the underlying event and contaminating the large-radius jet.

1. The jet formation steps can be undone (termed declustering), one step at a time. The last ingredients j_1 and j_2 from the original jet j at the penultimate stage can be labelled according to their masses, such that $m_{j_1} > m_{j_2}$.

2. Then the idea was to identify if the (final) clustering step involved subjets which came from the Higgs boson decay products, or from soft uncorrelated constituents. The latter would mean the subjets from the hard decay are rather unbalanced, both in terms of their mass and angular directions. This is tested by comparing masses and rapidities: (A) if $m_{j_1} < \mu m_j$, and (B) if

$$y = \frac{\text{Min}(p_{T_{j1}}^2, p_{T_{j2}}^2)}{m_j^2}(\Delta R_{j1,j2}^2) > y_{\text{cut}},$$ where μ and y_{cut} are two parameters of the

process, and are chosen so that maximum amount of signal is retained.

---

[2] Defining jets by their origin is inherently an ambiguous concept, as will be discussed in section 8.5.

This paper used $\mu = 0.67$ and $y_{cut} = 0.09$. If both of these conditions are satisfied, then the effect of soft uncorrelated constituents is considered negligible, and this is termed mass drop, as shown in figure 8.4. Otherwise, the initial jet j_1 is redefined as j, j_2 is rejected and the process is repeated again.

After this initial *cleaning* of the large-radius jet, the three hardest subjets from declustering the large-radius jet were picked as the Higgs boson decay products, with the requirement $R_{bb} > R_{filt}$, where R_{bb} is the distance between the two hardest subjets which were also required to be b-tagged. All constituents outside the three hardest subjets were discarded. This step is termed as filtering, which is shown in figure 8.4 as well. The R_{filt} is another parameter chosen to optimise the signal acceptance dynamically, with the value $\min(0.3, R_{bb}/2)$. The reason three subjets were considered is because there can be an extra hard gluon radiation from Higgs boson decay products.

There are several other selections, like lepton pair invariant mass, a certain missing energy threshold, number of charged leptons etc, which were applied as well depending on whether events with a W or Z boson were considered. At the end, the large-radius jet mass represented the Higgs boson candidate mass, and can be seen over the background in figure 8.5. The simulated Higgs boson mass was 115 GeV and an expected significance of 4.5σ was claimed for an integrated luminosity of 30 fb^{-1} combining all the channels.

While this was a rather promising proposal, there were a number of reasons why the technique did not work in this form in ATLAS or CMS initially. The most important reason being that b-tagging subjets inside a large-radius jet was not well established at that point. Also, the effect of pile-up was more severe than expected. However, there are still a number of important lessons we can take away from this:

- The large-radius jets need to be cleaned from soft uncorrelated radiation (mostly from pile-up and underlying event) before trying signal to background discrimination. This step has come to be known as *grooming*, with several techniques established to perform this step (section 8.2).
- The observables constructed from the large-radius jet constituents (mass in this case) need to defined, which would discriminate signal from background. In this case, a Higgs boson decaying to two bottom quarks suggests there will be two dominant directions of energy flow inside the large-radius jet, which is often referred to as *two-prong* substructure (top quark decay will result in a *three-prong* substructure). However, we should also account for hard gluon radiation from these prongs inside the large-radius jets, which is often the

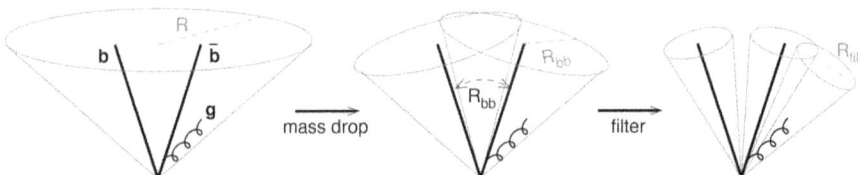

Figure 8.4. Illustration of the mass drop and filtering stages in the BDRS method [8].

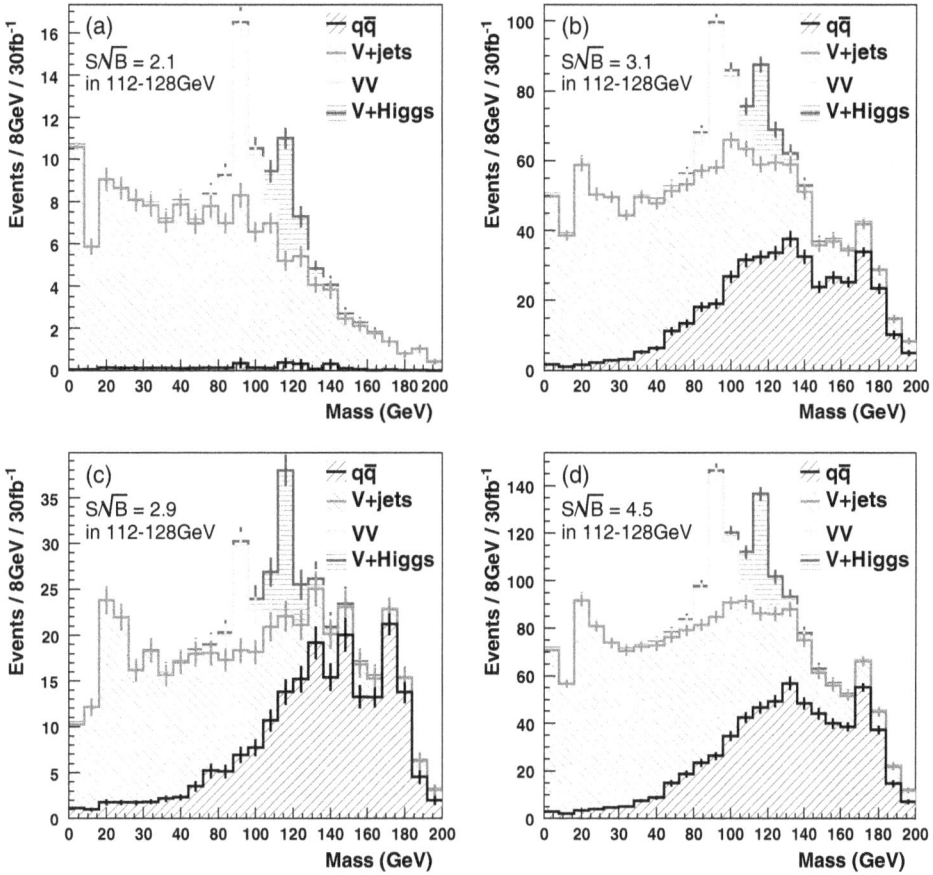

Figure 8.5. Results showing that the Higgs boson is reconstructed as the large-radius jet in four different associated production modes using simulation by BDRS method [8].

case. Many such *tagging observables* have been proposed. Additionally, *taggers* have also been constructed which combines several observables and/or grooming techniques (section 8.3).

- Experimental implementation of these algorithms often brings its own set of challenges, depending on how complex the algorithm is, what input objects it requires, whether the behaviour of those objects is well understood, and the calibration of large-radius jets. In other words, the suggested grooming and tagging techniques from theoretical consideration, like in this paper, may not translate to experimental success (section 8.5).

- It often helps to use final states which offer additional signal to background discriminating power along with the jet substructure techniques, as the presence of leptons(s) here.

Usually, jet substructure techniques prefer k_t or Cambridge–Aachen (CA) algorithms over more commonly used anti-k_t algorithm. In these techniques, as seen in the BDRS example, the last jet clustering steps are undone first, and checks are performed to see if those subjets correspond to the decay products of interest. The k_t algorithm prefers to cluster softer particles first, so at the last stage, the hardest subjets remain, which is ideal. With the CA algorithm only clusters based on distance measure, and the closest constituents are clustered first. So when reversing the clustering, the wide-angle ones are obtained first, and those mostly correspond to UE or PU. On the other hand, anti-k_t algorithm clusters the hardest particles first, so in order to find the hard subjets, it is necessary to dig deep through the clustering history, which is rather inefficient.

8.2 Grooming

The idea of grooming is to selectively eliminate soft particles coming from UE or PU from the large-radius jets, so that they do not obscure the interesting substructure of the jet. Wide-angle soft radiations for example can make measurable contributions to the mass. Usually large-radius jets from gluons or light quarks (or leptons) contain a single dense core of energy surrounded by soft radiation from the parton shower, hadronisation, and underlying events. On the other hand, large-radius jets containing the decay products of single massive particles, on the other hand, can be distinguished by hard, wide-angle components. The different grooming techniques, therefore, try to remove wide-angle or soft particles, but not the decay products from the heavy particle itself. Theoretically, grooming eliminates the kinematic region dominated by non-perturbative effects.

Mass drop/filtering: This technique was used in the BDRS paper. The idea is to isolate retain part of the large-radius jet in two stages by identifying relatively symmetric subjets, each with significantly smaller masses than that of the original jet, as seen in figure 8.6.

As an example, for a W boson of $p_T = 200$ GeV, mass of 80 GeV, if $\mu = 0.67$, that means mass of the first de-clustered subjet must be less than 61 GeV. Taking $\Delta R = 0.8$ and $y_{cut} = 0.09$, the second condition reduces to the lower p_T subjet having a minimum p_T of 30 GeV, thus forcing both subjets to carry some significant fraction of the momentum of the original jet. At the second filtering step, by dynamically re-clustering the jet at an appropriate angular scale, the sensitivity to highly collimated decays is maximised.

Trimming: Trimming [9] removes soft particles, assuming that those do not come from the heavy particle decay. It uses the k_t algorithm to create subjets of size R_{sub} from the constituents of the large-radius jet. Any subjets with $p_T^i/p_T^{jet} < f_{cut}$ are removed, where p_T^i is the p_T of the ith subjet, as shown in figure 8.7. The algorithm has two parameters f_{cut} and R_{sub}, which determine the severity of the rejection. In ATLAS they are taken to be 0.03–0.05 and 0.2–0.3, respectively for anti-k_t large-radius jet with $R = 1.0$. The remaining constituents then form the trimmed jet.

Pruning: The pruning algorithm [10] is similar to trimming in the sense that it removes constituents with a small relative p_T compared to the original jet p_T, but it additionally applies a veto on wide-angle radiation. It does not use subjets, but steps back in jet clustering history. At each merging step of the previous two merged

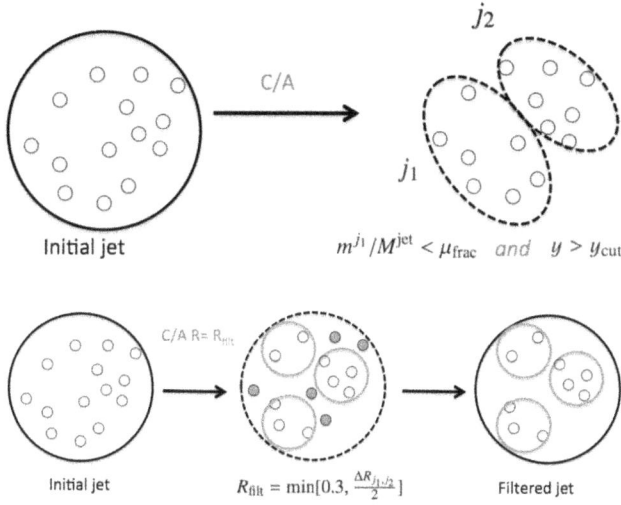

Figure 8.6. Schematic illustration of the mass drop (top) and filtering (bottom) procedures [5] (ATLAS Experiment © 2018 CERN). The small circles represent the inputs to jet forming, and the grey shaded ones are discarded.

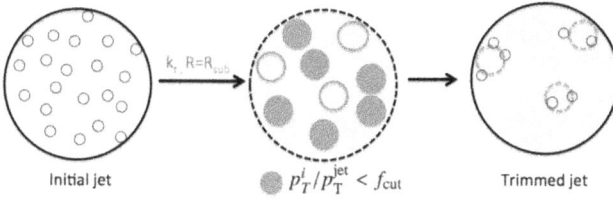

Figure 8.7. Schematic illustration of the trimming procedure [5] (ATLAS Experiment © 2018 CERN). The small circles represent the inputs to jet forming, and the grey shaded ones are discarded.

inputs, it selects whether that constituent will be part of the large-radius jet. The criteria for acceptance is either:

$$p_T^{j2}/\left(p_T^{j1}, +, p_T^{j2}\right) > z_{cut}$$

or

$$\Delta R_{j1,j2} < R_{cut}\frac{2m^{jet}}{p_T^{jet}}$$

where it is assumed the constituents $p_T^{j1} > p_T^{j2}$. Here z_{cut} and R_{cut} are parameters of the algorithm, CMS uses values of 0.1 and 0.25–0.5 respectively for CA large-radius jets with $R = 0.8$. The level of pruning tends to be determined by the less aggressive of the two parameters (since both requirements need to satisfied for a constituent to be removed). Figure 8.8 illustrates the pruning procedure.

Modified mass drop and soft drop: While the above grooming algorithms have been used extensively by experiments, a problematic feature of these algorithms is that

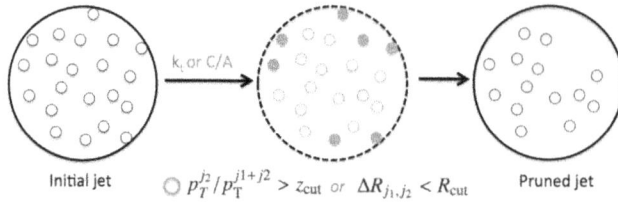

Figure 8.8. Schematic illustration of the pruning procedure [5] (ATLAS Experiment © 2018 CERN). The small circles represent inputs to jet forming, and the grey shaded ones are discarded.

theoretically their effect on jet substructure observables, most notably mass, is difficult to calculate. This is primarily caused by wide-angle gluon radiations emitting a soft gluon back inside the jet, which gives rise to logarithmically enhanced higher-order corrections in the calculation, often termed as non-global log (NGL).

In order to address this issue, the modified mass drop (mMDT) [11] and soft drop [12] algorithms were developed, which are insensitive to NGL corrections. Again, the last step of jet forming is undone, and j_1 and j_2 are the ingredients merged to form the original jet j.

In mMDT, rather than redefine the larger p_T j_1 as j and continue iterating, either j_1 or j_2 is redefined as j, depending on which has the larger transverse mass $(m^2 + p_T^2)$. Then the iteration is continued. This eliminates a theoretical ambiguity for three particle configurations.

The soft drop procedure generalises the mMDT procedure. For a large-radius jet with $R = R_0$, the soft drop procedure removes the softer of the last two subjets unless:

$$\frac{\min\left(p_T^1, p_T^2\right)}{p_T^1 + p_T^2} > z_{\text{cut}} \left(\frac{R_{12}}{R_0}\right)^\beta$$

where R_{12} is the distance between the subjets. This condition is checked at each splitting step by traversing backward in clustering history, as shown in figure 8.9. The two parameters of the algorithm are Z_{cut} and the exponent β. The former sets the scale of the energy removed by the algorithm, while the latter indicates the sensitivity to wide-angle radiation, larger values result in less soft radiation being removed. Thus $\beta \to \infty$ means the jet is ungroomed and $\beta = 0$ returns mMDT. Typical values used are $z_{\text{cut}} = 0.1$ and $\beta = 1.0$.

Comparison of the different grooming algorithms: The obvious question that can arise is: how do these grooming methods compare? The effect of filtering, trimming and pruning was compared in [1] for large-radius jets from top quark (signal) and from light quark or gluons (background). Figure 8.10 shows that all three grooming techniques reduce the jet mass for background events, and pruning is most aggressive. For the signal, the effect and the difference is much less pronounced (at least for this particular choice of grooming parameters), which is what we would expect.

The current preference of the experiments is to use the analytically calculable soft drop method. Historically CMS mostly used pruning while ATLAS mostly used

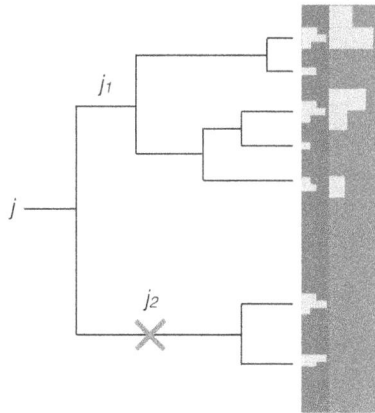

Figure 8.9. Schematic illustration of the soft drop procedure [13] (ATLAS Experiment © 2018 CERN). The yellow blocks represent jet inputs (with red and green areas representing calorimeters), and the softer merged input is discarded if it failed the soft drop condition.

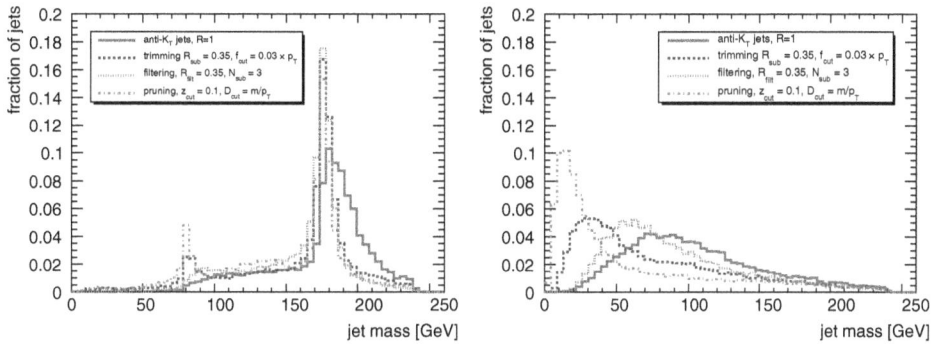

Figure 8.10. Comparison of grooming procedures illustrated by the effect on large-radius jet mass from a top quark (left) and from light quark or gluon (right) using simulated events [1]. © A Abdesselam, A Belyaev and E B Kuutmann.

trimming. Additionally the pile-up removal methods like CHS, PUPPI can also be used, often in conjunction with grooming.

8.3 Observables/taggers

Observables are constructed to characterise the large-radius jet substructure and distinguish massive boosted objects from gluons or light quarks. These observables, used for that purpose, are often referred to as tagging observables or taggers, as they are used to tag or identify a jet based on the original heavy particle (after grooming has been applied). Often a combination of observables are used as taggers, but there are also more complex algorithms, with the sole purpose of signal to background discrimination. We will describe a few representative observables which illustrate the substructure of jets, and some tagging algorithms. It must be noted, that in order to compare performance of different taggers, usually via ROC curves, the large-radius jets must be in a similar p_T range. The properties of jets, and tagging algorithms depend sensitively on their p_T.

Mass: The jet mass is defined as the invariant mass calculated from the four-momentum sum of all jet constituents. Even though the constituents are considered massless, the jet acquires its mass from the opening angle between all contributors. As a simple example, for two massless particles with (transverse) energies E_1 and E_2, the mass is generated by the opening angle, θ_{12}:

$$M_{12}^2 = E_1 E_2(1 - \cos\theta_{12}) = 2z(1 - z)E^2(1 - \cos\theta_{12})$$
$$= z(1 - z)E^2\theta_{12}^2$$

where z is the fraction of energy carried by one particle, and small angle approximation is used.

Mass is possibly the most obvious discriminating variable, as the mass of the large-radius jet corresponds to the mass of the heavy decaying particle for signal, assuming the decay products are contained in the large-radius jet. So for a large-radius jet coming from a top quark, or W/Z or Higgs boson, the mass is close to the mass of its originating particle. However, for large-radius jets coming from light quark or gluons, which are effectively massless, the jets still have non-zero mass because of wide-angle soft radiations. The average mass of the jet can be shown to be $\approx \alpha_s p_T R$, where p_T and R are the transverse momentum and the radius of the jet. This is referred to as the *Sudakov peak*. However, as large-radius jets get contaminated by pile-up and underlying events, an exact value of Sudakov peak is hard to determine.

In figure 8.11, a distribution of mass of trimmed large-radius jets from a hadronically decaying boosted top quark event is shown. While the clear peak around the top quark mass value can be seen, a peak at the W boson mass value can be seen as well as a peak at a much lower value. The W boson mass peak illustrates a feature called *(non-)containment*, where the large-radius jet only included decay products from the W boson, but missed the b-hadron decay products. The low mass peak corresponds to the Sudakov peak, where either all the decay products from the top quark are not included in the large-radius jet, or it is not from the top quark.

k_t splitting scale: If the k_t algorithm is employed to construct the large-radius jet, then the distance measure used for combining constituents can provide information about the structure of the jet. The distance measure (from chapter 3) is: $d_{ij} = \min[p_{T_i}^2, p_{T_j}^2]\frac{\Delta R_{ij}^2}{R^2}$, $d_{iB} = p_{T_i}^{2p}$. So at the final step, d_{12} will be the distance measure between two penultimate ingredients which are combined to form the large-radius jet. This variable in general is referred to as the splitting scale, with subscripts denoting which stage of clustering that distance measure corresponds to.

This observable is important to assess if the splitting is symmetric, which is usually the case for a heavy particle decay. The expected value for a two-body heavy particle decay is approximately $d_{12} \approx m_{\text{particle}}/2$, whereas jets from the parton shower of gluons and light quarks tend to have smaller values of the splitting scales and to exhibit a steeply falling spectrum for $\sqrt{d_{12}}$ and $\sqrt{d_{23}}$.

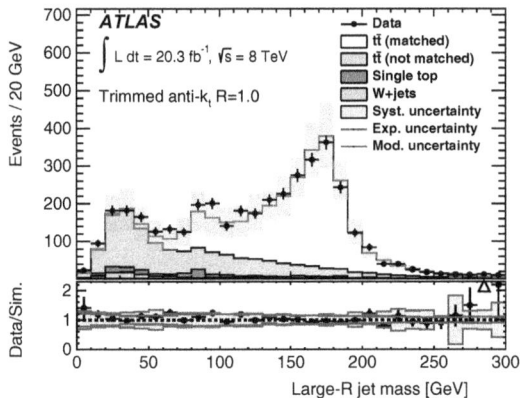

Figure 8.11. Mass of trimmed large-radius jet from hadronically decaying top quarks, passing a dedicated event selection [14] (ATLAS Experiment © 2018 CERN).

N-subjettiness: N-subjettiness [15, 16] is used to categorise how many dominant directions of energy deposition are inside the large-radius jet, or in other words how consistent it is with N or fewer subjets. Large-radius jets from top quarks are expected to have at least three prongs, and from $W/Z/H$ bosons have at least two prongs. Light quark/gluon large-radius jets have no definite structure, but most of them have one prong.

The τ_N variable is calculated by clustering the constituents of the jet with the k_t algorithm and requiring that exactly N subjets be found (this is termed an exclusive jet algorithm, where the clustering stops at a specified number of jets, not at a predetermined jet radius). The variables τ_N are then defined as the sum over all constituents k of the jet:

$$\tau_N = (1/d_0) \sum_{k=0}^{k=N} p_{T,k} \times \Delta R_{\min,k}$$

where $\Delta R_{\min,k}$ is the minimum distance of constituent k from the nearest subjet, and the normalisation factor d_0 is the sum of p_T all constituents, multiplied by the jet radius. This ensures that $0 < \tau_N < 1$.

Using this definition, τ_N describes how well jets can be described as containing N or fewer k_t subjets by assessing the degree to which constituents are aligned with the axes of these subjets for a given hypothesis N. The smaller values of τ_N correspond to N or fewer energy deposits (i.e. energy spread is close to the subjet axes), while the larger values indicate more than N energy deposits. So in general, for an N-prong substructure, it can be inferred that $\tau_{N-1} > \tau_N$.

In that sense τ_N can act as a discriminating variable, but some ambiguities arise. For example, for a two-prong substructure, τ_2 will have a smaller value than τ_1. But even for an one-prong substructure, that still can be the case, as τ_N does not strongly differentiate between substructures with N or fewer prongs. But we can exploit the fact that the τ_2/τ_1 ratio will be different in these two situations. For the two-prong case, the ratio will be a smaller number closer to zero, while for the one-prong case, the ratio will be closer to unity. The same argument holds for τ_{32} to identify three-prong decay over two- or one-prong decay.

The calculation of τ_N depends on the choice of axis, as the jet constituents are partitioned into N regions centred on the subjet axes, with respect to which the minimum distance is calculated. However, it was found that this definition is not robust enough, i.e. an additional constituent can change the directions of subjets. So a different *winner-takes-all* (WTA) [17] approach was proposed, which uses the direction of the hardest constituent in the exclusive k_t subjet instead of the subjet axis. While the previous definition can be thought of as taking a mean of jet constituent momenta to determine the axis, the WTA approach can be thought of as taking the median.

Recently it has been proposed that *dichroic* variants of N-subjettiness [18], where different subjets (i.e. from groomed and non-groomed large-radius jet) are used for the numerator and denominator of the $\tau_{N/N-1}$ ratio, can result in extra sensitivity in some cases.

Energy correlation functions: The angular distribution of subjets inside the large-radius jet is sensitive to how the jet was formed, i.e. to the internal radiation structure. Using the jet constituents, the energy correlation functions (ECF) [19, 20] are defined as follows:

$$\text{ECF}(1, \beta) = \sum p_{T_i}$$

$$\text{ECF}(2, \beta) = \sum_{i<j} p_{T_i} p_{T_j} (R_{ij})^\beta$$

$$\text{ECF}(3, \beta) = \sum_{i<j<k} p_{T_i} p_{T_j} p_{T_k} (R_{ij} R_{jk} R_{ki})^\beta$$

$$\cdots$$

$$\text{ECF}(N, \beta) = \sum_{allN} \left(\prod_{a=1}^{N} p_{T_a} \right) \left(\prod_{b=1}^{N-1} \prod_{c=b+1}^{N} R_{ib} R_{ic} \right)^\beta$$

The β parameter allows sensitivity to different angular scales.

The ECF variables behave in a very similar way to τ_N ones for the purpose of identifying N-prong substructure. If there are only N constituents, then $\text{ECF}(N + 1)$ will be zero. More generally, for N-prong substructure, $\text{ECF}(N + 1) < \text{ECF}(N)$.

So analogous to N-subjettiness ratios, ECF ratios can be defined:

$$r_N = \text{ECF}(N + 1)/\text{ECF}(N)$$

However, it has been found that double ratios of ECFs have more discriminating power:

$$\begin{aligned} C_N &= r_N/r_{N-1} \\ &= \frac{\text{ECF}(N + 1)\text{ECF}(N - 1)}{\text{ECF}(N)^2} \end{aligned}$$

for specific values of β.

The energy correlation double ratio C_N effectively discriminates higher-order radiation from leading order (LO) substructure. For a system with N subjets, the LO substructure consists of N hard prongs, so if C_N is small, then the higher-order radiation must be soft or collinear with respect to the LO structure. If C_N is large, then the higher-order radiation is not strongly-ordered with respect to the LO structure, so the system has more than N subjets. Thus, if C_N is small and C_{N-1} is large, then the lager-radius jet has at least N subjets. In this way, the energy correlation double ratios C_N, behave like N-subjettiness ratios. They are, however, more robust than N-subjettiness as they are less sensitive to multiple soft radiation.

The C_N observables are not boost-invariant; to address that D_N observables are constructed as:

$$D_N = \frac{\mathrm{ECF}(N+1)\mathrm{ECF}(N-1)^3}{\mathrm{ECF}(N)^3}$$

again for specific β values.

The constructions of C_N and D_N observables are examples of what in literature is referred to as power counting [21]. Using a set of ECF observables as a basis, they are constructed to amplify specific features of the large-radius jet. Other observables are also constructed [22], denoted by M (sensitive to groomed jets), N (similar to N-subjettiness ratios, but defined without considering the subjet axes), and U (sensitive multiple emissions within one-prong jets).

Les Houches angularity: Generalised angularities [23] for a jet are defined by:

$$\lambda_\beta^\kappa = \sum_{i \in jet} z_i^\kappa \theta_i^\beta$$

where z_i is the transverse momentum fraction, and $\theta_i = R_i/R_{jet}$ is the (normalised) angle of the ith constituents of the jet. The exponents κ and β probe different aspects of the jet fragmentation. While ($\kappa = 0$, $\beta = 0$) simply reduces to the multiplicity of the constituents, (2, 0) is referred to as p_T^D, as it is effectively the squared sum of the transverse momenta of the constituents. Further, (1.1) and (1.2) are analogous to jet shape observables like jet broadening and mass. The (1, 0.5) variant is termed Les Houches angularity [24]. It has been shown that angularities with $\kappa = 1$ are IRC safe.

Momentum fraction: The momentum fraction, z_g is defined from the soft drop condition [12]:

$$z_g = \frac{\min\left(p_T^1, p_T^2\right)}{p_T^1 + p_T^2}$$

which is essentially the transverse momentum fraction of the softer branch. Another quantity r_g is defined as the angular distance, or R_{12} between the last two subjets. These variables are sensitive to the parton splitting. For a symmetric splitting, the momentum fraction will be close to half.

HEPTopTagger: Heidelberg–Eugene–Paris Top Tagger (HEPTopTagger) [25] is an example of a multi-step algorithm to identify a large-radius jet coming from hadronic decay of a top quark. The following steps are performed, as shown in figure 8.12. CA algorithm is used in all steps.

1. Step 1: The large-radius jet is chosen, usually with $R = 1.5$ CA jets with $p_T > 200$ GeV are used. To find the hard substructure, a mass drop criteria is applied after undoing the clustering stages successively. If $m_{decay} < 0.8 M_{origin}$, then both decay products are kept, otherwise the softer one is discarded. This procedure stops when all substructure objects (i.e. subjets) have a mass below some m_{cut} (usually between 30–50 GeV).

2. Step 2: Then all combinations of three subjets among the subjets obtained in the previous step are considered.

3. Step 3: Filtering is performed.

4. Step 4: The five most energetic subjets are kept.

5. Step 5: The constituents of those five subjets are then re-clustered exclusively into three subjets.

6. Step 6: If the mass of the combined object is not within the top mass window of 150–200 GeV, then that combination is discarded. If a top quark candidate is found in more than one triplet, only the one with its mass closest to the measured top quark mass is used.

Additional kinematic cuts are used to increase the signal efficiency, and reject background. One of the di-subjet combinations needs to be close to W boson mass:

HEP Top Tagger details

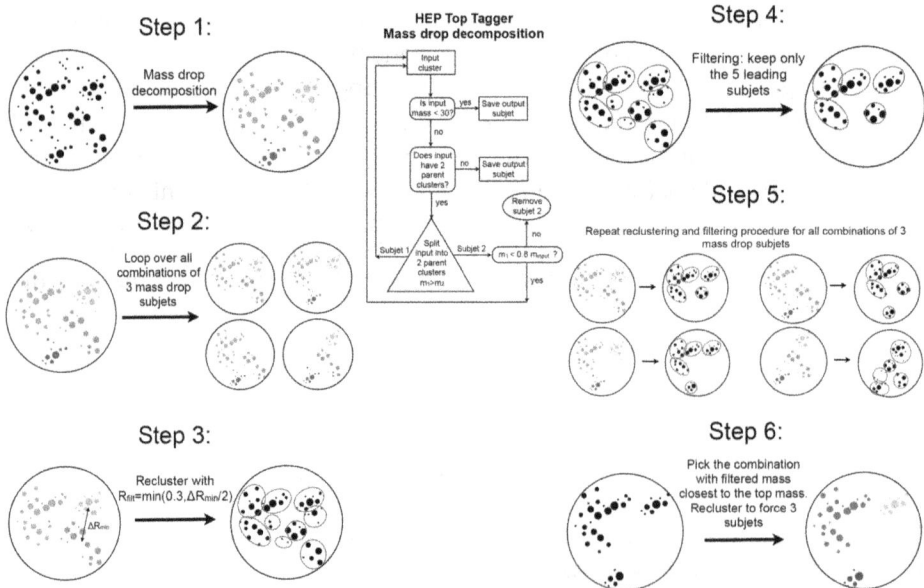

Figure 8.12. Step-by-step illustration HEPToptagger algorithm [26] (CMS Experiment © 2018 CERN).

$$0.85 \frac{m_W}{m_{top}} < \frac{m_{ij}}{m_{123}} < 0.85 \frac{m_W}{m_{top}}$$

Additionally if m_{23} is assumed to be closest to m_W, then $0.2 < \arctan m_{13}/m_{12} < 1.3$ or $m_{23}/m_{123} > 0.35$ is required. These internal parameters can be changed to optimise the performance.

Shower deconstruction: The idea of shower deconstruction [27, 28] is literally to revert the parton showering steps that has resulted in the observed large-radius jet, in order to determine its origin. The algorithm works by constructing all possible *shower histories*, or ways in which the jet under consideration could have been created from a specific object. Both signal and background hypotheses are considered, which is to say that shower histories are constructed assuming the jet originated from an object consistent with signal, and separately assuming it originated from an object consistent with background. In each case, multiple (running into hundreds) shower histories are possible, but with varying probabilities. The probabilities are calculated analytically from emission and decay probabilities at each vertex, colour connections, and kinematic requirements. This can be considered analogous to running a parton shower Monte Carlo generator in reverse. Then a likelihood ratio $\chi(p_N)$ is constructed,

$$\chi(p_N) = \frac{P(p_N|S)}{P(p_N|B)}$$

where $P(p_N|S)$ is the probability density that a jet configuration p_N is signal-like, and $P(p_N|B)$ is the probability density that the configuration p_N is consistent with background jets coming from other processes. As there are many shower histories that could lead to a given p_N, all those are summed. The $\log\chi(p_N)$ is used as the discriminating substructure variable.

Although the algorithm can be used for tagging top quarks [28], W bosons or Higgs bosons, or discriminating between quark and gluon initiated jets [29], the current discussion focusses on top quark tagging, for which it has been used predominantly in ATLAS [30]. Experimentally, the inputs to the algorithm are subjets (which were referred to as microjets in the original paper) from the large-radius jet, as they are expected to act as proxies for individual quarks and gluons coming from the decay of the particle initiating the jet. So the shower histories are constructed corresponding to subjet configuration inside the large-radius jet under consideration.

This can be best illustrated graphically, as in figure 8.13. The studies used a simulated sample of Z' events, where the Z' decays to a top–anti-top quark pair, in order to have a large sample of boosted top quark events. The mass of the Z' was set at 1.75 TeV. For a particular event, three out of more than 1500 shower histories with the largest signal probabilities are shown. The leading large-radius jet (anti-k_t, with $R = 1.0$) in this event was re-clustered into six subjets (CA $R = 0.2$). For a large-radius jet coming from a top quark decay, the subjets should correspond to hadronic decay products from a W boson, at least a bottom quark, gluon radiation

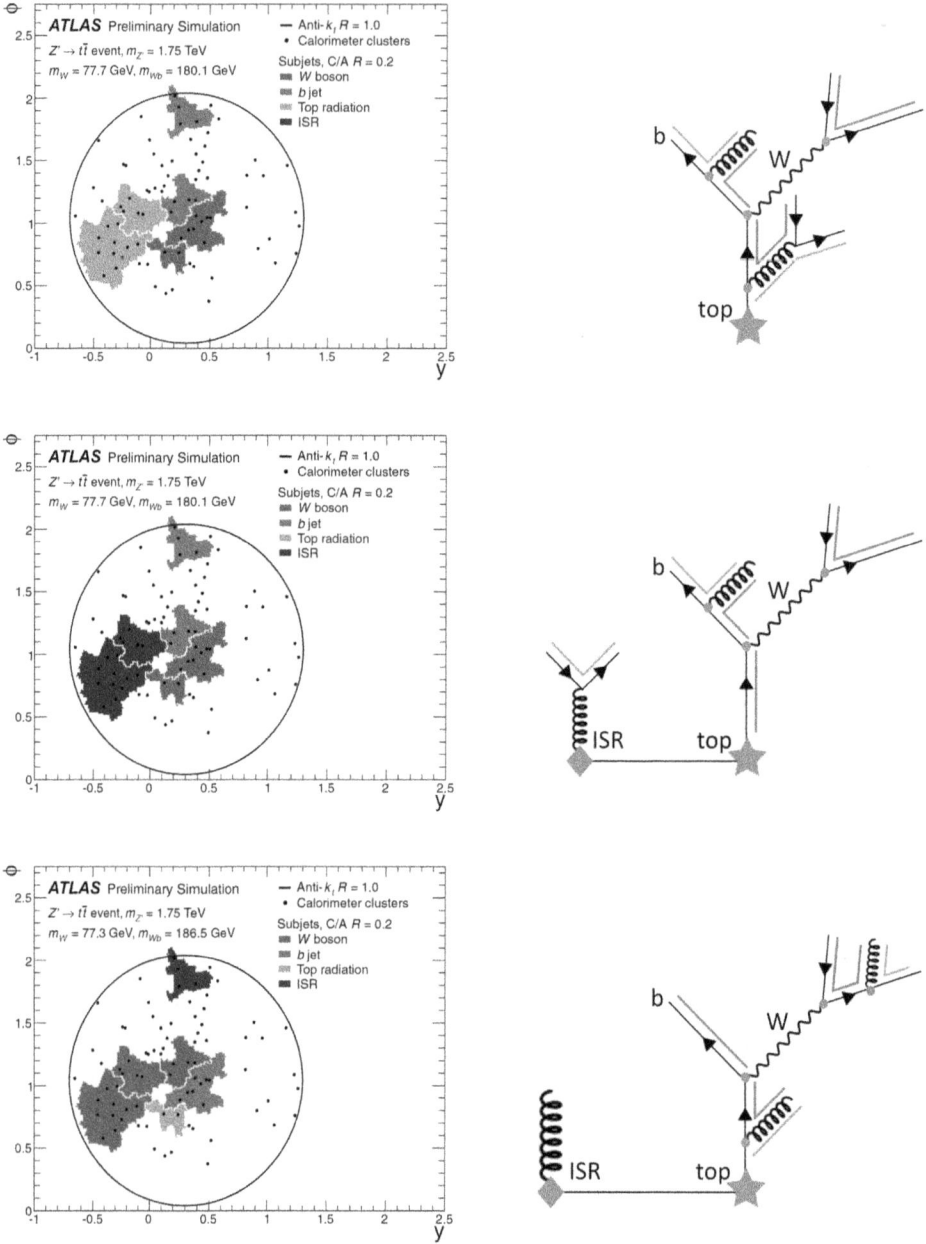

Figure 8.13. Illustrations of shower histories from a top quark initiated large-radius jet from a simulated event [30]. The left panel shows the event displays, with the right panel presenting the corresponding histories. On the left panel diagrams, subjets assigned to particular objects have the same fill colour and their size represents the subjet active catchment area. The red deposits correspond to jets from W boson decay, the green to b jets, the olive to jets from gluon radiation off the top quark, and the blue to jets from initial state radiation. On the corresponding right panel diagrams, jet constituents are shown as black dots, the hard scatter is indicated as the (red) star, initial state emissions by diamonds, and parton emissions are indicated by filled circles. Coloured straight lines represent the colour flow (ATLAS Experiment © 2018 CERN).

from the top quark, and unrelated initial state radiation (ISR). To obey the signal hypothesis, the W boson can only be constructed from two (or more, including parton radiation from the W boson) subjets, where the combined mass is close to the actual W boson mass. Similarly the combination of W boson and the bottom quark, plus any parton radiation associated with the top or the bottom quark should give a mass close to the actual top quark mass.

The left panel shows the energy deposits inside the large-radius jet assigned to specific origins (i.e. possible shower histories), with the right panel showing the corresponding histories themselves, satisfying the above constants from the signal hypothesis. The three histories shown (and many others) satisfy the above mentioned constraints, and represents how different subjets can be assigned to different objects. The same process is performed for background hypothesis as well. For a top quark tagging, which has a three body hadronic decay, at least three subjets are required to build the shower history. Also, a limit is applied on maximum number of subjets to be used, due to computational limitations. Subsequently, exclusively clustered subjets were also used as inputs to the algorithm [31], leading to an improvement in performance.

The algorithm requires setting some input parameters, which can be optimised. These are the top quark and W boson mass values, and their mass windows.

8.4 Experimental results using jet substructure

The discussion of techniques is incomplete without examples of their usage in actual analyses. As in the previous chapters, we do not intend this to be a summary of such results, rather present some results illustrating the techniques, including the examples mentioned at the beginning of the chapter. Although we focus on the searches, measurement of substructure observables is also a critical aspect, as that feeds into development of taggers, and improvement in modelling some aspects of MC generators (FSR, fragmentation and hadronisation) sensitive to such distributions.

The primary consideration for any search using large-radius jets is to decide which tagging algorithm to use. ROC curves are used to compare signal efficiency and background rejection of different tagging algorithms. The choice in a particular analysis is driven by the need to find a balance between high signal efficiency and high signal purity. Also, it is critical to have reasonable systematic uncertainties on the tagger in order for it to be used. So based on all these considerations, a specific tagger and a working point of the tagger (which would correspond to the cut on the tagging variable for signal to background discrimination in the analysis) is chosen. Now we must note that this approach is like a moving target, improvements and optimisation of tagging algorithms, introduction of newer methods, better estimation of systematic uncertainty on a tagger can lead to change of preference on what tagging algorithm to use. So rather than showing the state-of-the-art ROC curves, (which probably has become obsolete already!), we show ROC curves for top tagging from ATLAS Run 1 in figure 8.14, as an example. The ROC curves are made in different large-radius jet p_T ranges. Since we prefer higher signal efficiency

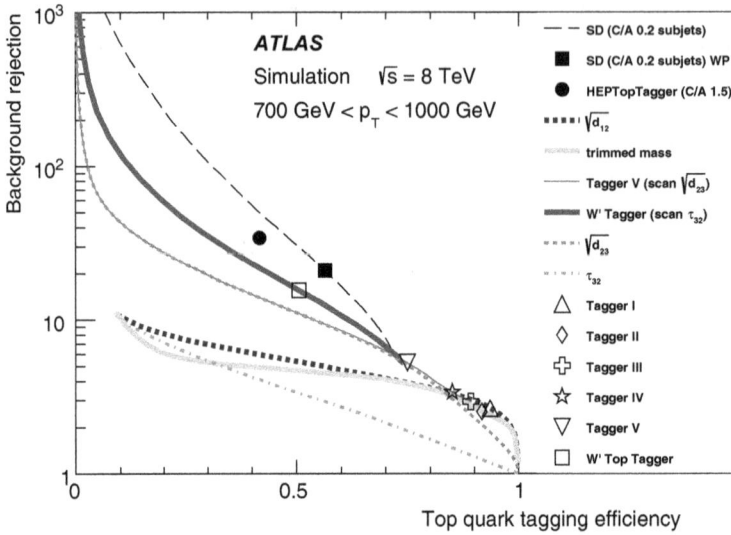

Figure 8.14. An example plot showing ROC curves for various algorithms showing the background rejection as a function of the tagging efficiency of trimmed large-radius jets using simulation [14] (ATLAS Experiment © 2018 CERN). The shower deconstruction line can be judged to have the best performance in this example.

(more to the right along the x-axis), and lower background (higher along the y-axis), the diagonally upward direction represents the best compromise.

Along with the ROC curves, it is imperative to see the discriminating power of the large-radius jet observables in signal and background samples. The use of large-radius jets in searches are predominantly in tagging hadronically decaying top quark and $W/Z/$ Higgs bosons. In figure 8.15, the mass of the large-radius jets, and D_2 and τ_{21} distributions are shown. The characteristic mass peaks for Higgs boson and top quark can be seen, as well as the lower mass peak from multijet background. The discrimination power increases with higher p_T. Individually D_2 and τ_{21} do not offer as much discriminating power as mass in this case, but observables can be combined.

In the previously mentioned all hadronic W' search in ATLAS, shower deconstruction algorithm is used for tagging boosted top quarks. Agreement between data and MC predictions for the relevant tagging observable is important, and in figure 8.16 (top left), $\log \chi$ distribution was seen to be well modelled in data. To improve the performance of SD algorithm, subjets were constructed with an exclusive k_t algorithm with a splitting scale cut-off of 15 GeV. The uncertainty was estimated by varying the p_T of the subjets by 2.5%, which was found to cover any data/simulation differences in the $\log \chi$ distribution. Then the invariant mass of the top and bottom quark system is shown in one of the signal regions (one b-tagged small-radius jet, one top-tagged large-radius jet with 80% efficiency WP) in figure 8.16 (top right).

Moving onto boosted boson tagging, diboson resonance is one of the early adopters of large-radius jets to reconstruct hadronically decaying WW, WZ or ZZ pairs. The dijet mass distribution shown in figure 8.16 (bottom left) was obtained by using D_2 to distinguish signal over multijet background. The value of D_2 cut and

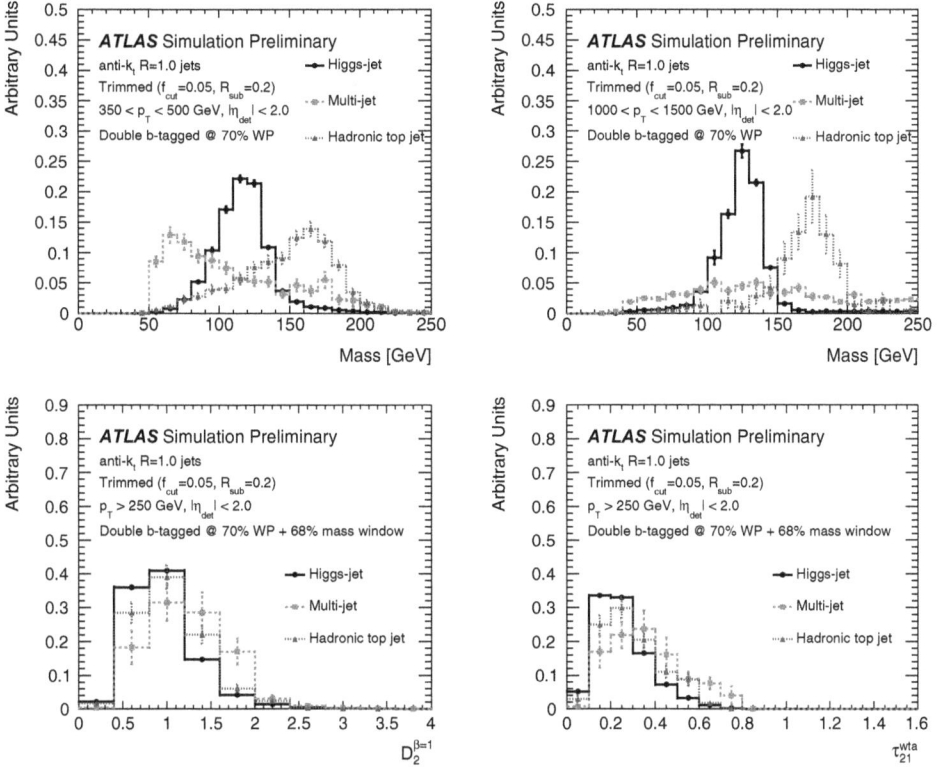

Figure 8.15. Comparisons of distributions of large-radius jet mass in low (top left) and high (top right) p_T ranges, D_2 (bottom left) and τ_{21} (bottom right) from top quark, Higgs boson, and multijet background events after event selection requirements using simulation [32] (ATLAS Experiment © 2018 CERN).

mass windows were simultaneously optimised, depending on large-radius jet p_T to achieve the maximal background-jet rejection for a fixed W or Z signal-jet efficiency of 50%. Additionally, to reject gluon initiated jets, the number of tracks associated with each jet were required to be less than 30. This is an example of an IRC unsafe observable that was nevertheless used.

For tagging Higgs bosons decaying to $b\bar{b}$, the large-radius jets need to be b-tagged. In ATLAS this is performed by requiring (one or two) b-tagged anti-k_t $R = 0.2$ track-jets to be associated with the large-radius jet. Here the signal events are chosen based on b-tag and the large-radius jets within mass window consistent with that of the Higgs boson. In figure 8.16 (bottom right), the correlation of leading and subleading large-radius jet masses are shown, where both the jets have exactly one b-tag. The signal region, shown by the inner (red) dashed line, is lightly populated with the current statistics. Increasing the required number of associated b-tagged track-jets in the event increases signal purity at the expense of lower signal efficiency.

Finally, two event displays are shown in figure 8.17, corresponding to a boosted top quark pair production in semileptonic decay mode, and boosted diboson

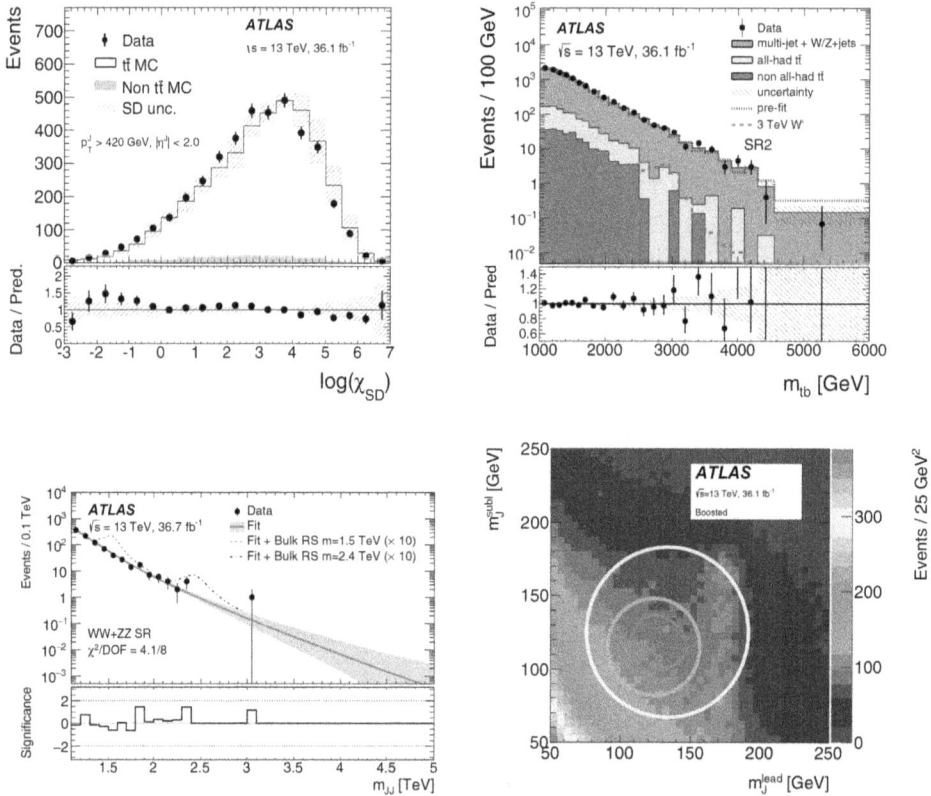

Figure 8.16. Distributions illustrating use of jet substructure techniques in new physics searches. Comparison of shower deconstruction $\log \chi$ between data and simulation (top left), and between data and SM background after top tagging the large-radius jet in a signal region (top right) in all hadronic W' search [31]. The dijet mass distribution using two boson-tagged large-radius jets in diboson resonance search [33] (bottom left) and a two-dimensional distribution showing the correlation between the masses of leading and subleading large-radius jets in di-Higgs to four bottom quarks search [34] (ATLAS Experiment © 2018 CERN). In the latter case, the circular rings indicate signal, control and side-band regions.

production. The hadronically decaying top quark (formed by re-clustering the three shown small-radius jets) had a transverse momentum of about 600 GeV and a mass of about 180 GeV. The leading and subleading jet mass and p_T in second case were 83.7 GeV, 1.26 TeV, 73.9 GeV and 1.15 TeV respectively, resulting in the dijet invariant mass of 2.4 TeV. Both the jets also had values of D_2 consistent with two-prong substructure.

8.5 Miscellaneous theoretical and experimental aspects

8.5.1 Ambiguity of jet identification based on origin

So far in this chapter, the focus has been hadronic decays of heavy particles. Another active area is trying to distinguish jets originating from light quarks and gluons. This is more relevant for small-radius jets, but many of the same observables (such as

Figure 8.17. Event display of a boosted top quark pair event (left) and a boosted dijet event (right) (ATLAS Experiment © 2018 CERN).

angularities) are used. Jets originating from gluons radiate more, so typically the jets have more softer particles spread wider, compared to a jet originating from a light quark. However, as was shown in [24], the definition of a quark or a gluon jet is fundamentally ambiguous, as shown in figure 8.18. While identifying the jet by the initiating parton is most intuitive, it ignores the effect of additional radiation. The definition of born-level process is similarly ambiguous, and tying the definition into parton shower restricts us to leading-logarithmic accuracy. Going further along, the penultimate definition ignores the effect of hadronisation, so the suggestion was to use an experimentally well-defined event selection.

This inherent ambiguity in defining a jet by its origin manifests in another situation. The comparison of different taggers, or tagging variables to discriminate large-radius jets from an intended signal against background is performed by drawing ROC curves. In most cases, they are constructed by using simulated samples for both signal and background (occasionally a pure sample of data multijet events is used for background), as it is almost impossible to obtain a pure sample of signal events in data. This means that to obtain signal efficiency, the denominator has to be the total number of large-radius jets originating from the heavy signal particle, or in other words, the origin of the large-radius jet needs to be uniquely ascertained. Even when detector level simulation is used to make the ROC curves, calculating the denominator involves what is referred to as *truth-matching*, where the corresponding particle level jet is located for each detector level jet, and the origin of the particle level jet is assigned as a *label* of the detector level jet.

8.5.2 Calibration and uncertainties

The large-radius jets need to be calibrated like regular jets, and JES and JER must be determined. Additionally, jet mass scale (JMS) and jet mass resolution (JMR), defined in the same spirit are important here, as large-radius jet mass is an important observable. Different methods are used, depending on the experimental setup and needs. The usual *in situ* calibrations using Z+jets and multijet events are used, as long as they go high enough in p_T (this is rather important as, in 2015, ATLAS initially saw a $>3\sigma$ excess at diboson invariant mass of 2 TeV using boson-tagged jets, which was later pinned down to using calibrations derived at lower p_T to higher

What is a Quark Jet?

From lunch/dinner discussions

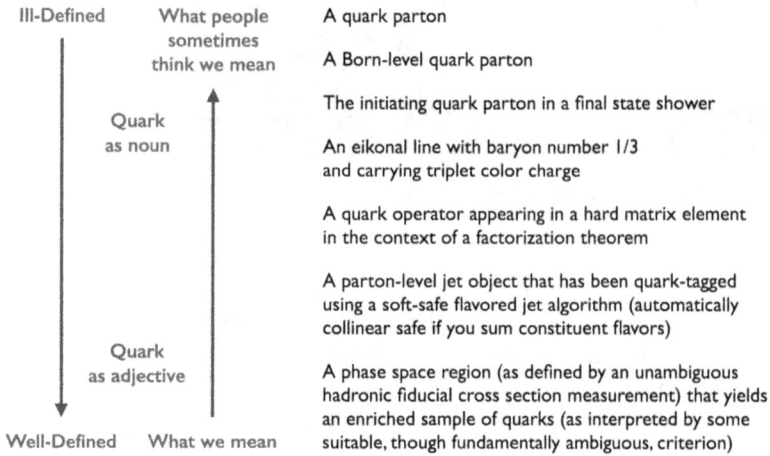

Ill-Defined   What people   A quark parton
sometimes
think we mean   A Born-level quark parton

The initiating quark parton in a final state shower

Quark
as noun   An eikonal line with baryon number 1/3
and carrying triplet color charge

A quark operator appearing in a hard matrix element
in the context of a factorization theorem

A parton-level jet object that has been quark-tagged
using a soft-safe flavored jet algorithm (automatically
collinear safe if you sum constituent flavors)

Quark
as adjective   A phase space region (as defined by an unambiguous
hadronic fiducial cross section measurement) that yields
an enriched sample of quarks (as interpreted by some
Well-Defined   What we mean   suitable, though fundamentally ambiguous, criterion)

Figure 8.18. A schematic diagram showing different possible definitions when classifying a jet based on its quark or a gluon origin [24]. This was the summary of an after dinner discussion in Les Houches workshop in 2015. © Philippe Gras, Stefan Höche, Deepak Kar, Andrew Larkoski, Leif Lönnblad, Simon Plätzer, Andrzej Siódmok, Peter Skands, Gregory Soyez and Jesse Thaler.

p_T. However, that was perhaps the first time jet substructure techniques in searches was discussed prominently).

In ATLAS, the R_{trk} method is used to derive uncertainties on mass and p_T calibrations [35]. The uncertainty is obtained based on double ratios between data and MC for that observable, calculated with clusters and tracks as inputs. For an observable X, the uncertainty R_{trk} is calculated as a function of the large-jet p_T.

$$R_{trk} = \left(\left(r_{trk}^X \right)^{data} \right) \Big/ \left(\left(r_{trk}^X \right)^{MC} \right)$$

$$r_{trk}^X = X_{calo} / X_{track}$$

A relatively new *bottom-up* method is being used in ATLAS to obtain uncertainties on the overall shape and scale of the substructure variables. Variations, corresponding to cluster reconstruction efficiency, cluster energy scale, cluster energy smearing, and cluster angular resolution are applied directly to input calorimeter clusters, based on [36]. This approach does not depend on tracks, potentially allowing use of large-radius jets beyond the tracker acceptance.

8.5.3 Mass sculpting and variable radius jets

The usual practice is to select large-radius jets within a certain range of values of a tagging variable, which is expected to be dominated by signal-like jets. However, no tagger is perfect, and that selection will also pick up a certain fraction of large-radius jets not coming from the signal. It has been observed, and theoretically explained, that the selected large-radius jets from background ends up having a similar (but artificial) peak-like mass distribution as the signal. This feature is referred to as *sculpting*, and it leads to difficulties in estimation of background under the signal mass peak. This sculpting depends strongly on the p_T range.

Among the approaches proposed to address this, designing decorrelated taggers [37, 38] (or DDT, the reference to professional wrestling not to be missed!) was the pioneer. By introducing a variable $\rho = \log(m^2/p_T^2)$, and identifying the ranges of ρ values, the tagging observables can be made uncorrelated with the groomed jet mass. A shift in the scale of ρ removed most of the p_T dependence.

Another concept worth mentioning is variable radius (VR) jets [39, 40]. Use of fixed values of the radius parameter of jets ignores the dependence of the jet radius on particular physics processes (as $R = 2\, m/p_T$). Shrinking the jet radius with increasing p_T for example can result in less contribution from pile-up and underlying event in the jet, which can improve the tagging performance. The VR jet reconstruction proceeds with a radius parameter R_{eff}, which is a function of the p_T, given by $R_{\text{eff}}(p_T) = \mu/p_T$, where μ is a scaling parameter, tuned to obtain maximum signal efficiency from simulation. The radius is restricted between R_{max} and R_{min}, which prevent the jets from becoming too large at lower p_T, and from shrinking below the detector resolution at high p_T. However, since jet calibrations in experiments are typically performed only for some fixed values of the radius parameter, the use of VR jets necessitates the use of track-jets, or some dedicated calibration procedures.

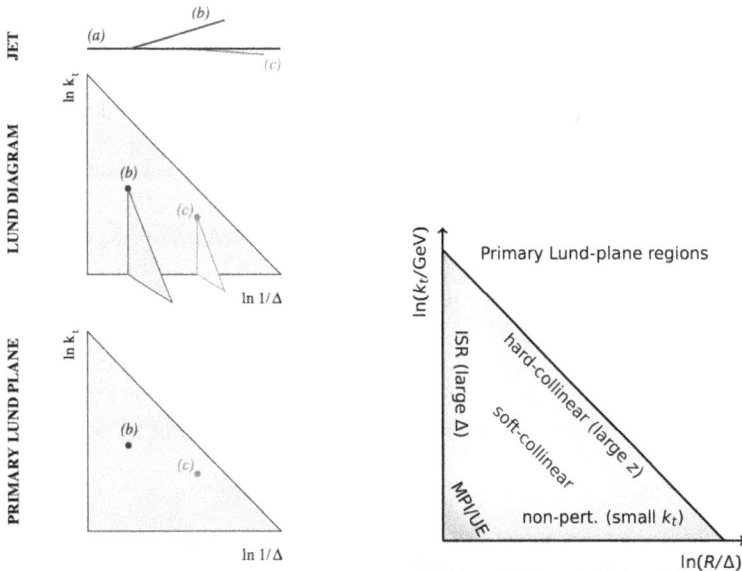

Figure 8.19. Schematic illustration showing construction of Lund plane for a jet (left), and of different regions of the Lund plane sensitive to different aspect (right). Figure from [39]. © F A Dreyer, G P Salam and G J Soyez.

8.5.4 Lund plane

Lund plane has recently been popularised [41] as a tool for representing the emissions inside a jet. The concept evolved from Lund diagrams [42], which show the transverse momentum and the angle of any given emission with respect to its emitter in a two-dimensional (logarithmic) plane. In figure 8.19 (left), a simple jet with two emissions (b, c) from the initial particle (a) is shown at the top, with the direction and length of each line segment schematically representing the direction and scalar momentum of the emitted particles. The middle figure is the corresponding Lund diagram(s), where the x-axis of each triangle represents the $\ln k_t$, and the y-axis represents $\ln 1/\Delta$ for that emission. The k_t and Δ are, respectively, the transverse momentum and the angle of the emitted particle. If any of the emitted particles emitted another particle, it would lead to a secondary Lund diagram. The Lund plane is constructed only from the primary radiation, as shown in the bottom figure. The triangular area is populated by a point corresponding to each primary radiation.

Different regions of the Lund plane of a jet are sensitive to different origins of jet constituents, as shown in figure 8.19 (right). Soft collinear radiations are emitted uniformly in the whole plane, whereas for example large angle but soft radiation from MPI will be concentrated near the origin of the triangle. Substructure based tagging algorithms can be interpreted as cuts in the Lund plane.

Constructing a Lund plane for each jet is not straightforward, as the final jet constituents do not directly give access to the angles and transverse momenta of each primary splitting. Working backward via the C/A jet clustering to identify the emissions is used to construct the plane.

References

[1] Abdesselam A *et al* 2011 Boosted objects: A probe of beyond the Standard Model physics *Eur. Phys. J.* C **71** 1661

[2] Altheimer A *et al* 2012 Jet substructure at the tevatron and LHC: new results, new tools, new benchmarks *J. Phys.* G **39** 063001

[3] Altheimer A *et al* 2014 Boosted objects and jet substructure at the LHC. Report of BOOST2012, held at IFIC Valencia, 23rd-27th of July 2012 *Eur. Phys. J.* C **74** 2792

[4] Asquith L *et al* 2018 Jet substructure at the Large Hadron Collider: experimental review arXiv: 1803.06991

[5] Aad G *et al* 2013 Performance of jet substructure techniques for large-R jets in proton-proton collisions at $\sqrt{s}$ = 7 TeV using the ATLAS detector *JHEP* **09** 076

[6] Seymour M H 1994 Searches for new particles using cone and cluster jet algorithms: A Comparative study *Z. Phys.* C **62** 127–38

[7] Butterworth J M, Cox B E and Forshaw J R 2002 W W scattering at the CERN LHC *Phys. Rev.* D **65** 096014

[8] Butterworth J M, Davison A R, Rubin M and Salam G P 2008 Jet substructure as a new Higgs search channel at the LHC *Phys. Rev. Lett.* **100** 242001

[9] Krohn D, Thaler J and Wang L-T 2010 Jet trimming *JHEP* **02** 084

[10] Ellis S D, Vermilion C K and Walsh J R 2010 Recombination algorithms and jet substructure: Pruning as a tool for heavy particle searches *Phys. Rev.* D **81** 094023

[11] Dasgupta M, Fregoso A, Marzani S and Salam G P 2013 Towards an understanding of jet substructure *JHEP* **09** 029

[12] Larkoski A J, Marzani S, Soyez G and Thaler J 2014 Soft drop *JHEP* **05** 146

[13] Aaboud M *et al* 2018 Measurement of the soft-drop jet mass in pp collisions at $\sqrt{s} = 13$ TeV with the ATLAS detector *Phys. Rev. Lett.* **121** 092001

[14] Aad G *et al* 2016 Identification of high transverse momentum top quarks in *pp* collisions at $\sqrt{s} = 8$ TeV with the ATLAS detector *JHEP* **06** 093

[15] Thaler J and Van Tilburg K 2011 Identifying boosted objects with N-subjettiness *JHEP* **03** 015

[16] Thaler J and Van Tilburg K 2012 Maximizing boosted top identification by minimizing N-subjettiness *JHEP* **02** 093

[17] Larkoski A J, Neill D and Thaler J 2014 Jet shapes with the broadening axis *JHEP* **04** 017

[18] Salam G P, Schunk L and Soyez G 2017 Dichroic subjettiness ratios to distinguish colour flows in boosted boson tagging *JHEP* **03** 022

[19] Jankowiak M and Larkoski A J 2012 Angular scaling in jets *JHEP* **04** 039

[20] Larkoski A J, Salam G P and Thaler J 2013 Energy correlation functions for jet substructure *JHEP* **06** 108

[21] Larkoski A J, Moult I and Neill D 2014 Power counting to better jet observables *JHEP* **12** 009

[22] Moult I, Necib L and Thaler J 2016 New angles on energy correlation functions *JHEP* **12** 153

[23] Larkoski A J, Thaler J and Waalewijn W J 2014 Gaining (Mutual) information about quark/gluon discrimination *JHEP* **11** 129

[24] Gras P, Höche S, Kar D, Larkoski A, Lönnblad L, Plätzer S, Siódmok A, Skands P, Soyez G and Thaler J 2017 Systematics of quark/gluon tagging *JHEP* **07** 091

[25] Plehn T and Spannowsky M 2012 Top Tagging *J. Phys.* G **39** 083001

[26] HEPTopTagger: *one slide description* https://twiki.cern.ch/twiki/pub/CMSPublic/PhysicsResultsJME13007/HEP_one_slide_description.pdf (accessed: 2018-12-30)

[27] Soper D E and Spannowsky M 2011 Finding physics signals with shower deconstruction *Phys. Rev.* D **84** 074002

[28] Soper D E and Spannowsky M 2013 Finding top quarks with shower deconstruction *Phys. Rev.* D **87** 054012

[29] de Lima D F, Petrov P, Soper D and Spannowsky M 2017 Quark-gluon tagging with shower deconstruction: unearthing dark matter and Higgs couplings *Phys. Rev.* D **95** 034001

[30] ATLAS Collaboration 2014 Performance of shower deconstruction in ATLAS 2014 *Technical Report* ATLAS-CONF-2014-003, CERN, Geneva http://cdsweb.cern.ch/record/1648661

[31] Aaboud M *et al* 2018 Search for $W' \to tb$ decays in the hadronic final state using pp collisions at $\sqrt{s} = 13$ TeV with the ATLAS detector *Phys. Lett.* B **781** 327–48

[32] ATLAS Collaboration 2015 Expected Performance of Boosted Higgs ($\to b\bar{b}$) Boson Identification with the ATLAS Detector at $\sqrt{s} = 13$ TeV *Technical Report* ATL-PHYS-PUB-2015-035, CERN, Geneva http://cdsweb.cern.ch/record/2042155

[33] Aaboud M *et al* 2018 Search for diboson resonances with boson-tagged jets in *pp* collisions at $\sqrt{s} = 13$ TeV with the ATLAS detector *Phys. Lett.* B **777** 91–113

[34] Aaboud M *et al* 2019 Search for pair production of Higgs bosons in the $b\bar{b}b\bar{b}$ final state using proton-proton collisions at $\sqrt{s} = 13$ TeV with the ATLAS detector *JHEP* **01** 030

[35] ATLAS Collaboration 2019 In situ calibration of large-radius jet energy and mass in 13 TeV proton–proton collisions with the ATLAS detector *Eur. Phys. J. C* **79** 135

[36] ATLAS Collaboration 2017 A measurement of the calorimeter response to single hadrons and determination of the jet energy scale uncertainty using LHC Run-1 *pp-collision* data with the ATLAS detector *Eur. Phys. J. C* **77** 26

[37] Kasieczka G, Plehn T, Schell T, Strebler T and Salam G P 2015 Resonance searches with an updated top tagger *JHEP* **06** 203

[38] Dolen J, Harris P, Marzani S, Rappoccio S and Tran N 2016 Thinking outside the ROCs: Designing decorrelated taggers (DDT) for jet substructure *JHEP* **05** 156

[39] Krohn D, Thaler J and Wang L-T 2009 Jets with variable R *JHEP* **06** 059

[40] Nachman B, Nef P, Schwartzman A, Swiatlowski M and Wanotayaroj C 2015 Jets from jets: Re-clustering as a tool for large radius jet reconstruction and grooming at the LHC *JHEP* **02** 075

[41] Dreyer F A, Salam G P and Soyez G 2018 The Lund jet plane *JHEP* **12** 064

[42] Andersson B, Gustafson G, Lonnblad L and Pettersson U 1989 Coherence effects in deep inelastic scattering *Z. Phys. C* **43** 625

Chapter 9

Advanced topic: machine learning

Deep Learning inspires Deep Thinking.

—Jesse Thaler

Remarkable advances in computing power in recent decades have opened up the field of particle physics to the use of so-called machine learning (ML) methods. ML is an emerging and active area of research in its own right, and the applications span diverse academic and commercial domains. This chapter is intended to give a general overview of ML in experimental particle physics, but for an in-depth discussion of the methods, the reader is referred to the dedicated resources [1, 2].

We start by introducing multivariate analysis and neural networks in sections 9.1 and 9.2, and conclude with a brief overview of the application of ML methods in particle physics in section 9.3.

9.1 Precursor: multivariate analyses

As we have discussed before, a main focus in the analyses is signal to background discrimination, or in more general terms, classification of events or objects according to their origin. The simplest approach is based on applying *selection cuts* on different variables, such as on four-momentum of particles, or on the kinematic quantities obtained from them, and on observables indicative of global event characteristics. This is often referred to as *rectangular cuts*, as thresholds applied on each variable are independent of each other.

This is easy to visualise for a two variable case as in figure 9.1, where fixed cut values for the two variables lead to a rectangular area delimiting signal and background points. However, this ignores correlation between the variables, which implies the classification is not fully optimal. Rather than rectangular cuts, a linear boundary, or even a non-linear one accounting for the correlations can result in a better classification performance. While for the two-dimensional case shown here, it

doi:10.1088/2053-2563/ab1be6ch9

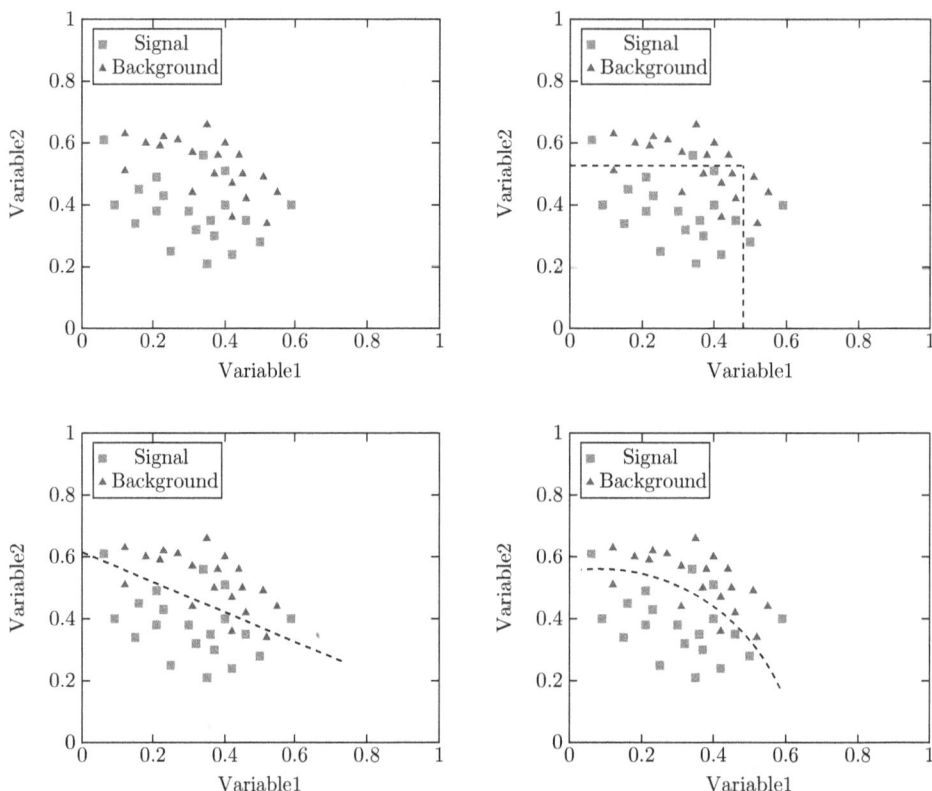

Figure 9.1. Illustration of separating signal and background (top left) with rectangular cuts (top right), a linear decision boundary (bottom left), and a non-linear decision boundary (bottom right). Our target is to include as many of the 20 signal events, and as few of the 20 background events in an area. The rectangular area includes 16 signal and 7 background events, the linear cut improves it to 17 and 4, and the non-linear cut further to 17 and 3.

does not look too daunting, actual analyses involve many more correlated variables, so selecting the optimal *decision boundary* is the main challenge.

Multivariate analysis (MVA) [3–5] is a general class of techniques designed for this purpose. The N variables which are used in classification are referred to as *feature variables*, which form the N-dimensional feature space. Correlations between the variables effectively reduce the dimensionality of the problem. The specific objects or events of interest (i.e. signal) are expected to occupy specific continuous regions in this feature space. Then MVA algorithms construct a constant surface in this N-dimensional feature space, accounting for the correlations, corresponding to the decision boundary. This represents a mapping from N- to one-dimensional output variable, then a cut on this new variable is applied. The actual cut threshold will of course be determined by the usual compromise of efficiency and misidentification rates preferred in that particular analysis.

In most cases this is preceded by a preprocessing step, often referred to as *feature extraction*. The initial set of variables may not always reflect the distinguishing features of the signal well, so combining them to construct physics-motivated

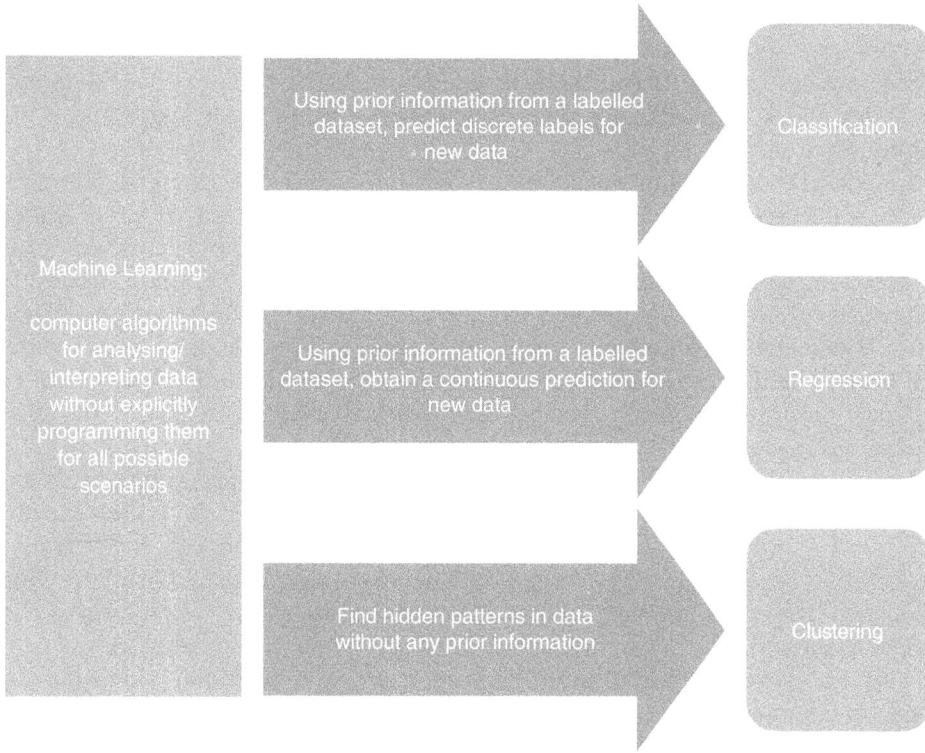

Figure 9.2. A schematic diagram illustrating the basic features of commonly used machine learning methods.

variables is a common approach. The construction of transverse mass in W boson events can be considered as an example. Other transformations such as scaling certain variables (for example by $\sqrt{s}$) are also common. Optimal discrimination, most simply, is a procedure that minimises the probability of misclassification.

9.2 Machine learning in a nutshell

Machine learning in some sense is a generalisation of the MVA technique described above, where the *learning* happens *automatically* (i.e. without being explicitly programmed) from *data*. The term data is used loosely here, to include simulation as well, as will become clear as we go along. The origin and development of ML methods is intricately tied to the field of artificial intelligence (AI).

The aim here is to start with the data, and guess an approximate function $f(x) \to y$ relating the input variables x and the output variables y. The functional form, or whether or not an underlying functional form exists at all is not *a priori* known. Once that form is inferred (which represents the *learning* component in machine learning), it can be used to predict *future* behaviour for similar datasets by estimating the output variable y' for new input data x'.

Machine learning methods are broadly classified into supervised, unsupervised, and semi-supervised learning, shown schematically in figure 9.2. which will be discussed again in section 9.2.4. Classification and regression are two common approaches, which are categorised under the supervised machine learning, where predictions are made utilising known datasets (referred to as training datasets). Classification is applicable for discrete output variables, while regression is for continuous output variables.

Classification algorithms predict the category or *label* for a given input. When the output category is restricted to two choices, it is termed a binary classification problem, and multi-class classification problem otherwise. Marking emails as spam or not-spam may be thought of as an example of binary classification, where the training involved scanning through many emails already marked as spam or not-spam. The structure of the email, or the occurrence and frequency of certain words will be used by the algorithm in this case to make the classification. A classification accuracy can be defined as the percentage of correctly classified examples out of all predictions made.

Regression algorithms are designed to output continuous real values, with real valued or discrete input variables. For multiple input variables, it is often called a multivariate regression algorithm. An example of regression can be predicting temperature based on historical weather data. Root mean squared error (RMSE) calculated from each output point can be taken as the measure of performance or *skill* of the algorithm.

There is, however, some overlap between them. A classification algorithm can predict a continuous value, but the continuous value then may be interpreted as a probability, which can be translated to a class label. For example in particle physics, identifying jets based on their origin, or categorising events into signal and background almost always results in a continuous output variable, with ranges corresponding to desired labels. When we use simulation, we automatically have labelled data. A regression algorithm may predict a discrete value in the form of an integer quantity.

Unsupervised learning is where there are no outputs y, so the idea is to find patterns in the input x, and then it is up to the user to interpret the patterns. For example, real data events do not come with labels, so picking a potential new physics signal event from a large number of background events will need unsupervised learning. Clustering is a commonly used method, where the data are portioned into groups based on some characteristics.

9.2.1 Decision trees

Decision trees [6] are used in classification problems. The idea is essentially to apply a sequence of criteria to evaluate features in data in order to arrive at a final decision corresponding to a selection of a class of events, thereby minimising contamination from any other classes. Let us start with a simple example, we want to determine from a set of events with two electrons, which events are coming from a decay of a Z boson, as shown in figure 9.3. Our first test can be to check if the two electrons have opposite charge, which is represented by a diamond-shaped decision node. Two possible outcomes are represented by two lines or branches. Since the case of the same sign electrons automatically means they cannot come from decay of a Z boson,

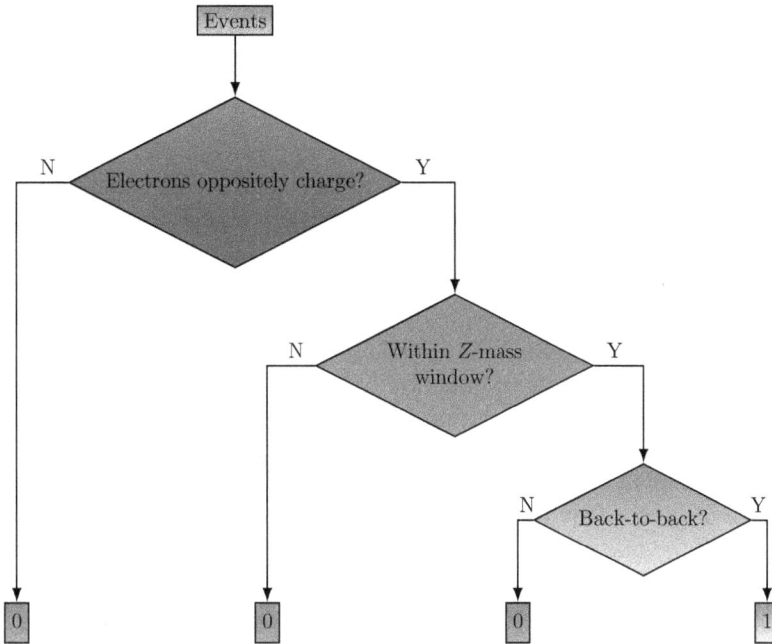

Figure 9.3. Illustration of a simple decision tree. The input node is shown in brown, the decisions nodes in blue, with progressively lighter shades indicating weaker classifiers. The final leaves are indicated in green and red depending on desired or rejected outputs.

we have a decision, represented by a square leaf with zero. The other branch leads to another decision node, where the test can be on the invariant mass of the two leptons. Let us say we require the invariant mass to be within 66–111 GeV for a Z boson, then again we have two branches going out from this node. In case we have another requirement, for example, in an inclusive production mode, the electrons need to balance each other, then we will have another decision node. If this is our final test, then we will have two final leaves, one of them corresponding to the desired output, marked one. This can be thought of as an upside down tree.

However, the obvious question is how do we know we should test for charge first followed by invariant mass? Also, while the charge criteria is obvious, how do we know our selected mass window is the best choice, and indeed how do we determine what choice is the best? While in this overtly simple example, these can be answered rather easily (charge is a stronger classifier than mass, and maybe this mass range only rejects a negligible fraction of simulated $Z \to \ell^+\ell^-$ events), in general for a dataset with more features and not-so-intuitive distinctions between categories, the answers are not *a priori* obvious. So in essence *growing* a tree involves deciding which features to choose, and what conditions to use for branching, along with knowing when to stop. This part is considered *training* the tree. Every decision node corresponds to an analysis cut. Trying different orders is equivalent to keeping events rejected by one criterion and see whether other criteria could help classify them properly.

We will give one example to illustrate the training algorithm, but many such methods exist. If we have N features in data, and subdivide the training data into N groups, starting with the different features as the first node. Now at each node, loss of accuracy due to the branching criteria can be ascertained. This is done using *cost functions*, which essentially test different branching possibilities in terms of which one leads to best separation. We will come back to the cost function in the next section, but this is a key concept in machine learning algorithms, and the aim is to always minimise the cost function.

In our example, for the invariant mass node, this can simply be how many Z boson events were missed, or if we are doing a signal to background discrimination, how many background events were accidentally selected. Then a simple squared difference can represent the cost function, or more commonly Gini score is used: $G = p * (1 - p)$, where p is the fraction of correctly categorised inputs by a decision node. So for $p = 1$ or $p = 0$, we get a perfect score, $G = 0$, whereas for $p = 0.5$, the worst decision, the score is $G = 0.25$. So for each of these N trees, the sum of cost functions for all nodes (for all combinations, varying the branching criteria) can be calculated, and the one giving the least value can be picked. Essentially Gini scores indicate how the separation power of the branching is. This is commonly referred to as the Classification and Regression Trees (CART) approach.

Another way is to use entropy: $S = \sum_{possibilities} - p \log_2 p$, which would be zero for a perfectly classified case. Then information gain is defined as the difference of entropy after a branching into n outcomes by: $IG = S - \sum_n p(n)S(n)$, where $p(n)$ denotes the fraction of events in the nth branch as a fraction of total events, and $S(n)$ is its entropy. The tree corresponding to the most entropy gain will be the desired tree. This is referred to as the Iterative Dichotomiser 3 (ID3) approach.

In both cases, the higher up nodes have the stronger classification power, with the first node (root) as the best classifier. These algorithms are recursive in nature as the groups formed can be sub-divided using the same strategy. As the overwhelming focus is lowering the cost, these algorithms are also known as the greedy algorithm, The stopping criteria can be determined by requiring that a minimum fraction of inputs are classified as a certain outcome in a decision leaf, or a pre-set depth (i.e. number of nodes). Apart from binary classification as described here, multi-class classification trees can also be formed, as well as trees outputting continuous values.

Overfitting, i.e. creating complex trees overtly specific to the training data, but not best performing for a more general dataset is a problem for this approach. This is opposite to underfitting, where the training is stopped before reaching the best possible separation structure. In figure 9.4, the concepts are illustrated using the previous cut optimisation figure. While the exact details are different from different algorithms, the basic concept remains the same. Another problem is variance or instability, which means small variations in the data causing a completely different tree being generated. A few approaches are used to mitigate these problems:

- Pruning: the branches which make use of features with low importance are cut out, essentially restricting the tree to a certain depth.

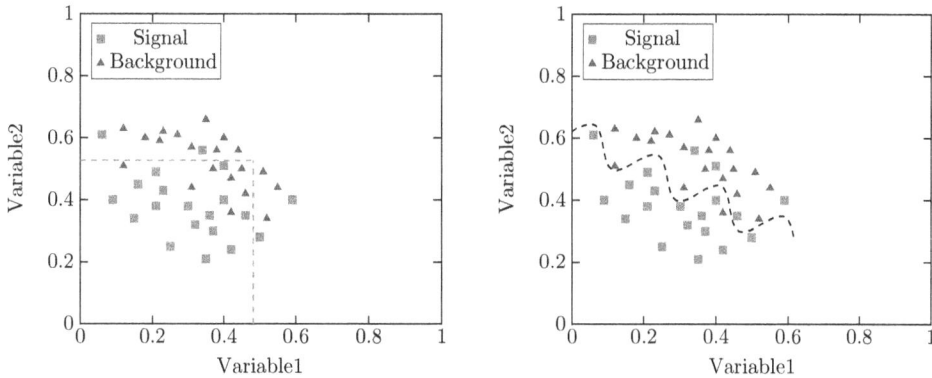

Figure 9.4. Illustration of underfiitting and overfitting using the example of figure 9.1. The linear boundary (left) can be considered a non-optimal signal to background separation, while the sinusoidal curve (right) probably has the best performance for this particular case, but it is so much tuned to this particular dataset, it may fail badly for a similar but non-identical dataset.

- Bagging (Bootstrap AGGregatING): the training is performed by *bootstrapping*, on multiple random sub-samples of the data with possible overlap. This results in individual trees having large variance, but the average is used as estimated prediction for the data. This has the advantage over using a single tree which may not address overlapping feature spaces.
- Random forest: this is an extension of bagging. Apart from a random subset of data, it also uses random selection of features rather than using all features. For each tree, again the feature among those selected giving the best classification is the first node.
- Boosting: here also an ensemble of trees is used, but sequentially. The first sets of trees are relatively simple, based only on a few features, which are not powerful discriminators individually. They are termed *weak learners*. Then their output is combined, trying to minimise the net error from the previous step. This can be achieved by increasing the weight of a misclassified input in the next step. This approach, which can be iterative, is known as Adaptive Boosting (AdaBoost). Another approach, termed Gradient Boosting, which involves fitting the residuals (i.e. the difference in cost functions).

9.2.2 Neural networks

To explain neural networks (NN), let us start with an overtly simplistic example of a dataset, where a linear relationship exists between two variables (x, y). If we already know of this linear relationship, then it is simple to fit the data with a functional form $y = \alpha x$, and determine the value of the fit parameter α by some fit convergence criteria. While for such a dataset, it is probably not difficult to determine an approximate functional form, for a dataset with large dimensionality, it is practically impossible. We will still illustrate the basic feature with this simple example, which will introduce the most salient concepts and terms.

Table 9.1. Example linear function input/output and loss function.

Input	Output	Target	Squared difference
0	0	0	0
1	2	3	1
2	4	6	4
3	6	9	9
4	8	12	16
Sum			30

Let us assume that the five pairs of data points have been generated with the functional form $y = 3x$, but *a priori* we do not know the functional form. We need to start with some functional form $y = f(w, x)$, where w represents a set of parameters of the function, referred to as *weights*, the reason for the terminology will become clear later. These are the parameters the algorithm is going to learn. These functions are referred to as *activation functions*. Usually a few types of activation functions are used, which will be discussed later. For the sake of simplicity, let us start with a linear activation function, $y = wx$. The weight w can be initialised randomly. Then the input x_i value is passed through this activation function (termed *forward propagation*), and the output y_i^w is obtained, which must be compared with the actual y_i values. The comparison can be done simply by summing the differences $y_i - y_i^w$, or of their absolute values, but usually the squared difference is taken, in order to amplify the effect of larger differences. This is termed as a *loss function* (analogous to the cost function defined earlier), as it quantifies how much we lose by using the current approximation. Then the value of w which would minimise the loss function will give us the desired functional form. While for our simple example, a brute-force scan of w values will give us the correct result, for functions with many weights, a general optimisation method is needed. Table 9.1 depicts the situation for our example, if we start with $w = 2$.

As the derivative of a function at a certain point gives the rate at which this function is changing its values at that point, a derivative of the loss function indicates what modification of the weight will help. For example, a change of $\delta w = 0.01$ will result in loss function of $(0.99)^2 + (1.98)^2 + (2.97)^2 + (3.96)^2 = 29.403$, corresponding to a rate of $\delta y / \delta w = -0.597/0.01 \approx 60$ (where $30 - 29.403 = 0.597$). Since the derivative is negative, it means the error decreases if we increase the weights, which is of course true in this case. If we decreased the weight by the same value, then the loss function will be the 30.603, but the derivative will be the same amount but positive, meaning the error increases if we increase the weights. This is shown in figure 9.5. A convex loss function like the mean squared one will always have a minima. The optimisation or *learning/training* therefore involves re-running the activation function again with the modified weight (termed *back propagation*), until a set minimisation criteria is reached.

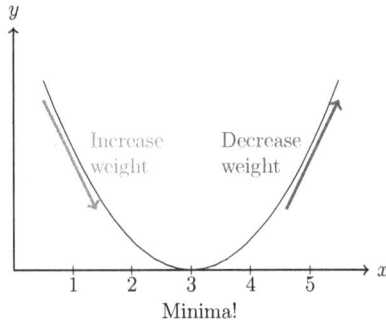

Figure 9.5. Illustration of how the loss function derivative is minimised for our simple example.

While for this case, the rate of change of loss function clearly indicates what change (increase 0.5 times) is needed to reach a zero value of loss function corresponding to $w = 3$, such a large change in the loss function is usually avoided to not encounter numerical instability (a derivative is only local at the point where we are calculating the derivative). The weights are updated by introducing a constant learning rate parameter, L: $w_{new} = w_{current} - L\frac{\delta J(w)}{\delta w}$, where $J(w)$ is the loss function. The learning rate determines how much we are adjusting the weights of our network with respect to the loss gradient, while approaching a minimum. If this rate is too large, the optimiser may keep overshooting a minimum and never converge. If it is too small, the network may take a very long time to converge or may select a local minima. Learning rate is an example of a *hyperparameter*, which are the parameters set in the algorithm (as opposed to the weights) independent of the training process.

This network represents what is termed a *neural network* (NN), with the line representing this application of the weight w, termed a *neuron*. In NN terminology, a neuron is essentially something which translates a set of inputs (using the activation with weights) to a single output.

In this example, we used only one neuron connecting the input and the output. Also, we had a one-dimensional input, corresponding to the input layer consisting of only one node. We also had a single node in the output layer. In general, for an N-dimensional input (i.e. with N features), we will have N input nodes in the input layer. Some configurations add one additional node for a bias term, which essentially corresponds to an intercept in the linear functions. Figure 9.6 shows an NN with three input nodes, and a bias node, with one output node, connected by four neurons.

The output can be succinctly represented by matrix multiplication:

$$[w_1 \quad w_2 \quad w_3]\begin{bmatrix} x_1 \\ x_2 \\ x_3 \end{bmatrix} + b$$

where w_1, w_2 and w_3 are the weights applied along each neuron by the chosen activation function, and b is the bias term.

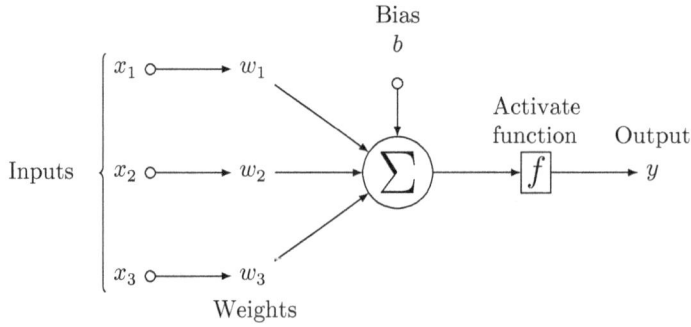

Figure 9.6. Simple neural network with three input nodes x_i, and one output node y, with an additional bias node b. The weights are denoted by w_i, and the activation function by f.

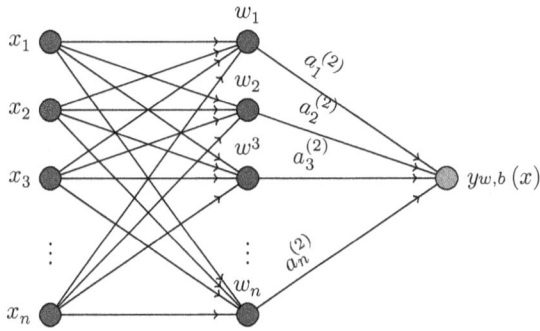

Figure 9.7. Illustration of a more complicated neural network.

A more complicated relationship between the input and output may not be represented by a simple linear mapping, rather by a series of functions, each represented by an additional layer in the neural network. These are termed *hidden layers* (because its values are not observed in the training set). One hidden layer is usually sufficient for the large majority of problems.

Determining the number of nodes n in the hidden layer is a non-trivial aspect, with no fixed rule. It is usually between the size of the input and size of the output layers. Using too few modes causes *underfitting*, where all the attributes in the data may not be captured. Using too many neurons results in *overfitting*, where the network effectively becomes a memory bank that can recall the training set to perfection, but does not perform well on samples that were not part of the training set. Also, it results in a prohibitively large training time. Figure 9.7 shows a generalised version of NN, with N input nodes, and one hidden layer with n nodes. The neurons connect each node with one another, such that the output of a node can be the input of another. The nodes are also referred to as neurons.

For this generalised case, let us represent the activation values (output of the activation functions f) of node i in layer l by a_i^l, which are of the form (w_{ij}^l, b_i^l), where the indices i and j correspond to the nodes in layer $l + 1$ and l respectively. In this notation, w_{ij}^l applies the weight corresponding to the forward propagation from node j in layer l and node i in layer $l + 1$. Bias nodes would not have input connections going into them, since they always output a constant value (but their weights are updated like other weights). Let us represent b_i^l as the bias associated with node i in layer $l + 1$. We can use x_i to denote the ith input, corresponding to node i in the input layer. Then mathematically for the output of the hidden layer:

$$a_1^{(2)} = f\left(w_{11}^{(1)}x_1 + w_{12}^{(1)}x_2 + w_{13}^{(1)}x_3 + \cdots + w_{1n}x_n + b_1^{(1)}\right) \tag{9.1}$$

$$a_2^{(2)} = f\left(w_{21}^{(1)}x_1 + w_{22}^{(1)}x_2 + w_{23}^{(1)}x_3 + \cdots + w_{2n}x_n + b_2^{(1)}\right) \tag{9.2}$$

$$a_3^{(2)} = f\left(w_{31}^{(1)}x_1 + w_{32}^{(1)}x_2 + w_{33}^{(1)}x_3 + \cdots + w_{3n}x_n + b_3^{(1)}\right) \tag{9.3}$$

$$\cdots$$

$$a_N^{(2)} = f\left(w_{N1}^{(1)}x_1 + w_{N2}^{(1)}x_2 + w_{N3}^{(1)}x_3 + \cdots + w_{Nn}x_n + b_N^{(1)}\right) \tag{9.4}$$

And the final output:

$$y_{w,b}(x) = f'\left(w_{11}^{(2)}a_1^{(2)} + w_{12}^{(2)}a_2^{(2)} + w_{13}^{(2)}a_3^{(2)} + \cdots + w_{1n}^{(2)}a_N^{(2)} + b_1^{(2)}\right)$$

where the activation function applied at the nodes of the hidden layer is denoted by f, and f' is the activation function of the output. These are matrices, so essentially the whole operation boils down to a series of linear operations.

The choice of activation functions is of critical importance. The activation function makes the neural network non-linear. They must be evaluated simply for forward propagation, and be analytically differentiable in order for back propagation. The commonly used ones are (shown in figure 9.8):

- Sigmoid, where $f(z) = \frac{1}{1 - \exp(-z)}$, with output range between 0 to 1.
- Hyperbolic tangent (tanh), where $f(z) = \tanh(z) = \frac{\exp(z) - \exp(-z)}{\exp(z) + \exp(-z)}$, with output range between -1 to 1.
- Rectified linear unit (ReLU), where $f(z) = \max(0, z)$. While it is not bounded, it is piecewise linear and saturates at exactly 0 whenever the input $z < 0$.

Creating the NN, therefore, means coming up with values for the number of layers of each type and the number of nodes in each of these layers, along with the choice of activation functions.

Let us consider another non-linear example, the so-called exclusive or (or XOR) logic gate, shown in table 9.2. It gives a binary output corresponding to two binary inputs.

We want to predict the output of this logic gate. It can be represented by a NN with one hidden layer,

$$w_1 = \begin{bmatrix} 1 & 1 \\ 1 & 1 \end{bmatrix} \qquad w_2 = \begin{bmatrix} 1 & -2 \end{bmatrix} \qquad b_1 = \begin{bmatrix} 0 \\ -1 \end{bmatrix} \qquad b_2 = \begin{bmatrix} 0 & 0 \end{bmatrix}$$

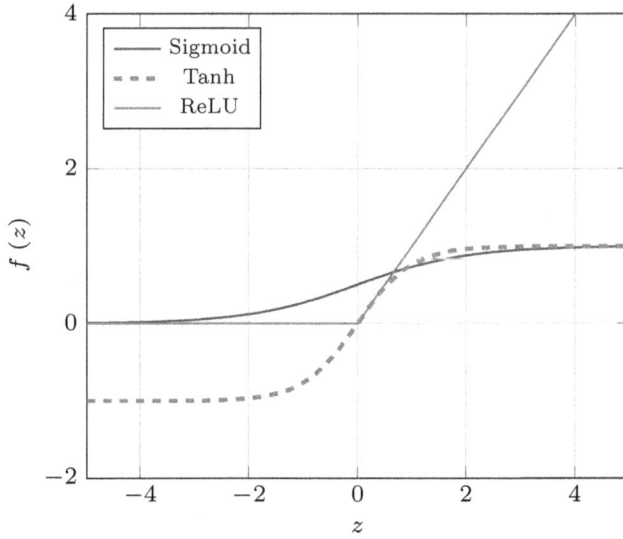

Figure 9.8. Examples of commonly used activation functions.

Table 9.2. Inputs and corresponding outputs for an XOR gate.

Input		Output
x_1	x_2	y
0	0	0
0	1	1
1	0	1
1	1	1

The activation function is actually a ReLU. This will act on any set of input value pairs, and give the corresponding output value. As an example, If we take the input as $x \equiv (0, 0)$, then applying the first activation function results in:

$$a_1 = w_1 x + b_1 = \begin{bmatrix} 1 & 1 \\ 1 & 1 \end{bmatrix} \begin{bmatrix} 0 \\ 0 \end{bmatrix} + \begin{bmatrix} 0 \\ -1 \end{bmatrix} = \begin{bmatrix} 0 \\ 0 \end{bmatrix}$$

Then the second activation function, after the hidden layer gives:

$$w_2 a_1 + b_2 = \begin{bmatrix} 1 & -2 \end{bmatrix} \begin{bmatrix} 0 \\ 0 \end{bmatrix} + \begin{bmatrix} 0 & 0 \end{bmatrix} = 0$$

Let us consider another case, where the input is $x \equiv (1, 0)$, then applying the first activation function results in:

$$a_1 = w_1 x + b_1 = \begin{bmatrix} 1 & 1 \\ 1 & 1 \end{bmatrix} \begin{bmatrix} 1 \\ 0 \end{bmatrix} + \begin{bmatrix} 0 \\ -1 \end{bmatrix} = \begin{bmatrix} 1 \\ 0 \end{bmatrix}$$

Then the second activation function in this case, after the hidden layer gives:

$$w_2 a_1 + b_2 = \begin{bmatrix} 1 & -2 \end{bmatrix} \begin{bmatrix} 01 \\ 0 \end{bmatrix} + \begin{bmatrix} 0 & 0 \end{bmatrix} = 1$$

Similarly, the other two cases can be shown to give the expected result. This also introduces us to another useful concept. Activation functions here can result in input being 0 or 1, which can also be considered *ON* and *OFF*. This is often designated as whether one neuron fired or not.

9.2.3 Types of neural networks

The previous architecture is part of a broad category called artificial neural networks (ANN). The particular example is often referred to as multi-layer perceptron (MLP) in literature. The depth of the network is determined by the number of hidden layers. While a network with no hidden layers can be categorised as a shallow neural network, neural networks with one or multiple hidden layers are termed deep neural networks (DNN). Here we will talk about two other types of neural networks, which are often used in particle physics. Again, this list is by no means exhaustive.

CNN: Convolutional neural networks (CNN) are used in image recognition and classification, a description of which is beyond the scope of this book. However, in the next section, we will mention how jets can be represented as images, so CNNs have application in particle physics. It derives its name from the convolution operator, which we can illustrate with a simple example. We consider that an image is represented very simply in a 4 × 4 matrix with each element representing a pixel, with binary entries (1 corresponding to that pixel being black, 0 corresponding to it being white):

$$\begin{bmatrix} 1 & 1 & 0 & 0 \\ 1 & 1 & 0 & 0 \\ 1 & 1 & 0 & 0 \\ 1 & 1 & 0 & 0 \end{bmatrix}$$

Then if we have a 3 × 3 matrix of the form:

$$\begin{bmatrix} 0 & 1 & 0 \\ 1 & -4 & -1 \\ 0 & 1 & 0 \end{bmatrix}$$

We can overlay this second matrix on the four possible 3 × 3 blocks of the first matrix, starting from top left corner, moving to top right corner, then to bottom left corner, and finally to bottom right corner. In each case we can do an element-wise multiplication, and then sum up the nine numbers, and use the resulting four numbers to form a 2 × 2 matrix:

$$\begin{bmatrix} -1 & 1 \\ -1 & 1 \end{bmatrix}$$

This new matrix amplifies the vertical edge feature of the first matrix. Of course, in this case, it is visually obvious, but for orders of magnitude higher dimensional matrices with even grey-scale images (where numbers 0–10 can indicate different shades of darkness), this operation can extract the vertical edges of the image. This operation is termed convolution, with the second matrix referred to as *filter* or *kernel*, as it helps to enhance certain features of the first image. Different values of filter matrix can produce edge detection, sharpening, blurring of the image. Training of a CNN involves finding optimised values of these filters for a specific feature extraction, eventually leading to recognition of patterns in new images. The term depth is used to denote number of filters used, and stride indicates the number of pixels by which the filter matrix is moved at a step. Occasionally *zero-padding*, i.e adding extra columns or rows of zeros around the border is employed to apply the filter more effectively to bordering elements.

GAN: Generative adversarial networks (GANs) are combinations of two neural networks, competing against one another, giving rise to the *adversarial* part of the name. The two components are generator neural network and discriminator neural network, the first of which generates data mimicking some required behaviour, while the latter tries to predict whether the input is real or generated (like a classifier). In that sense, this continuous competition fine tunes both the networks, resulting in an optimal performance in generating a specific output. In perfect equilibrium, the generator will capture all the features of training data, rendering the discriminator idle. The usual considerations for designing neural networks apply to both of them.

9.2.4 Types of learning

Different training models are used based on the aim, and type of available data.

- Supervised learning: is employed when the training data is labelled, i.e. we have access to correct input–output pairs. The learning proceeds by comparing the answer of the NN with the correct answer. So in order to classify a new input, the NN effectively compares it to the training examples to predict the correct label. Apart from classification problems, this is used also in regression problems.
- Unsupervised learning: is employed where the available data is not labelled, or a clear labelling is not possible. The NN then attempts to extract patterns and features in the data. A few terms are used. *Clustering* is where the NN looks for training data that are similar to each other and groups them together. *Anomaly detection* flags outliers in the dataset. *Association* looks for correlation in data. *Autoencoders* take input data, compress it into a code, then try to recreate the input data from that summarised code. This can be useful to remove noise from data. As there is no reference to compare to, it is difficult to measure the accuracy of an algorithm trained with unsupervised learning.

- Semi/weakly supervised learning: is somewhat like the middle ground between supervised and unsupervised learning. The training dataset has both labelled and unlabelled data, which improves its accuracy compared to a fully unsupervised model. GAN architecture mentioned earlier is an example of this.
- Reinforcement learning: is employed where there is no training set, but the NN (termed an *agent* in this case) learns from experience. This makes it an iterative process, where at each step, the NN performs a trial-and-error as it attempts its task, but it is incentivised to remember the successes. So in the long run, it learns the optimal way to perform the task.

9.3 Applications in data analysis

Historically the use of machine learning methods, starting with multivariate techniques, has been in identifying physics objects, and discriminating signal against background [7]. These categories of tasks are essentially pattern-recognition problems. The BABAR collaboration used it as early as the mid-2000s for suppressing background to identify B hadrons [8]. The TMVA package [9] (Toolkit for Multivariate Analysis) in ROOT came about at the same time (which has been merged into ROOT subsequently, and expanded to include BDT and NN). In the last few years, it has expanded into all aspects of particle physics [10, 11]. A complete survey is beyond the scope of this book (and impossible, since this is an evolving area of research), instead we mention some examples of the use of machine learning techniques in this field, not in any particular order.

- Object reconstruction and identification [12–18]: Finding a charged particle track is essentially a pattern-recognition problem, as hits in detector elements are fitted to make the tracks. In the dense environment at the LHC, this comes with inherent ambiguity. Identification and reconstruction of particles in detectors is a binary classification problem. For example, electron identification in ATLAS is based on a likelihood (LH) method using a host of variables corresponding to the amount and pattern of energy deposits in different layers of calorimeters, isolation, and matching track and cluster. One-dimensional probability density functions for each of the variables are used,

  Another prominent example is identification of the τ lepton, which has been the primary ML test candidate in ATLAS and CMS. They result in one or three charged hadrons (low-track multiplicity and narrow hadronic showers) potentially accompanied by neutral pions (EM showers), which need to be distinguished against similar events originating from quark- and gluon-initiated jets. The number of associated tracks, track variables, energy density and the width and depth of the shower are the usual discriminating variables used in BDTs and MVAs. Calibrations of objects, like jet mass, energy and transverse momentum can be performed by NNs, using many variables, and properly accounting for the correlations between them.

- Jet tagging and jet images [19–38]: Classifying the jets based on their origin has been a test-bed of ML applications. This covers flavour tagging (discriminating between b, c, and light quarks), parton tagging (quark or gluon initiated), and substructure tagging (mostly for large-radius jets, discriminates between jets from W, Z, top and Higgs against light-quark and gluon jets). These mostly involve supervised learning. While many methods have been proposed, experimentally both ATLAS and CMS have started using BDT and DNNs.

In figure 9.9, an example is shown how BDT and DNN was used to tag a large-radius jet coming from a top quark. The relative background rejection with different sets of variables are shown for BDT and DNN. For the BDT case, more variables were added successively, leading to a performance saturation at a particular point. For DNN, different groups of variables, as indicated in the table at the middle row are tested once at a time. Some of these variables were introduced in chapter 8, but the details are not important for understanding the working of BDT/DNN taggers. The bottom row shows the signal and background discriminating power of the classifier obtained from BDT and DNN respectively. In this particular instance, their performance was almost identical. Additionally both experiments investigated using jet constituent level observables, such as the TopoDNN tagger in ATLAS using four vectors topological energy clusters, and DeepJet in CMS using track, cluster and vertex information along with substructure and flavour tagging information.

The energy deposit in calorimeter cells, which are the inputs to jet forming can be represented as pixels in a two-dimensional y–ϕ plane, treating the jet as an image. This allows using CNN, as employed in image recognition and classification problems, to be used in jet identification. Since the actual detector geometry is not perfectly regular, some preprocessing is required to represent the jet as an image, and standardise the representation.

1. Translation: the leading p_T subjet (i.e. with the highest pixel intensity) is placed in the centre of the image.
2. Rotation: the subleading p_T subjet is aligned along the vertical axis of the jet-image.
3. Parity flip: the images are flipped over the vertical axis such that the right side of the image has a higher energy than the left.
4. Normalisation: each pixel in the image is multiplied by a normalisation factor such that the sum of the squared pixel information is equal to unity. This is required so that the algorithm is not misled by large shifts in intensity between two training images.

In figure 9.10, the effect of preprocessing, as well as the difference in images between a signal (simulated W') and background (simulated dijet) jet is shown. The left column shows the jet images after translation, but before rotation or parity flip, while the right column shows them after the full preprocessing. The signal jet at the

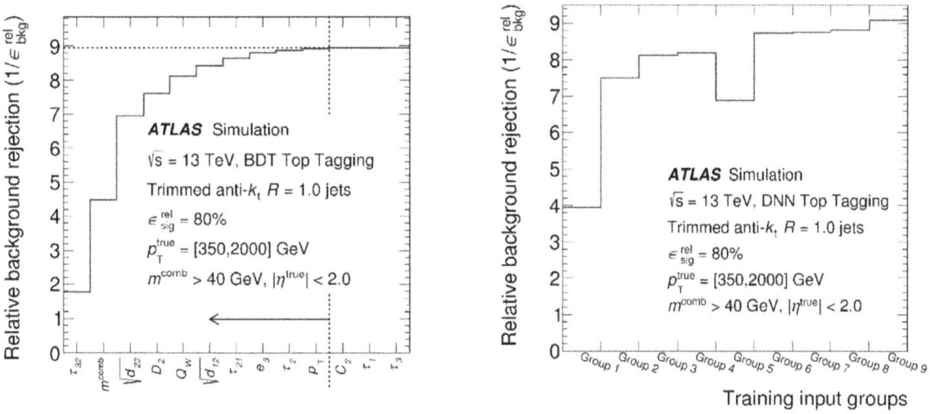

Relative background rejection ($1/\epsilon^{rel}_{bkg}$)
ATLAS Simulation
$\sqrt{s}$ = 13 TeV, BDT Top Tagging
Trimmed anti-k_t R = 1.0 jets
ϵ^{rel}_{sig} = 80%
p_T^{true} = [350,2000] GeV
m^{comb} > 40 GeV, $|\eta^{true}|$ < 2.0
τ_{32} m^{comb} $\sqrt{d_{23}}$ D_2 Q_W $\sqrt{d_{12}}$ τ_{21} e_3 τ_2 p_T C_2 τ_1 τ_3
ATLAS Simulation
$\sqrt{s}$ = 13 TeV, DNN Top Tagging
Trimmed anti-k_t R = 1.0 jets
ϵ^{rel}_{sig} = 80%
p_T^{true} = [350,2000] GeV
m^{comb} > 40 GeV, $|\eta^{true}|$ < 2.0
Group 1 Group 2 Group 3 Group 4 Group 5 Group 6 Group 7 Group 8 Group 9
Training input groups

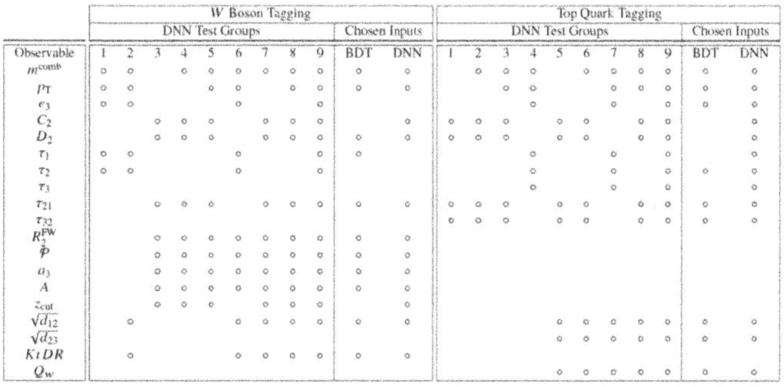

	W Boson Tagging											Top Quark Tagging										
	DNN Test Groups									Chosen Inputs		DNN Test Groups									Chosen Inputs	
Observable	1	2	3	4	5	6	7	8	9	BDT	DNN	1	2	3	4	5	6	7	8	9	BDT	DNN
m^{comb}	o	o		o	o	o	o	o	o	o	o		o	o	o		o	o	o	o	o	o
p_T	o	o			o	o		o	o	o	o			o	o			o	o	o	o	o
e_3	o	o				o			o						o			o		o	o	o
C_2			o	o	o		o	o	o		o	o	o	o		o	o		o	o		o
D_2			o	o	o		o	o	o	o	o	o	o	o		o	o		o	o		o
τ_1	o	o				o			o	o					o			o		o		o
τ_2	o	o				o			o						o			o		o	o	o
τ_3															o			o		o		o
τ_{21}			o	o	o		o	o	o	o	o	o	o	o		o	o		o	o	o	o
τ_{32}												o	o	o		o	o		o	o	o	o
R_2^{FW}			o	o	o	o	o	o	o	o	o											
$\mathcal{P}$			o	o	o	o	o	o	o	o	o											
a_3			o	o	o	o	o	o	o	o	o											
A			o	o	o	o	o	o	o	o	o											
z_{cut}			o	o	o		o	o	o		o											
$\sqrt{d_{12}}$		o				o	o	o	o	o	o					o	o	o	o	o	o	o
$\sqrt{d_{23}}$																o	o	o	o	o	o	o
$KtDR$		o				o	o	o	o	o	o											
Q_w																o	o	o	o	o	o	o

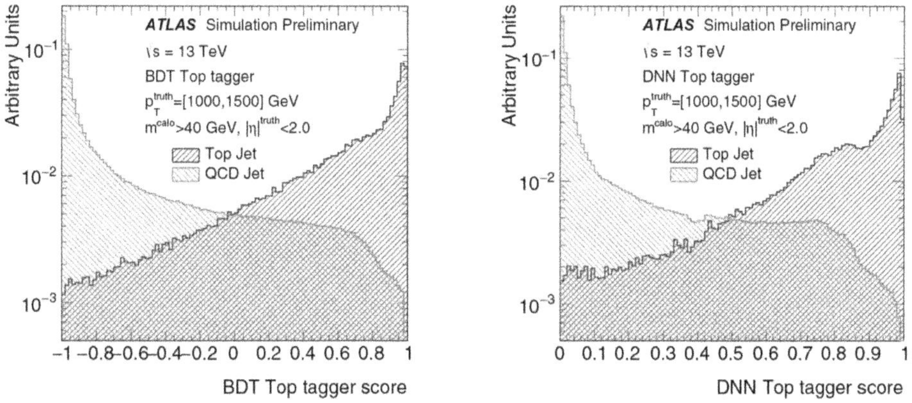

Arbitrary Units
ATLAS Simulation Preliminary
$\sqrt{s}$ = 13 TeV
BDT Top tagger
p_T^{truth}=[1000,1500] GeV
m^{calo}>40 GeV, $|\eta|^{truth}$<2.0
Top Jet
QCD Jet
BDT Top tagger score
ATLAS Simulation Preliminary
$\sqrt{s}$ = 13 TeV
DNN Top tagger
p_T^{truth}=[1000,1500] GeV
m^{calo}>40 GeV, $|\eta|^{truth}$<2.0
Top Jet
QCD Jet
DNN Top tagger score

Figure 9.9. Relative background rejection with successively adding variables for top tagging using BDT (top left), using DNN (top right), the grouping of variables for DNN [37] (middle row), performance of the obtained classifiers [26] for BDT (bottom left) and DNN (bottom right) case (ATLAS Experiment © 2018 CERN).

top has a secondary deposition due to the two-prong hadronic decay of a high-p_T W boson, while the background jet at the bottom has a large central core of deposited energy from a single hard hadronic parton. This approach also helps in identifying pileup contamination in jets.

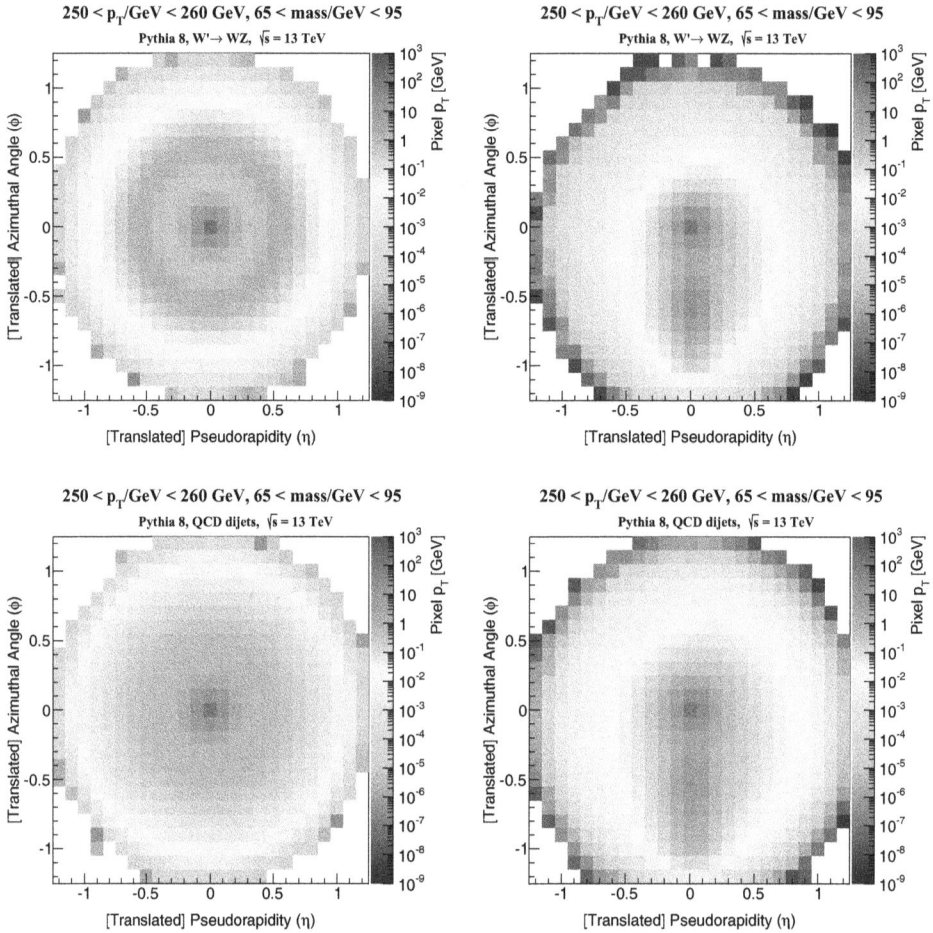

Figure 9.10. Examples of jet images, from [20]. © Luke de Oliveira, Michael Kagan, Lester Mackey, Benjamin Nachman and Ariel Schwartzman.

- Triggering [39–42]: LHCb experiment pioneered the use of NN/BDTs in triggering, resulting in better fake track rejection and better particle identification.
- New physics searches [43–52]: The first papers to demonstrate the ability of deep networks to significantly increase classification performance and discovery potential were for new physics particle searches, and many such methods have been proposed subsequently. Classification without labels (CWoLa) is perhaps worth mentioning, where the two mixed samples are the signal region and Standard Model background.
- Calorimeter simulation [53–55]: GAN is being used to model continuous high-dimensional distributions, such as detector simulation. Full simulation programme Geant4 is computationally intensive, and parametrised fast

simulation programmes do not yield comparable results, as discussed in chapter 4. In CaloGAN approach the GAN is trained using an initial set of fully simulated events, and then used to replicate showers in the calorimeter, resulting in large computational speed-ups while achieving reasonable modelling of the energy deposition.

- Event generation [56–58]: Deep learning networks or GANs are being used to generate events and simulate parton showers. The important aspect here is to have a realistic sampling.
- Parton distribution functions [59]: NNs are used to parametrise experimental distributions to obtain parton distribution functions by the NNPDF group.
- General analysis techniques [50, 54, 60–70]: This covers a wide spectrum, from event classifications based on topology using both supervised, weakly supervised and unsupervised methods, to better estimation of uncertainties.

Historically, the development in the field was based on (supervised) learning from simulated samples. This comes with a concern if the ML algorithms are learning features of simulation programmes, especially in the case they do not describe the data perfectly. Another aspect is the propagation of systematic uncertainties through these algorithms, especially assigning experimental systematic uncertainty on the new classifier variable obtained. Training directly on data corresponds to weakly supervised or unsupervised learning. In this approach, using multiple sub-samples with known class proportions, the algorithms can be trained on proportions instead of labels.

The eventual goal is not only to have these ML methods give a better result than the previous methods, but also to extract physics information from what the algorithms are learning. So rather than focussing solely on what we teach the machine, what the machine teaches us is getting more attention. In this context, we can refer to what a lot of particle physicists thought at the beginning of the ML era, that *deep thinking triumphs deep learning*. Now the paradigm has shifted to *deep thinking and deep learning together*!

References

[1] Goodfellow I, Bengio Y and Courville A 2016 *Deep Learning* (Cambridge, MA: MIT Press)
[2] Burkov A 2018 *The Hundred-Page Machine Learning Book* (Andriy Burkov)
[3] Cranmer K 2003 Multivariate analysis from a statistical point of view *Statistical Problems in Particle Physics, Astrophysics, and Cosmology* ed L Lyons, R Mount and R Reitmeyer (Singapore: World Scientific) p 211
[4] Wolter M 2007 Multivariate analysis methods in physics *Phys. Part. Nucl.* **38** 255–68
[5] Bhat P C 2011 Multivariate analysis methods in particle physics *Annu. Rev. Nucl. Part. Sci.* **61** 281–309
[6] Moore D H II 1987 Classification and regression trees, by Leo Breiman, Jerome H Friedman, Richard A Olshen, and Charles J Stone. Brooks/Cole Publishing, Monterey, 1984 *Cytometry* **8** 534–5
[7] Denby B H 1988 Neural networks and cellular automata in experimental high-energy physics *Comput. Phys. Commun.* **49** 429–48

[8] Boutigny D *et al* 1998 The BABAR physics book: Physics at an asymmetric *B* factory *Workshop on Physics at an Asymmetric B Factory (BaBar Collaboration Meeting) (Pasadena, California, September 22–24, 1997)*

[9] Hoecker A, Speckmayer P, Stelzer J, Therhaag J, von Toerne E and Voss H 2007 TMVA: toolkit for multivariate data analysis *PoS* **ACAT** 040

[10] Guest D, Cranmer K and Whiteson D 2018 Deep learning and its application to LHC physics *Annu. Rev. Nucl. Part. Sci.* **68** 161–81

[11] Albertsson K *et al* 2018 Machine learning in high energy physics community white paper *J. Phys. Conf. Ser.* **1085** 022008

[12] Farrell S *et al* 2017 Particle track reconstruction with deep learning *Proc. of the Deep Learning for Physical Sciences Workshop at NIPS (2017)*

[13] Farrell S *et al* 2018 Novel deep learning methods for track reconstruction *4th Int. Workshop Connecting The Dots 2018 (CTD2018) (Seattle, Washington, USA, March 20–22, 2018)*

[14] Stoye M 2017 *DeepFlavour in CMS. IML Machine Learning Workshop* (Geneva: CERN)

[15] ATLAS Collaboration 2017 Measurement of the tau lepton reconstruction and identification performance in the ATLAS experiment using pp collisions at $\sqrt{s} = 13$ TeV *Technical Report* ATLAS-CONF-2017-029, CERN, Geneva http://cdsweb.cern.ch/record/2261772

[16] Khachatryan V *et al* 2016 Reconstruction and identification of τ lepton decays to hadrons and ν_{τ} at CMS *JINST* **11** P01019

[17] Aaboud M *et al* 2017 Electron efficiency measurements with the ATLAS detector using 2012 LHC proton-proton collision data *Eur. Phys. J.* C **77** 195

[18] ATLAS Collaboration 2018 Generalized numerical inversion: a neural network approach to jet calibration *Technical Report* ATL-PHYS-PUB-2018-013, CERN, Geneva http://cdsweb.cern.ch/record/2630972

[19] Cogan J, Kagan M, Strauss E and Schwarztman A 2015 Jet-images: Computer vision inspired techniques for jet tagging *JHEP* **02** 118

[20] de Oliveira L, Kagan M, Mackey L, Nachman B and Schwartzman A 2016 Jet-images— deep learning edition *JHEP* **07** 069

[21] Almeida L G, Backović M, Cliche M, Lee S J and Perelstein M 2015 Playing tag with ANN: boosted top identification with pattern recognition *J. High Energy Phys.* **2015** 86

[22] Baldi P, Bauer K, Eng C, Sadowski P and Whiteson D 2016 Jet substructure classification in high-energy physics with deep neural networks *Phys. Rev.* D **93** 094034

[23] Komiske P T, Metodiev E M and Schwartz M D 2017 Deep learning in color: towards automated quark/gluon jet discrimination *JHEP* **01** 110

[24] Kagan M, de Oliveira L, Mackey L, Nachman B and Schwartzman A 2016 Boosted jet tagging with jet-images and deep neural networks *EPJ Web Conf.* **127** 00009

[25] CMS Collaboration 2018 Identification of heavy-flavour jets with the CMS detector in pp collisions at 13 *TeV J. Inst.* **13** P05011

[26] ATLAS Collaboration 2017 Identification of hadronically-decaying W bosons and top quarks using high-level features as input to boosted decision trees and deep neural networks in ATLAS at $\sqrt{s} = 13$ TeV *Technical Report* ATL-PHYS-PUB-2017-004, CERN, Geneva http://cdsweb.cern.ch/record/2259646

[27] Shimmin C, Sadowski P, Baldi P, Weik E, Whiteson D, Goul E and Søgaard A 2017 Decorrelated jet substructure tagging using adversarial neural networks *Phys. Rev.* D **96** 074034

[28] Datta K and Larkoski A J 2018 Novel jet observables from machine learning *JHEP* **03** 086

[29] Datta K and Larkoski A 2017 How much information is in a jet? *JHEP* **06** 073

[30] Louppe G, Cho K, Becot C and Cranmer K 2017 QCD-aware recursive neural networks for jet physics arXiv: 1702.00748

[31] Stoye M, Kieseler J, Verzetti M, Qu H, Gouskos L, Stakia A and CMS Collaboration 2017 DeepJet: Generic physics object based jet multiclass classification for LHC experiments *Proc. of the Deep Learning for Physical Sciences Workshop at NIPS (2017)*

[32] Pearkes J, Fedorko W, Lister A and Gay C 2017 Jet constituents for deep neural network based top quark tagging arXiv:1704.02124v2

[33] Kasieczka G, Plehn T, Russell M and Schell T 2017 Deep-learning top taggers or the end of QCD? *JHEP* **05** 006

[34] Komiske P T, Metodiev E M and Thaler J 2019 Energy flow networks: deep sets for particle jets *JHEP* **01** 121

[35] Heimel T, Kasieczka G, Plehn T and Thompson J M 2018 QCD or what? *SciPost Phys.* **6** 030

[36] Kasieczka G, Kiefer N, Plehn T and Thompson J M 2018 Quark-gluon tagging: Machine learning vs detector *SciPost Phys.* **6** 069

[37] ATLAS Collaboration 2018 Performance of top-quark and W-boson tagging with ATLAS in Run 2 of the LHC *Eur. Phys. J. C* **79** 375

[38] Butter A *et al* 2019 The machine learning landscape of top taggers arXiv:1902.09914

[39] Gligorov V V and Williams M 2013 Efficient, reliable and fast high-level triggering using a bonsai boosted decision tree *JINST* **8** P02013

[40] Likhomanenko T, Ilten P, Khairullin E, Rogozhnikov A, Ustyuzhanin A and Williams M 2015 LHCb topological trigger reoptimization *J. Phys. Conf. Ser.* **664** 082025

[41] Weitekamp D III, Nguyen T Q, Anderson D, Castello R, Pierini M, Spiropulu M and Vlimant J-R 2017 Deep topology classifiers for a more efficient trigger selection at the LHC *Proc. of the Deep Learning for Physical Sciences Workshop at NIPS (2017)*

[42] Bourgeois D, Fitzpatrick C and Stahl S 2018 Using holistic event information in the trigger arXiv:1808.00711

[43] Baldi P, Sadowski P and Whiteson D 2015 Enhanced Higgs Boson to $\tau^+\tau^-$ search with deep learning *Phys. Rev. Lett.* **114** 111801

[44] Baldi P, Sadowski P and Whiteson D 2014 Searching for exotic particles in high-energy physics with deep learning *Nat. Commun.* **5** 4308

[45] Searcy J, Huang L, Pleier M-A and Zhu J 2016 Determination of the WW polarization fractions in $pp \rightarrow W^\pm W^\pm jj$ using a deep machine learning technique *Phys. Rev. D* **93** 094033

[46] Bertone G, Deisenroth M P, Kim J S, Liem S, de Austri R R and Welling M 2016 Accelerating the BSM interpretation of LHC data with machine learning arXiv:1611.02704

[47] Renner J *et al* 2017 Background rejection in NEXT using deep neural networks *JINST* **12** T01004

[48] Caron S, Kim J S, Rolbiecki K, de Austri R R and Stienen B 2017 The BSM-AI project: SUSY-AIgeneralizing LHC limits on supersymmetry with machine learning *Eur. Phys. J. C* **77** 257

[49] Frate M, Cranmer K, Kalia S, Vandenberg-Rodes A and Whiteson D 2017 Modeling smooth backgrounds and generic localized signals with Gaussian processes arXiv:1709.05681

[50] Metodiev E M, Nachman B and Thaler J 2017 Classification without labels: learning from mixed samples in high energy physics *JHEP* **2017** 174

[51] Lin J, Freytsis M, Moult I and Nachman B 2018 Boosting $H \to b\bar{b}$ with machine learning *JHEP* **2018** 101

[52] Brehmer J, Cranmer K, Louppe G and Pavez J 2018 Constraining effective field theories with machine learning *Phys. Rev. Lett.* **121** 111801

[53] Paganini M, de Oliveira L and Nachman B 2018 Accelerating science with generative adversarial networks: an application to 3D particle showers in multilayer calorimeters *Phys. Rev. Lett.* **120** 042003

[54] de Oliveira L, Paganini M and Nachman B 2017 Learning particle physics by example: location-aware generative adversarial networks for physics synthesis *Softw. Big Sci.* **1** 4

[55] Hooberman B, Farbin A, Khattak G, Pacela V, Pierini M, Vlimant J-R, Spiropulu M, Wei W, Zhang M and Vallecorsa S 2017 Calorimetry with deep learning: particle classification, energy regression, and simulation for high-energy physics *Proc. of the Deep Learning for Physical Sciences Workshop at NIPS (2017)*

[56] Bendavid J 2017 Efficient Monte Carlo integration using boosted decision trees and generative deep neural networks arXiv:1707.00028v1

[57] Monk J W 2018 Deep learning as a parton shower *JHEP* **12** 021

[58] Otten S, Caron S, de Swart W, van Beekveld M, Hendriks L, van Leeuwen C, Podareanu D, de Austri R R and Verheyen R 2019 Event generation and statistical sampling for physics with deep generative models and a density information buffer arXiv:190100875

[59] Rojo J C 2006 The neural network approach to parton distribution functions *PhD thesis* Barcelona University

[60] Stevens J and Williams M 2013 uBoost: A boosting method for producing uniform selection efficiencies from multivariate classifiers *JINST* **8** P12013

[61] Baldi P, Cranmer K, Faucett T, Sadowski P and Whiteson D 2016 Parameterized neural networks for high-energy physics *Eur. Phys. J.* C **76** 235

[62] Barnard J, Dawe E N, Dolan M J and Rajcic N 2017 Parton shower uncertainties in jet substructure analyses with deep neural networks *Phys. Rev.* D **95** 014018

[63] Dery L M, Nachman B, Rubbo F and Schwartzman A 2017 Weakly supervised classification in high energy physics *JHEP* **05** 145

[64] Cohen T, Freytsis M and Ostdiek B 2017 (Machine) learning to do more with less *JHEP* **2018** 34

[65] Chang S, Cohen T and Ostdiek B 2018 What is the machine learning? *Phys. Rev.* D **97** 056009

[66] Estrade V, Germain C, Guyon I and Rousseau D 2017 Adversarial learning to eliminate systematic errors: a case study in High Energy Physics *Proc. of the Deep Learning for Physical Sciences Workshop at NIPS (2017)*

[67] Andrews M, Paulini M, Gleyzer S and Poczos B 2018 End-to-end physics event classification with the CMS open data: applying image-based deep learning on detector data to directly classify collision events at the LHC arXiv:1807.11916

[68] D'Agnolo R T and Wulzer A 2019 Learning new physics from a machine *Phys. Rev.* D **99** 015014

[69] Andreassen A, Feige I, Frye C and Schwartz M D 2019 JUNIPR: a framework for unsupervised machine learning in particle physics *Eur. Phys. J.* C **79** 102

[70] Englert C, Galler P, Harris P and Spannowsky M 2019 Machine learning uncertainties with adversarial neural networks *Eur. Phys. J.* C **79** 4

Experimental Particle Physics
Understanding the measurements and searches at the Large Hadron Collider
Deepak Kar

Appendix A

Solutions

Chapter 1

1. $[E/\hbar] = 1/T$, which cannot be expressed as power of only c. Other powers of $\hbar$ would leave an unbalanced mass dimension. $[L] = \hbar c/\text{eV}$ and $[M] = \text{eV}/c^2$.

2. Essentially to cross-check each other's results. It also results in intense competition, and occasionally a wrong (but stunning) result from one collaboration potentially driving the other to *reproduce* that.

3. Dark matter is not antimatter, because we do not see the unique gamma rays that are produced when antimatter annihilates with matter.

4. $M_{\text{inv}} = \sqrt{(E_1 + E_2)^2 - (p_{1x} + p_{2x})^2 - (p_{1y} + p_{2y})^2 - (p_{1z} + p_{2z})^2} = 125.4$ GeV, which is the Higgs boson.
   For future reference: if the four-momenta were expressed in terms of collider coordinates: $M_{\text{inv}} = 2p_T^1 p_T^2 (\cosh(\eta_1 - \eta_2) - \cos(\phi_1 - \phi_2))$

5. $\gamma = E/m = 6500/0.9838 \approx 6927$, leads to $v/c = \sqrt{1 - 1/\gamma^2} \approx 99\,999\,998\,958$.

6. $\tau = 1/2.5$ GeV $= 0.4 \times \hbar = 2.5 \times 10^{-25}$ s.

7. There is no frame where a photon has zero momentum. However, there exists a frame, the center-of-mass frame, in which the electron–positron system has zero momentum. This incompatibly makes it kinematically impossible. However, *virtual photons* can temporarily violate conservation laws, leading to pair production.

8. We have, $m_e^2 = m_e^2 + 0 + 2E_\gamma(E_e - p_e \cos\theta)$ So, $E_e = p_e \cos\theta < p_e$. Whereas, $E_e = \sqrt{m_e^2 + p^2} > p_e$.

9. For conservation of strangeness, $-1 + 0 = X + 1 + 1 \Rightarrow X = -3$. Checking the mass: 1672.45 MeV < 1321.31 MeV + 498 MeV.

doi:10.1088/2053-2563/ab1be6ch10

10. No, as they respectively violate momentum and charge-conjugation. Neutral λ decay to a proton and pion violate parity, but can proceed via weak interaction.

Comparison of linear and circular colliders.

	Linear collider	Circular collider
Energy ramp up:	Restricted by the length	Multiple *kicks* while going around
Luminosity:	Low, as beams cross only once	Beams cross many times
Energy loss due to synchrotron radiation:	Less	More
Cost:	Less	Expensive magnets are needed

Comparison of lepton and hadron colliders.

	Lepton collider	Hadron collider
Synchrotron radiation energy loss:	More	Less
Focussing beams:	Easier with large *e/m* ratio	Harder with large *e/m* ratio
Cross-section:	Smaller, only electroweak, much cleaner environment	Much larger, hadronic, messier
Final states:	Only charge neutral	No such restrictions
Probed energies:	Fixed	Protons are composite, range of energies
Physics aim:	Tune c.m. energy to a discovered particle being probed	Scan an energy range for discovery

Chapter 2

1. The energy lost by a particle going through turn in a circular accelerator is proportional to $E^4 R^{-1} m^{-4}$, where E, m are the energy and mass of the particle, and R is the radius. That explains why energy loss for leptons is much larger, and all proposd future electron/positron colliders are linear. Also, anti-protons are harder to make than protons, which is why LHC moved to pp. Muon colliders have the advantage that muons are heavier than electrons, so less energy loss, but muons can decay.

2. Protons move at speeds closer to speed of light, so they will traverse the 27 lm LHC ring in $\approx 10^{-8}$ s. We clearly cannot keep injecting protons at the rate.

3. Field strength of the bending magnets is the limiting factor. The protons are accelerated by RF electric fields, and the dipole magnets bend them into a circular orbit, and quadrupole magnets focus them.

4. When calculating $\Delta\phi$, it must be within 0 to 2π. In order to achieve that, if $\Delta\phi < 0$, 2π is added to it.

5. As the detectors cover the full $\phi = 4\pi$ area, so particle with any value of ϕ can be measured.

Chapter 3

1. Although hadron colliders can probe a wide range of energies, the main target was the Higgs boson. The promising decay modes in the mass range of interest were two photons, two τ leptons, two W bosons and two Z bosons. So the focus was on good lepton and photon identification capacity, as much detector coverage as possible, good momentum mass and angular resolution. Also efficient b-jet and τ identification needed pixel detectors close to the beamline, and well segmented calorimeters for jet and MET construction.

2. The most prevalent final state particle is the photon. This is because the majority of the produced hadrons are mesons, and most prevalent mesons are the light pions, which decay to two photons. Next are the hadrons.

3. There can be multiple reasons. The collision may not have happened at the centre of the detector, leading to what is termed a *displaced beam spot*. It can also be coming from pile-up interaction.

4. The electron charge is measured from the curvature of the associated tracks. The higher p_T they have, the straighter the tracks become, so misidentification probability increases. The tag electron is required to have *tight* characteristics that ensure that its charge is very likely to be correctly reconstructed, while the probe electrons will be the electrons used in the analysis. The probability of wrong charge assignment is then the number of events where the probe has the same reconstructed charge as the tag (which is assumed to be correct), divided by the total number of events passing tag-and-probe mass requirement.

5. All the objects used in the construction of MET have some threshold p_T requirement.

6. The JVF depends on tracks, while for forward jets beyond $|\eta| > 2.5$, the tracker does not exist.

7. Calculation of differential cross-sections in perturbative QCD needs IRC safety. All experimental observables are IRC safe by definition, there are no infinities experimentally because of the finite size of the particles.

8. We are limited by calorimeter resolution, as typical cell resolution in η and ϕ is 0.1×0.1.

9. Jets are *smaller* in the forward region.

10. Energy deposits from electrons and photons are typically included in jet construction, as deposits in both electromagnetic and hadronic calorimeters

are considered. Muons deposit their energy only in muon chambers, so they are typically not included. Even though the jet reconstruction is democratic, electron reconstruction is affected if there is real hadronic activity close-by.

Chapter 4

1. (a) For a hadronic collision, we have three possible hard scattering processes, qq, qg and gg. With the higher energy of the proton at the LHC, there are more gluons, making gg dominant, as opposed to qq in Tevatron.
    (b) The ratio of energies (which is termed Björken-x) of partons with respect to the proton (or anti-proton) required for formation of a $t\bar{t}$ pair is much higher in Tevatron than in LHC, as the beam energy is lower. The PDFs show that higher x values are dominated by valence quarks, while lower x values are dominated by gluons.
    (c) The s channel production is initiated at LO by $q\bar{q}$ annihilation, which is more abundantly produced in $p\bar{p}$ collisions.
2. We can generate a large number of random, uniformly distributed points inside the square, and test.
3. Parton shower can be thought of as the mechanism which gives rise to jets from partons. However, it must be realised that we never really have a single quark, rather a quark–anti-quark pair to start off with. As they move apart, they produce more anti-quark pairs from vacuum. Then hadronisation involves forming colour neutral hadrons from combination of these quarks. All the conservation laws hold for initial partons and final hadrons, but not for the jets.
4. Running of the strong coupling constant hints at what value it should have for higher energy jets. However, the determination is dependent on scale choice (should we use leading jet p_T or the average p_T or the H_T for multijets?) and non-perturbative corrections.
5. To obtain the correct shape for the leading jet p_T spectrum, one must apply the event weight, then normalize each slice to the corresponding cross-section, filter efficiency, sum-of-weights (or total number of events), and luminosity. The usual problems are not having all the slices, not having all the events in a slice, or not using the correct weight. If the event weights are calculated by the generator, then sum-of-weights should be used, whereas if event weights are only added to flatten the individual slice p_T distribution after the generation (such as by throwing out some events, and reweighting the rest), then total number of events should be used.
6. When calculating this process at NLO, the $t\bar{t}$ contribution has to be subtracted. Several method are used, one of them is to reject the events where invariant mass of the $W^-\bar{b}$ system is close to mass of the top quark.

7. (a) Not all generators, such as HERWIG or SHERPA keep the intermediate Z bosons in the event records.

    (b) With increasing p_T more and more b quarks are produced in the PS via gluon splitting. So requiring four b quarks from ME will ignore as an example two from shower and two from ME. However, the distinction between ME and PS is operational at best, and depends on the generator set-up. So these two samples will result in double counting, as the two b quark at the ME sample with extra b quarks from showering will cover some of the same phase space as the four b quark ME sample.

8. The tunes of the generators are performed inclusively, not just for specific sub-processes. Turning of MPI in PYTHIA for example will result in the initial state shower increasing its activity to compensate, since both the MPI and shower processes are competing with each other in the evolution of the interleaved shower and MPI equation.

9. If the generation uses a LO PDF, we expect the distribution to be narrower than in data (and than using a NLO PDF).

10. Usually the hadronic calorimeter shower shape variables are found to be least well modelled, so that tends to affect internal structure of jets the most.

Chapter 5

1. If we do create a miniature black hole, that will have at most have a mass equal to the energy of the pp collision, or up to 13 TeV. That corresponds, via $E = mc^2$, to a mass of just 5×10^{-23} kg, and most likely less, and a decay (via Hawking radiation) time of 10^{-83} s, too tiny to have any effect.

2. If the pile-up vertex is very close to hard scatter vertex, they can appear merged. Tracks from pile-up tend to have much less p_T, so the effect on sum p_T will be negligible, but track multiplicity may be affected.

3. We define μ as the number of pp pairs interacting in a bunch crossing. We aim to correct for this by scaling the selection efficiency of an event so that on average the predicted μ distribution matches that in data, which is known as μ reweighting. It depends on how good the modelling of pile-up is in the generator used to produce the overlay distributions.

4. We usually reject negatively weighted events during the analysis, but use the total cross-section generated. So we then need more events to attain the same statistical accuracy as would be in an MC with only positive weights. For multileg generators, there will always be a distribution of weights, leading to weights greater than unity for most *common* events. Sometimes in order to avoid spikes in distributions, these are set to unity.

5. This can be due to non-application of the (same) trigger in the simulated sample. If we do not have all the objects which fire the trigger at the trigger plateau, then this would lead to such a behaviour.

6. The staggered cuts are preferred, because for theoretical (soft emission) and experimental (resolution) reasons, objects are not perfectly balanced.

7. At the LHC, more positively charged W bosons are produced. W^+ predominately has $u\bar{d}$, while W^- predominately has $d\bar{u}$. Since protons are uud, we have more probability of u quarks participating in W formation. In Tevatron $p\bar{p}$, there was no such difference.

8. There will be no change in the first case. However, in the second case, if we have a three jet event, with one jet close to signal lepton, then step 5 will remove that jet, and the event will pass exact two jets criteria in 6. However if we swap, step 5 will kill the event.

9. The implicit assumption here is that all the background processes have the same shape over the entire range. Otherwise, we may ignore the contribution of a background which may, for example, start having a flatter shape toward the higher values.

10. It must be dominated by one single process.

11. We will have more fake electrons from jets.

12.   (a) A single jet and missing energy. Largest background is W+jet events. Can be reduced by vetoing lepton.

    (b) Four leptons. Largest background is SM ZZ production. Can be reduced by opposite charge and Z mass window requirements.

    (c) Four jets. Largest background is multijet and combinatorial Can be reduced by b tagging and top-mass requirement.

    (d) A single jet and missing energy. Largest background is W+jet events. Can be reduced by vetoing lepton and also by ISR tagging.

    (e) Two leptons and two jets. Largest backgrounds are $t\bar{t}$ and Z+jets. Can be reduced by b veto, Z-mass window requirement.

    (f) Diphoton. Largest backgrounds are SM diphoton production, diphoton decay of π^0 in jets. The latter can be reduced by requiring isolated photons.

13. We get 25 signal events. Assuming a 40% signal efficiency, we get 10 events. We definitely need about 10 events for the analysis to be feasible.

14. Human bias!

15. The b tagging efficiency is not 100%, so one b-tagged CR can have signal contamination. A further MET requirement may help.

Chapter 6

1. Yes, the range given by mean value and uncertainty are overlapping. No, larger uncertainty.

2. The default in ROOT is to do simple errors, $\sigma = 1/\sqrt{N}$. This will get the errors wrong in a weighted distribution. In that case, sum of squares of weights need to be used (which in ROOT terms mean using *Sumw2* function).

3. If the background is purely data-driven, then there can be uncertainty assessed on the method used, but it can be minimal. Then the systematic uncertainty on signal yield, which is from simulation, will be a contributing factor.

4. There are systematic uncertainties, which are determined using data. For example jet energy scale and resolutions uncertainties are often determined using *in situ* or data-driven methods. In those cases, more data can help constrain the systematic uncertainty better.

5. The top-down approach demands that we know of all possible sources of potential systematic effects *a priori*. However, then assessing the uncertainties is more straightforward. The bottom-up approach is more *democratic*, but it has its own challenges. It is sometimes difficult to estimate what variations of the input objects are reasonable (in terms of their effect on the final observable), and it is often also difficult to separate statistical fluctuations from real systematic effects.

Chapter 7

1. Tracks correspond to charged particles. Two (or any even number of) charged particles cannot make a singly charged τ lepton, we need one or three. We get one in about half the cases.

2. τ leptons decay to single pion or three pion jets, as opposite to *usual* jets with large hadron multiplicity. This is because the τ lepton is light, so the decay proceeds via a process called weak charged current interaction, where the quark pairs from the virtual W form the pions instantaneously. The branching fraction for this decay mode is several orders of magnitude smaller than the usual dijet decay mode of on-shell W bosons.

3. The leptonic branching fraction for both the τ is roughly 0.35×0.35, and needing two same flavour leptons further reduces it by a factor of two, so we are at about 6%. Then we have to consider the kinematic cuts and acceptance, which usually reduces the τ contamination in signal to sub-percent levels.

4. Semi-leptonic $t\bar{t}$ implies presence of one neutrino, leading to real missing energy. In the hadronic decay mode, the missing energy results from mismeasured objects, and threshold requirements.

5. If we produce a DY pair with $p_T^{\ell\ell} \approx 0$, then $M_{\ell\ell} \approx 2p_T^{\ell\ell}$.

6. This is because when the measurement is done with an identified hard scatter (as in the case of UE), preferentially events with a higher activity are selected.

7. As the events were required to have a larger radius jet, we preferentially selected events with higher overall activity.

8. If we just merge the neighbouring bins, then obviously each bin will go up. To avoid this, usually bin entries are divided by bin width.

9. There are several options:
   - Divide each bin content by bin width.
   - Divide each bin content by bin unity.
   - Divide each bin content by sum of entries.
   - Divide each bin content by integral.

- Divide each bin content by (integral/bin width), If already divided by bin width.

Histograms with fixed and variable bin size behave differently when bin width is included in normalisation.

10. Smooth distributions from simulation are needed to decide analysis strategy.

11. *This problem is motivated by https://cds.cern.ch/record/2320419.* The number of events from a specific process X, with a cross-section of σ_X, is given by, $N_X = L\sigma_X$, where L is the luminosity. Assuming average number of pp collisions in a bunch crossing is given by μ, this implies that the corresponding average number of inelastic collisions is $L\sigma_X\mu$. The probability that an inelastic interaction gives a hard scatter process Y with a cross-section of σ_Y is $P_B = \sigma_Y/\sigma_{\text{inel}}$. Putting all together, the number of events which have both processes X and Y from multiple pp interactions is: $N_{XY} = L\sigma_X\sigma_Y\mu/\sigma_{\text{inel}} = L\sigma_{t\bar{t}}^2\mu/2\sigma_{\text{inel}}$, where we have used $t\bar{t}$ process for both X and Y, and the factor of 2 in the denominator avoids double counting of identical processes. Taking $\sigma_{\text{inel}} = 80$ mb and $\sigma_{t\bar{t}} = 832$ pb (both at $\sqrt{s} = 13$ TeV), and using $t\bar{t}$ semileptonic branching fraction of 44%, we get: $N = 150 \times 10^3 \times (832 \times 0.44)^2 \times 50/(2 \times 80 \times 10^9) \approx 1400$. This number is the expected number of events before acceptance and efficiency has been taken into account.

12. Expected events 15. So statistical error is $\sqrt{15} = 3.8$. We saw 20, so roughly that is 1.3σ away. This actually corresponds to measuring σ with the more conservative estimate of $\sigma = S/\sqrt{B}$, where S is the excess of signal events. Alternatively, by using $\sigma = S/\sqrt{S + B}$, we get 1.12σ. In order to use $S/\sqrt{S + B} = 5$, if N is the number of background events needed for discovery, then the excess will be $S = N/3$, assuming proportionality. Then we can write $N/3 = 5\sqrt{N(1 + 1/3)}$, or $\sqrt{N} = 17$, or $N = 290$. To have a $290/15 = 20$ times increase in event yields, $200\ \text{fb}^{-1}$ luminosity is needed.

13. Observing zero events in the data means $N_{\text{bg}} = N_{\text{sig}} = 0$. If we expected S signal events, using Poisson distribution, the probability to observe zero events is e^{-S}. So we can set an upper limit on the expected signal yield S with $e^{-S} < \alpha$, where $\alpha = 5\%$ for 95% CL. That gives $S \leqslant -\ln 0.05 \approx 3$.

14. (a) This was an example of a missing systematic uncertainty. The jet energy corrections needed to be slightly different for quark and gluon initiated jets, as the latter are broader and contain a larger number of hadrons. The overcorrection of gluon jets, which do not uniformly populate the phase space, caused mismodelling of the background distribution.

 (b) The absence of jets rules out the gluon–gluon fusion production mechanism of this new resonance, and the only remaining alternative of photon–photon fusion cannot give such a high cross-section. The result should not have been insensitive to photon isolation, as new resonance will produce well isolated photons.

15. They are usually the papers which introduced widely used programmes or techniques. The original PYTHIA manual features prominently (it crossed 10 000 citations as of April 2019), and GEANT4 manual is up there as well. Other prominent ones include papers describing CT* series of PDFs, and anti-k_t clustering paper.

www.ingramcontent.com/pod-product-compliance
Lightning Source LLC
Chambersburg PA
CBHW080521220326
41599CB00032B/6160